REWARD
Pre-intermediate
Teacher's Book

REWARD

Pre-intermediate

Teacher's Book

Simon Greenall

MACMILLAN
HEINEMANN
English Language Teaching

Macmillan Education
Between Towns Road, Oxford OX4 3PP
A division of Macmillan Publishers Limited
Companies and representatives throughout the world

ISBN 0 435 242466

© Simon Greenall 1994
Heinemann is a registered trademark of Reed Educational & Professional
Publishing Limited
First published 1994

Designed by Stafford & Stafford
Cover design by Stafford & Stafford

Illustrations by:
Adrian Barclay (Beehive Illustration), pp26, 50, 70, 85;
Rowan Barnes-Murphy, pp62/63; Hardlines, pp41, 53, 87
Martin Sanders, pp12, 14/15, 18, 28, 29, 32, 37, 48, 51, 56/57, 58,
64/65, 75, 77, 89, 101; Jeannette Slater (Beehive Illustration), pp74,
92; Simon Stafford, p25.

Commissioned photography by:
Paul Freestone pp66, 67; Chris Honeywell pp38, 39, 42, 43, 60, 61,
78, 82, 83; Simon Stafford pp68, 73.

Author's acknowledgements
I am very grateful to all the people who have contributed towards the
creation of this book. My thanks are due to:
- All the teachers I have had the privilege to meet on seminars in
  many different countries and the various people who have
  influenced my work.
- James Richardson for producing the tapes, and the actors for their
  voices.
- The various schools who piloted the material in Brazil, France, Italy,
  Spain and the UK: Cultura Inglesa, Belo Horizonte; IBI, Brazilia; CEL
  Chalon sur Saone; CEL Melun; CEL Chartres; CEL Troyes et Aube;
  CLM Bell, Bolzano; English Institute, Valencia; San Pablo CEU
  Monteprincipe, Madrid; Eurocentre, Cambridge; International
  House, London; King's School, Bournemouth.
- All the readers, especially David Newbold, Peter McCabe and Liz
  Driscoll.
- Simon Stafford for the stunning design of the book.
- Sue Kay, Janet Bianchini and all the staff and students at the Lake
  School, Oxford for their enthusiastic trialling of the material, their
  feedback, support and encouragement.
- Jill, Jack and Alex for providing the most supportive home
  environment in which to write. Nothing would have been possible
  without them.
- Jacqueline Watson for researching the photos.
- Karen Jamieson for shaping the project at its early stages.
- Chris Hartley for doing everything a good publisher can do to help.
- Catherine Smith for her kind but very thorough style of editing. Her
  contribution towards shaping the book, her advice and her attention
  to detail have played a central role in making this book possible.

Acknowledgements
The authors and publishers would like to thank the following for their
kind permission to reproduce material in this book:
British Railways Board and Roald Dahl for extracts from *Roald Dahl's
Guide to Railway Safety*; James Ferguson for *Highway to the Andes*;
Hamish Hamilton Ltd for *The Kingdom by the Sea* by Paul Theroux;
HMSO for an extract from *Belt up*; Jonathan Cape for *The Book of
Ages* by Desmond Morris; Martin Secker & Warburg for an extract
from *The Return of Heroic Failures* by Stephen Pile; Michael Joseph
Limited and Penguin Books Limited for *The Umbrella Man* by Roald
Dahl; *The Independent on Sunday* for 'Flying is bad for your health'
by Jenny Bryan and 'A brief discussion about the environment'; Virgin
Publishing for two extracts from *Urban Myths* by Rick Glanvill and
Phil Healey.

Photographs by: Ace Photo Agency/Mauritius p90(b); Adams Picture
Library p46; The Bridgeman Art Library p40; The Detroit Institute of
Arts, Michigan/Bridgeman p40(m); Britstock IFA pp8(t),68, 69; The J
Allan Cash Photolibrary pp32(b), 96; Collections/Paul Watts pp14,
15; ©1992 Comstock/Julian Nieman/SGC p21; © Robert
Doisneau/Rapho p10; Eye Ubiquitous/J Stephens p86; Fotoccompli
p32(t); The Illustrated London News Picture Library pp16, 17, 24;
The Hulton Deutsch Collection/Steve Eason p52; The Image Bank
p4(t); Images Colour Library pp6, 45, 54(b) Images/Charlie Waite
pp22, 23, 34; The International Stock Exchange Library, London
p67(tr); Mexicolore/Tito Zauala p80; Courtesy of The National
Portrait Gallery, London p40(b); Paramount, courtesy of The Kobal
Collection p82; The Photographer's Library p55; Pictor International,
London p90; Picturepoint, London pp2, 3, 54; Popperfoto pp3, 94;
Reproduced by permission of Royal Mail p78; © Estate of Stanley
Spencer 1994 All rights reserved DACS p41; Tony Stone Images pp23,
30, 31, 53; Telegraph Colour Library pp4(b), 8(b); Zefa p44.

The publishers would also like to thank Ella Dennison, Bord Fáilte
Irish Tourist Board, Balvinder Gill, Mrs Jones, Peter Mays, Emma and
Kate Simpson-Wells, Mr and Mrs Youd and Zoë Youd.

Printed and bound in Spain by Edelvives

2003 2002

13 12 11 10 9 8

Contents

Introduction

Course organisation

Reward is a general English course which will take adult and young adult learners of English from pre-intermediate level to upper-intermediate level. British English is used as the model for grammar, vocabulary, spelling and pronunciation, but other varieties of English are included for listening and reading practice. The course components for each level are as follows:

For the student	For the teacher
Student's Book	Teacher's Book
Practice Book	Class cassettes
Practice Book cassette	Resource Pack
	Business Resource Pack

The Student's Book has forty teaching lessons and eight Progress check lessons. After every five teaching lessons there is a Progress check lesson to review the language covered in the preceding teaching lessons and to present new language work relevant to the grammar, functions and topics covered so far. Within the teaching lessons, the main grammar or language functions and the most useful vocabulary are presented in boxes which allow easy access to the principal language of the lesson. This makes the focus of the lesson clearly accessible for purposes of presentation and revision. Each lesson will take about 90 minutes.

The two **Class cassettes** contain all the listening and sounds work in the Student's Book.

The Practice Book has forty practice lessons corresponding to the forty teaching lessons in the Student's Book. The Practice Book extends work done in class with the Student's Book by providing further practice in grammar, vocabulary, reading, writing, listening and sounds work. The activities are designed for self-access work and can be used either in the class or as self-study material. Each lesson will take between 45 and 60 minutes. An answer key is included for self-checking.

The **Practice Book cassette** contains all the listening and sounds work in the Practice Book.

The Teacher's Book contains a presentation of the course design, methodological principles, as well as detailed teaching notes interleaved with pages from the Student's Book. It also includes four photocopiable tests. The teaching notes for each lesson include a step-by-step guide to teaching the lesson, a discussion of some of the difficulties the learners may encounter, and more detailed methodological issues arising from the material presented.

The Resource Packs provide additional teaching material to practise the main language points of the teaching lessons. *Reward* is designed to be very flexible in order to meet the very different requirements of learners. There is a Resource Pack for learners of general English and a Business Resource Pack for learners with language requirements of a more professional nature.

Each pack contains a wide variety of communicative practice activities in the form of photocopiable worksheets with step-by-step Teacher's Notes on the back. There is at least one activity for each lesson in the Student's Book and the activities can be used to extend a core teaching lesson of 90 minutes from the Student's Book with an average of 30 minutes of extra material for use in the classroom. They can also be used to revise specific structures, language or vocabulary later in the course.

As well as step-by-step Teacher's Notes for each activity, each Resource Pack includes an introduction which explains how to use the worksheets and offers tips on how to get the most out of the activities.

Course design

The course design is based on a broad and integrated multi-syllabus approach. It is broad in the sense that it covers grammar and language functions, vocabulary, reading, listening, speaking, writing and sounds explicitly, and topics, learner training and socio-cultural competence implicitly. It is integrated in that each strand of the course design forms the overall theme of each lesson. The lessons always include activities focusing on grammar and language functions, and vocabulary. They also include reading, listening, speaking, writing and sounds. The inclusion of each strand of the syllabus is justified by its communicative purpose within the activity sequence. The methodological principles and approaches to each strand of course design are discussed below.

Methodological principles

Here is an outline of the methodological principles for each strand of the course design.

Grammar and language functions

Many teachers and learners feel safe with grammar and language functions. Some learners may claim that they want or need grammar, although at the same time suggest that they don't enjoy it. Some teachers feel that their learners' knowledge of grammar is demonstrable proof of language acquisition. But this is only partly true. Mistakes of grammar are more easily tolerated than mistakes of vocabulary, as far as comprehension is concerned, and may be more acceptable than mistakes of socio-cultural competence, as far as behaviour and effective communication is concerned. *Reward* attempts to establish grammar and language functions in their pivotal position but without neglecting the other strands of the multi-syllabus design.

Vocabulary

There are two important criteria for the inclusion of words in the vocabulary boxes. Firstly, they are words which the pre-intermediate learner should acquire in order to communicate successfully in a number of social or transactional situations. The revised Threshold Level (1990)[1] has been used to arbitrate on the final selection. Secondly, they may also be words which, although not in Threshold Level, are generated by the reading or listening material and are considered suitable for the pre-intermediate level. However, an overriding principle operates: there is usually an activity which allows learners to focus on and, one hopes, acquire the words which are personally relevant to them. This involves a process of personal selection or grouping of words according to personal categories. It is hard to acquire words which one doesn't need, so this approach responds to the learner's individual requirements and personal motivation. Pre-intermediate *Reward* presents approximately 950 words in the vocabulary boxes for the learner's active attention, but each learner must decide which words to focus on. The *Wordbank* in the Practice Book encourages students to store the words they need in categories which are relevant to them.

Reading

The reading passages are generally at a higher level than one might expect for learners at pre-intermediate level. Foreign language users who are not of near-native speaker competence are constantly confronted with difficult language, and to expose the learners to examples of real-life English in the reassuring context of the classroom is to help prepare them for the conditions of real life. There is always an activity or two which encourages the learner to respond to the passage either on a personal level or to focus on its main ideas. *Reward* attempts to avoid a purely pedagogical approach to reading, and encourages the learner to respond to the reading passage in an personal and genuine way before using it for other purposes.

Listening

Listening is based on a similar approach to reading in *Reward*. Learners are often exposed to examples of natural, authentic English in order to prepare them for real-life situations in which they will have to listen to ungraded English. But the tasks are always graded for the learners' particular level. Learners at pre-intermediate level are often pleased by how much they understand. Learners at higher levels are often disappointed by how little they understand. A number of different native and non-native accents are used in the listening passages to reflect the fact that in real life very few speakers using English speak with standard British or American pronunciation.

Speaking

Many opportunities are given for speaking, particularly in pairwork and groupwork. Learners are encouraged to work in pairs and groups, because the number of learners in most classes does not allow the teacher to give undivided attention to each learner's English. In these circumstances, it is important for the teacher to evaluate whether fluency or accuracy is the most important criterion. On most occasions in Pre-intermediate *Reward* speaking practice in the Grammar sections is concerned with accuracy, and in the Speaking sections with fluency. In the latter sections, it is better not to interrupt and correct the learners until after the activity has finished.

Writing

The writing activities in *Reward* are based on guided paragraph writing with work on making notes, turning notes into sentences, and joining sentences into paragraphs with various linking devices. The activities are quite tightly controlled. This is not to suggest that more creative work is not valid, but it is one of the responsibilities of a coursebook to provide a systematic grounding in the skill. More creative writing is covered in the Practice Book. Work is also done on punctuation, and most of the writing activities are based on real-life tasks, such as writing letters and cards.

---

1 The Threshold Level 1990, J A van Ek and J L M Trim, Council of Europe Press, 1991.

Sounds

Pronunciation, stress and intonation work tends to interrupt the communicative flow of a lesson, and there is a temptation to leave it out in the interests of maintaining the momentum of an activity sequence. In *Reward* there is work on sounds in most lessons, usually just before the stage where the learners have to use the new structures orally in pairwork or groupwork. At this level, it seems suitable to introduce work beyond the straightforward system of English phonemes, most of which the learners will be able to reproduce accurately because the same phonemes exist in their own language. So activities which focus on stress in words and sentences, and on the implied meaning of certain intonation patterns are included. The model for pronunciation is standard British.

Topics

The main topics proposed by the Threshold Level are covered in *Reward* Pre-intermediate. These include personal identification, house and home, daily life, leisure activities, travel, relations with other people, health, education, shopping, food and drink, geographical location and the environment. On many occasions the words presented in the vocabulary box all belong to a particular word field or topic.

Learner training

Implicit in the overall approach is the development of learner training to encourage learners to take responsibility for their own learning. Examples of this are regular opportunities to use monolingual and bilingual dictionaries, ways of organising vocabulary according to personal categories and inductive grammar work.

Cross-cultural training

Much of the material and activities in *Reward* creates the opportunity for cross-cultural training. Most learners will be using English as a medium of communication with other non-native speakers, and certainly with people of different cultures. Errors of socio-cultural competence are likely to be less easily tolerated than errors of grammar or lexical insufficiency. But it is impossible to give the learners enough specific information about a culture, because it is impossible to predict all the cultural circumstances in which they will use their newly acquired language competence. Information about *sample* cultures, such as Britain and America, as well as non-English speaking ones, is given to allow the learners to compare their own culture with another. This creates opportunities for learners to reflect on their own culture in order to become more aware of the possibility of different attitudes, behaviour, customs,

traditions and beliefs in other cultures. In this spirit, cross-cultural training is possible even with groups where the learners all come from the same cultural background. There are interesting and revealing differences between people from the same region or town, or even between friends and members of the same family. Exploring these will help the learners become not merely proficient in the language but competent in the overall aim of communication.

Level and progress

One important principle behind *Reward* is that the learners arrive at pre-intermediate level with very different language abilities and requirements. Some may find the early lessons very easy and will be able to move quickly on to later lessons. The way *Reward* is structured, with individual lessons of approximately 90 minutes, means that these learners can confirm that they have acquired a certain area of grammar, language function and vocabulary, consolidate this competence with activities giving practice in the other aspects of the course design, and then move on. Others may find that their previous language competence needs to be reactivated more carefully and slowly. The core teaching lesson in the Student's Book may not provide them with enough practice material to ensure that the given grammar, language functions and vocabulary have been firmly acquired. For these learners, extra practice may be needed and is provided in both the Practice Book (for self-study work) and by the Resource Packs (for classroom work). If learners return to language training at pre-intermediate level after a long period of little or no practice, it is hard to predict quite what they still know. *Reward* is designed to help this kind of learner as much as those who need to confirm that they have already acquired a basic knowledge of English.

Correction

You may need to tell your students your policy on correction. Some may expect you to correct every mistake; others will be hesitant to join in if they are nervous of correction. You need to decide when and how often you want to correct people. Of course, this will depend on the person and the activity, but it might be worth making the distinction between activities which encourage accuracy, in which it is very suitable to provide a certain amount of correction, and activities which focus on fluency, in which it may be better to note down mistakes and give them to the student at a later stage. Another approach may be to encourage accuracy at the beginning of a practice sequence and fluency towards the end. Let them know the general principles. It will create a positive impression even for those who may, at first, disagree with it.

Interest and motivation

Another important principle in the course design has been the intrinsic interest of the materials. Interesting material motivates the learners, and motivated learners acquire the language more effectively. The topics have been carefully selected so that they are interesting to adults and young adults, with a focus on areas which would engage their general leisure-time interests. This is designed to generate what might be described as authentic motivation, the kind of motivation we have when we read a newspaper or watch a television programme. But it is obvious that we cannot motivate all learners all of the time. They may arrive at a potentially motivating lesson with little desire to learn on this particular occasion, perhaps for reasons that have nothing to do with the teacher, the course or the material. It is therefore necessary to introduce tasks which attract what might be described as pedagogic or artificial motivation, tasks which would not usually be performed in real life, but which engage the learner in an artificial but no less effective way.

Variety of material and language

Despite the enormous amount of research done on language acquisition, no one has come up with a definitive description of how we acquire either our native language or a foreign language which takes account of every language learner or the teaching style of every teacher. Every learner has different interests and different requirements, and every teacher has a different style and approach to what they teach. *Reward* attempts to adopt an approach which appeals to differing styles of learning and teaching. The pivotal role of grammar and vocabulary is reflected in the material, but not at the expense of the development of the skills or pronunciation. An integrated multi-syllabus course design, designed to respond to the broad variety of learners' requirements and teachers' objectives, is at the heart of *Reward's* approach.

RESEARCH

Heinemann ELT is committed to continuing research into coursebook development. Many teachers contributed to the evolution of *Reward* through piloting and reports, and we now want to continue this process of feedback by inviting users of *Reward* – both teachers and students – to tell us about their experience of working with the course. If you or your colleagues have any comments, queries or suggestions, please address them to the Publishing Director, Adult Group, Heinemann ELT, Halley Court, Jordan Hill, Oxford OX2 8EJ or contact your local Heinemann representative.

Map of the book

Lesson	Grammar and functions	Vocabulary	Skills and sounds
1 *Welcome!* Hospitality customs in different countries	Present simple (1) for customs and habits Questions Adverbs of frequency	Hospitality: verbs	**Listening:** listening for main ideas **Speaking:** talking about suitable questions to ask people **Reading:** reading for main ideas **Writing:** writing simple sentences with adverbs of frequency
2 *A day in the life of the USA* Typical daily routines	Present simple (2) for routines: third person singular Expressions of time	Routine activities Meals Times of the day Matching verbs and nouns	**Reading:** reading for main ideas **Listening:** listening for main ideas; listening for specific information **Sounds:** third person singular present simple endings: /s/, /z/, /ɪz/ **Writing:** writing connected sentences about daily routines using *and* and *then*
3 *Home rules* Different ideas of 'home' in different cultures	Articles Plurals	Types of housing Rooms Furniture and equipment	**Reading:** reacting to a passage and comparing information in a passage with personal opinion **Speaking:** talking about typical homes and ideas of 'home'
4 *First impressions* Impressions of Britain and the USA	Verb patterns (1): *-ing* form verbs Talking about likes and dislikes	Positive and negative adjectives Matching nouns and adjectives	**Reading:** reading for main ideas; inferring **Sounds:** strong intonation for likes and dislikes **Speaking:** talking about likes and dislikes
5 *Take a closer look* Describing a situation	Present simple and present continuous	Words for everyday transactions, such as shopping Identifying verbs and nouns; grouping words according to subject	**Speaking:** talking about people's lifestyles; describing a situation **Sounds:** /n/ and /ŋ/ **Listening:** listening for main ideas; listening for context
Progress check lessons 1 – 5	Revision	Words with more than one part of speech Choosing useful words Starting a *Wordbank*	**Sounds:** third person singular present simple endings: /s/, /z/, /ɪz/; /ɪ/ and /iː/; interested intonation **Speaking:** talking about customs and traditions of hospitality **Writing:** punctuation of sentences; writing about customs of hospitality
6 *Surprising behaviour* An extract from *The Kingdom by the Sea*, by Paul Theroux	Past simple (1): regular and irregular verbs	New words from a passage by Paul Theroux Adjectives and nouns which go together	**Reading:** reading for main ideas; inferring **Sounds:** past simple endings: /t/, /d/ and /ɪd/ **Speaking:** talking about surprising behaviour
7 *The world's first package tours* The life of Thomas Cook	Past simple (2): questions and short answers	Travel and tourism Nouns, adjectives and verbs which go together	**Reading:** reading for main ideas; reading for specific information **Speaking:** talking about important events in Thomas Cook's life; talking about experiences as a tourist
8 *Something went wrong* Travel situations when things went wrong	Expressions of past time *So, because*	Journeys by train, boat and plane Hotel accommodation	**Listening:** predicting, listening for specific information **Speaking:** talking about a situation where something went wrong **Writing:** writing a story about something that went wrong using *and, but, so* and *because*
9 *Family life* Portrait of a French family; typical families from different countries	Possessive *'s* Possessive adjectives	Members of the family	**Listening:** listening for specific information **Sounds:** weak syllables /ə/; contrastive stress **Reading:** reading for main ideas **Speaking:** giving general and/or specific information about families in your country
10 *The town where I live* Venice; talking about the advantages and disadvantages of life in your home town	*Have got*	Town features and facilities Adjectives to describe towns	**Listening:** listening for specific information; listening for main ideas **Writing:** writing a description of the town where you live using *and* and *but*

Lesson	Grammar and functions	Vocabulary	Skills and sounds
Progress check lessons 6 – 10	Revision	Word chains Compound nouns Categorising vocabulary	**Sounds:** syllable stress in words; /ð/ and /θ/; /ɒ/ and /əʊ/; friendly intonation **Writing:** predicting a story from questions **Speaking:** talking about past events; families
11 *How ambitious are you?* Talking about ambitions	Verb patterns (2): *to* + infinitive; *going to* for intentions, *would like to* for ambitions	Ambitions Verbs and nouns which go together	**Reading:** reading and answering a questionnaire **Writing:** writing a paragraph describing your ambitions using *because* and *so*
12 *English in the future* The role of the English language in the future of your country	*Will* for predictions	Jobs School subjects	**Listening:** listening for main ideas **Sounds:** syllable stress in words; /e/ and /eɪ/ **Speaking:** talking about the future of English **Writing:** writing a paragraph about what people think about the future of English
13 *Foreign travels* Planning a trip to South America	*Going to* for plans and *will* for decisions Expressions of future time	Equipment for travellers	**Listening:** listening for specific information **Speaking:** planning a trip
14 *In Dublin's fair city* Finding your way around town	Prepositions of place Asking for and giving directions	Town features Adjectives to describe bars	**Reading:** reacting to a passage **Listening:** listening for specific information **Sounds:** /ɑː/, /æ/ and /ʌ/ **Speaking:** giving directions around town
15 *An apple a day* Typical meals in different countries	Expressions of quantity (1): countable and uncountable nouns, *some* and *any*, *much* and *many*	Food and drink Meals	**Listening:** listening for specific information **Speaking:** talking about typical meals and food in different countries
Progress check lessons 11 – 15	Revision	Word maps Nouns from verbs and nouns from other nouns Noun suffixes for jobs	**Sounds:** weak syllables /ə/ ;/tʃ/ and /ʃ/; contrastive stress; polite intonation in questions **Speaking:** planning a lunch party for friends
16 *What's on?* Typical entertainment in different countries	Prepositions of time and place Making invitations and suggestions	Types and places of entertainment and related words	**Listening:** listening for specific information **Speaking:** talking about typical entertainment **Writing:** writing and replying to invitations
17 *Famous faces* Describing people	Describing appearance and character: *look like, be like*	Words to describe height, age, looks, build and character	**Listening:** listening for main ideas **Speaking:** describing people **Writing:** writing a letter describing your appearance
18 *Average age* Personal qualities at different ages	Making comparisons (1): comparative and superlative adjectives	Adjectives of character	**Reading:** reacting to a passage and comparing information in a passage with your own experience **Speaking:** talking about exceptional people **Writing:** writing sentences describing exceptional people
19 *Dressing up* Typical clothes in different countries	Making comparisons (2): *more than, less than, as...as*	Clothes Colours Personal categories for organising new vocabulary	**Reading:** reading for specific information **Sounds:** weak syllables /ə/ and /ɪ/; weak forms /ðən/, /əz/ and /frəm/; stress for disagreement **Listening:** listening for main ideas **Speaking:** talking about clothing
20 *Memorable journeys* A car journey across the USA	Talking about journey time, distance, speed and prices	Numbers Words to describe a long-distance journey by car	**Listening:** listening for specific information **Sounds:** syllable stress in numbers **Speaking:** talking about a memorable journey
Progress check lessons 16 – 20	Revision	International words Adjective suffixes Male and female words	**Sounds:** /ʊ/ and /uː/; /dʒ/; polite and friendly intonation **Speaking:** mutual dictation **Writing:** mutual dictation to recreate a story

Lesson	Grammar and functions	Vocabulary	Skills and sounds
21 *How are you keeping?* Your body and your health	Present perfect simple (1) for experiences	Parts of the body	**Reading:** reading and answering a questionnaire **Speaking:** talking about experiences
22 *What's new with you?* Talking about changes in your life	Present perfect simple (2) for past actions with present results	Political and social conditions	**Listening:** predicting; listening for specific information **Sounds:** linking of /v/ and /s/ endings before certain verbs **Writing:** writing a letter describing recent changes in your life
23 *It's a holiday* Important national or local events and festivals	Present perfect simple (3): *for* and *since*	Words to describe important events and festivals	**Listening:** listening for specific information **Sounds:** weak form /fə/ **Writing:** writing a paragraph describing an important national occasion
24 *Divided by a common language?* A comparison of British and American English	Defining relative clauses: *who, which/that* and *where*	American English words with different meanings in British English	**Speaking:** talking about useful types of English **Reading:** inferring **Sounds:** comparing American and British standard pronunciation; difference in specific phonemes **Listening:** listening for specific information
25 *What's it called in English?* Describing objects	Describing things when you don't know the word	Adjectives for shape, material, size etc. Words to describe something if you don't know the English word Everyday objects	**Listening:** listening for main ideas **Sounds:** consonant clusters; word linking in sentences **Speaking:** describing everyday objects
Progress check lessons 21 – 25	Revision	Verbs from nouns and nouns from verbs Noun suffixes Meanings of *get*	**Sounds:** /ɜː/ and /ə/; /ɔː/ and /ʌ/ **Speaking:** Talking about your travel experiences **Writing:** writing a paragraph about other students travel experiences.
26 *Safety first* Safety instructions and rules	Modal verbs *Must* for obligation; *mustn't* for prohibition	Words to describe situations where safety instructions apply: on a motorway, in a train, at the border, in the street	**Speaking:** talking about safety instructions **Sounds:** linking of /mʌst/ and /mʌsnt/; insistent intonation **Reading:** reading about safety instructions **Listening:** listening for main ideas
27 *The Skylight* A short story by Penelope Mortimer	*Can, could* (1) for ability	New words from a story *The Skylight*	**Speaking:** talking about what you can or can't do; predicting what happens next in a story **Listening:** listening for main ideas; listening for specific information
28 *Breaking the rules?* Rules and customs in everyday situations in different countries	*Can, can't* (2) for permission and prohibition	Words to describe rules in everyday situations	**Reading:** reading and answering a questionnaire **Sounds:** strong and weak forms of *can*; American English *can* and *can't* **Listening:** listening for main ideas **Speaking:** talking about rules
29 *Warning: flying is bad for your health* Advice on staying healthy	*Should* and *shouldn't* for advice	Medical complaints Parts of the body	**Reading:** reading for specific information **Listening:** listening for specific information **Speaking:** talking about advice for staying healthy
30 *Doing things the right way* Behaviour and manners in different social situations	Asking for permission Asking people to do things Offering	Words from a questionnaire about behaviour in social situations	**Reading:** reading and answering a questionnaire **Sounds:** polite intonation in questions **Listening:** listening for specific information
Progress check lessons 26 – 30	Revision	Adjectives and nouns which go together Words with more than one meaning and part of speech Techniques for dealing with words you don't understand Word association	**Sounds:** /əʊ/ and /ɔɪ/; /ʃ/, /tʃ/ and /dʒ/; polite and friendly intonation **Speaking:** talking about advice and rules for foreign visitors to your country **Writing:** writing some advice and rules for foreign visitors

Lesson	Grammar and functions	Vocabulary	Skills and sounds
31 *My strangest dream* An English woman talks about her dream in which the Queen came to tea	Past continuous (1) for interrupted actions *When*	Verbs and prepositions which go together Adjective and noun or noun and noun combinations Words for story telling	**Speaking:** predicting what happens next in a story **Listening:** listening for specific information **Writing:** writing a story using *suddenly, fortunately, unfortunately, to my surprise, finally*
32 *Time travellers* A true story about two women who travelled back in time	Past continuous (2): *while* and *when*	New words from a passage called *Time travellers*	**Reading:** predicting; reading for main ideas **Speaking:** talking about travelling in time
33 *Is there a future for us?* Two children give their views on the environment in the future	Expressions of quantity (2): *too much/many, not enough, fewer, less* and *more*	Geographical features and location	**Reading:** reading for specific information; inferring **Speaking:** talking about the geography of your country; talking about the environment
34 *The day of the dead* An article about Mexico's Day of the Dead	Present simple passive	Religion Rituals and festivals	**Reading:** predicting; reading for main ideas; reacting to a passage **Speaking:** talking about a ritual or festival in your country **Writing:** writing about a ritual or festival
35 *Mind your manners!* Table manners and behaviour in social situations	Making comparisons (3): *but, although, however*	Food Plates, cutlery etc. Cooking utensils	**Reading:** reading and answering a questionnaire **Listening:** listening for specific information **Sounds:** stress and intonation in sentences with *but, however, although* **Speaking:** talking and writing about table manners and social occasions in your country
Progress check lessons 31 – 35	Revision	Multi-part verbs	**Sounds:** /v/ and /w/; /h/; stress in multi-part verbs **Writing:** punctuating a story; inserting words into a story
36 *Lovely weather* The best times to visit different countries	*Might* and *may* for possibility	Weather	**Reading:** reading for specific information **Writing:** writing a letter giving advice about the best time to visit your country
37 *Help!* Emergency situations	First conditional	Words to describe emergency situations	**Speaking:** talking about emergency situations; predicting the end of a story **Sounds:** /l/; word linking in sentences **Listening:** listening for main ideas, listening for specific information
38 *My perfect weekend* Two people describe their perfect weekend	*Would* for imaginary situations	Luxuries and necessities New vocabulary from a passage called *My perfect weekend*	**Speaking:** talking about luxuries and necessities; talking about your perfect weekend **Reading:** reacting to a passage **Sounds:** linking of /d/ ending before verbs beginning in /t/ or /d/ **Listening:** listening for main ideas
39 *The umbrella man* A short story by Roald Dahl	Second conditional	New vocabulary from a story called *The umbrella man*	**Reading:** predicting; reading for main ideas; reading for specific information **Listening:** listening for specific information **Writing:** rewriting a story from a different point of view
40 *How unlucky can you get?* Unlucky experiences	Past perfect: *after, when* and *because*	New vocabulary from a story called *How unlucky can you get?*	**Listening:** listening for specific information **Sounds:** linking of /d/ in past perfect sentences **Speaking:** predicting the end of a story **Writing:** writing sentences using *after, because* and *when*
Progress check lessons 36 – 40	Revision	*Make* and *do* Formation of adverbs	**Sounds:** /w/, /r/; /ɔː/, /aʊ/; stressed words **Speaking:** talking about difficult situations; preparing and acting out a dialogue

Communication activities   Grammar section   Irregular verbs   Phonetic alphabet

1 | *Welcome!*

Present simple (1) for customs and habits: questions; adverbs of frequency

SPEAKING AND LISTENING

1 Work in pairs. Where do you hear these words and phrases?
– in a bar – in a shop – in a hotel – at home – in class

Hello. Goodbye. Come in. How do you do. Pleased to meet you.
Can I help you? How much is this? Thank you. Fine, thanks.
Sorry! I'd like a Coca Cola. I don't understand. How are you?
Could you repeat that? This is my friend, Rosario.

Which ones go together?
Hello. How are you?

2 🔲 What are the situations? Listen and find out.

3 Put the words in the right order and make questions. Then underline the stressed words.

1 first your what's name	5 are married you
2 are old you how	6 do do you what
3 much earn how you do	7 and you brothers sisters do any have
4 do live you where	8 from where come do you

What's your first name?

🔲 Now listen and check. Say the sentences aloud.

4 Look at the questions in 3. Which ones are suitable questions for:
– someone you know well – someone you don't know well?

5 Go round the class greeting people and asking suitable questions.

READING

1 *Make yourself at home* is about hospitality in Germany, Saudi Arabia, Britain and Japan. Read it and match these headings with the paragraphs.

a type of clothes b length of stay c refreshments d special customs
e gifts f topics of conversation g time of arrival

2 Work in pairs. Match the countries and the paragraphs.
I think paragraph 1 is Japan because they sit on the floor in Japan.

3 Which paragraphs are true for your country?

Make yourself at home

1 'In my country, men usually go to restaurants on their own. They always take their shoes off before they go in. Then they usually sit on the floor around a small, low table. In the evening they often sing songs.'

2 'You usually take chocolates or flowers. But you always take an odd number of flowers, and you remove the paper before you give them to the hostess. You can also send flowers before you arrive. You don't usually take wine except when you visit very close friends.'

3 'We always offer our guests something to drink when they arrive, tea, coffee or perhaps water or soft drinks. We think it is polite to accept a drink even if you're not thirsty. If you visit someone you always stay for a few drinks. When you have had enough to drink, you tap your cup or put your hand over it. If you say no, your host will insist that you have more to drink.'

1

GENERAL COMMENTS

Beginning the course

It may be that at the start of the course, the students need to get to know each other. In *Speaking and listening* there are a number of activities which give the students the necessary language and opportunities to introduce themselves to each other. At this stage, most people will be unfamiliar with the coursebook and may be unsure of the reasons why they are doing certain activities. Be ready to explain anything which the students may be unsure about.

Top photo: possible questions

Who is the woman in the photo? Where do you think she is?
Is she the hostess or the guest? Are the flowers a gift to or from her? What other gifts can you take people when they invite you to their homes?

Bottom photo: possible questions

Who are the people? Where are they from?
Where are they? What else can you see in the photo?

SPEAKING AND LISTENING

1 Aim: to present some common words and phrases for greetings and simple transactions; to make a simple distinction between transactional and social situations.

● Write the situations on the board and ask the students to write the expressions under the situations.

> **Possible answers**
> **in a bar:** Can I help you? How much is this?
> Thank you. I'd like a Coca Cola.
> **in a shop:** Hello. Goodbye. Can I help you?
> How much is this? Thank you. I'd like a Coca Cola.
> **in a hotel:** Hello. Goodbye. Can I help you?
> Thank you.
> **at home:** Hello. Goodbye. Come in. Pleased to meet you. Thank you. Fine, thanks. Sorry!
> How are you? This is my friend, Rosario.
> **in class:** Hello. Goodbye. Pleased to meet you.
> Thank you. I don't understand. Could you repeat that?

2 Aim: to practise listening for main ideas.
● Explain that the students are going to listen to two dialogues. They should guess the situation from the five given in activity 1. Tell them not worry if they don't understand every word.

● 📟 Play the cassette.

> **Answers**
> Situation 1: at home, Situation 2: in a bar

● Elicit the correct answers. Try to resist explaining the meaning of individual words. Explain that the aim of the activity was to listen for main ideas.

3 Aim: to focus on word order in questions; to focus on word stress in questions.

● Explain that the words are in the wrong order and the task is to put them in the right order and make questions.

● Elicit the correct word order and word stress. You may like to explain that usually the words which are stressed are those which the speaker considers to be important. Wait until after the students have identified the stressed words before saying if their answers are correct.

● 📟 Play the cassette.

> **Answers**
> 1 first 2 old 3 much/earn 4 Where 5 married
> 6 What/do 7 brothers/sisters 8 come

4 Aim: to help the students become aware that personal questions may cause offence.
The questions might be linguistically simple, but they are quite complex on a socio-cultural level. You will find that there will be minor differences even among people from the same cultural background and these will serve as a source of discussion. Use this opportunity to help the students become aware that people from different cultural backgrounds may have different reactions to these questions.

● Ask the students to decide which questions are suitable for someone they know well and which are unsuitable for someone they don't know well.

● Explain that, for example, people in Korea and Japan don't use their first names with both types of people outside their family, talking about your family is not really acceptable in Arab countries, and asking how much people earn may be more acceptable in parts of Asia and in certain circumstances in the USA.

5 Aim: to practise the language presented so far in this lesson.
● Ask the students to go round the class using the words and phrases in *Speaking and listening* activity 1.

READING

1 Aim: to encourage the students to read for main ideas; to discourage them from trying to understand every word.

● Explain that the paragraphs give some very general information about customs of hospitality in different countries. The information is offered as a guide, not as a set of rules.

● Ask the students to read and match the headings with the paragraphs. Try not to explain too much vocabulary. If necessary, tell the students that you will explain the meaning of six words only, and they must choose these words carefully.

- When they have read the passage once on their own, ask them to work in pairs and to check their answers. Encourage them to help each other with any difficult vocabulary.

2 Aim: to practise reading for main ideas; to use outside knowledge to understand a passage.

- Ask the students to work in pairs and to decide which country each paragraph describes. In this task they are bringing their own general knowledge to find the answer, as they do in real-life problem-solving tasks.

3 Aim: to compare other people's cultures with the student's own culture.

- Ask the students to compare the situations they have read about with similar situations in their own culture. It may be that they belong to one of the countries mentioned. If so, ask them if they think the information given in the passage is accurate. If they think it isn't accurate, ask them to correct it.

GRAMMAR

1 Aim: to focus on verbs in the present simple.

- Tell the students that the grammar focus of this lesson is the present simple. Ask them to look at the passage again and find some verbs in the present simple. Ask them to make new sentences with these verbs. Suggest that they talk about customs and traditions of hospitality in their own country.

- Ask the students to guess what the missing verb might be. Then suggest that they check to see if the verb is in the passage. Take this opportunity of pointing out that they can often guess the general sense of a difficult word by looking for clues in the context.

- Ask if anyone knows when to use the present simple.

- Ask them to read the explanation about the present simple in the grammar box.

Answers
1 sit   2 arrive   3 drink / leave   4 wear / go

2 Aim: to focus on present simple questions.

- Write *who, what, where* on the board and ask the students to suggest other question words. Ask the students to use the words in questions.

- Ask the students to do the exercise.

3 Aim: to focus on adverbs of frequency.

- Ask the students to look at the adverbs of frequency in the grammar box and to make sentences using each one.

- Ask them to read the explanation about adverbs of frequency in the grammar box.

WRITING AND VOCABULARY

1 Aim: to focus on some key vocabulary from this lesson; to practise writing sentences in the present simple.

- Explain that the vocabulary boxes in *Reward* contain the most important vocabulary in the lesson.

- Ask the students to use the vocabulary to write seven or eight sentences about customs of hospitality in their country. They may like to do this in pairs.

2 Aim: to focus on the position of adverbs in sentences.

- Ask the students **either** to insert adverbs of frequency into their own sentences, **or** to pass their sentences onto another student/pair and insert adverbs into the new sentences, if they can.

3 Aim: to practise joining sentences into a paragraph.

- Ask the students to write a single sentence in answer to the questions they wrote in *Grammar* activity 2.

- Write up two sentences on the board, side by side, and link them with *and* or *but* to make a paragraph.

- Ask the students to join their sentences and make paragraphs. You may like to ask them to do this for homework.

4 'People's private lives are very important so they never ask you personal questions about your family or where you live or your job. They never talk about religion or matters of finance, education or politics, but usually stay with safe subjects like the weather, films, plays, books and restaurants.'

5 'It's difficult to know when to leave, but an evening meal usually lasts about three or four hours. When the host serves coffee, this is sometimes a sign that the evening is nearly over, but you can have as much coffee as you want.'

6 'If the invitation says eight o'clock then we arrive exactly at eight. With friends we know well, we sometimes arrive about fifteen minutes before.'

7 'Obviously it depends on the occasion, but most dinner parties are informal. The men don't usually wear a suit, but they may wear a jacket and tie. Women are usually smart but casual.'

GRAMMAR

> **Present simple (1) for customs and habits**
> **You use the present simple to talk about customs and habits.**
> *In my country men **go** to restaurants on their own. They **take** their shoes off.*
>
> **Negatives**
> *He **doesn't** live here.   You **don't** take wine.   We **don't** ask personal questions.*
>
> **Questions**
> **There are two types of questions:**
> – with question words: **who**, **what**, etc.
> ***What's** your first name?   **How are** you?*
> – without a question word
> ***Are** you married?   **Do** you **have** any brothers and sisters?*
> **You can answer this type of question with *yes* or *no*.**
>
> **Adverbs of frequency**
> **You can use adverbs of frequency to say how often things happen.**
> *They **always** take their shoes off.*   *We **sometimes** arrive early.*
> *We **usually** take chocolates or flowers.*   *We **never** ask personal questions.*
> *We **often** wear jeans and sweaters.*

1 Complete these sentences with verbs from the passage.

1 In my country we ___ at a table for our meals.
2 We usually ___ ten or fifteen minutes after the time on the invitation.
3 People ___ coffee at the end of the dinner party and then they ___ .
4 Men often ___ a suit when they ___ to a restaurant.

2 Write a question for each heading in *Reading* activity 1.
a type of clothes: What do you wear to a dinner party?

3 Look at *Make yourself at home* again. Underline the verbs which are with an adverb of frequency.

WRITING AND VOCABULARY

1 Choose seven or eight verbs from the box and write sentences about hospitality in your country.

> accept   arrive   ask   answer   come from   drink   earn   give   go   know   live
> offer   put   say   send   sing   sit   stay   take   take off   talk about   think   visit
> want   wear

We offer a cup of tea when guests arrive.

2 Now add a suitable adverb of frequency to each sentence.
We usually offer a cup of tea when guests arrive.

3 Look back at the questions you wrote in *Grammar* activity 2, and write answers for your country.

 **2** *A day in the life of the USA*

Present simple (2) for routines: third person singular; expressions of time

READING AND VOCABULARY

1 Read *A day in the life of the USA* and decide who you can see in the photos.

2 Work in pairs. Say what you usually do at the times mentioned in the passage.
At 6.30am I'm still asleep.
I also have lunch at 12.45pm.

3 Are there any differences from life in your country ?
In my country we usually have lunch at 3pm.

4 Match the verbs with suitable nouns in the box.

start	television	come	dinner
leave	·have	stop	lunch	watch
breakfast	home	get	school
finish	work		

start school, start work...

5 Say what time you do these things.

get up	get dressed	go shopping
go to sleep	have a shower/bath	
wake up	wash	wash up

I get up at seven thirty.

A day in the life of the USA

6.30am, **Poughkeepsie, New York.** Norman Davies, 37, gets up and washes. After breakfast, he goes to the station. He works in New York City and the journey takes an hour, so he hurries to catch his train.

7.15am, **Roanoke, Virginia.** A tired Annie Laurence, 10, wakes up and gets ready for school. An hour later she leaves home. She has lunch at school, usually sandwiches and an apple. It's a long day for Annie. She doesn't get home again until 5pm at the end of the afternoon.

10.30am, **Long Beach, California.** Tony de Valera takes a coffee break between meetings. He works for the Disney corporation as an imagineer, a job that is somewhere between an artist, an engineer and a science-fiction writer.

12.45pm, **Evanston, Illinois.** Thirty-four-year-old Amelia Noriega, head of public relations for a major car manufacturer, stops work and goes shopping. Then she has lunch with a friend. 'There aren't many women at my job level,' she says. 'But there are more every day.'

4.30pm, **Tampa, Florida.** George Markopoulos, 65, comes home after his daily swim. Then he joins his wife at the community centre, where she teaches physical education. 'I feel twenty years younger than I am.'

6.15pm, **Seattle, Washington.** Jo-Ann Rosenthal leaves work after a long day as a telephonist at a downtown bank. It's Friday night so she walks to her local bar and meets her friends.

7.45pm, **Lubbock, Texas.** Cliff Renton III, 61, meets Walter Avery, 62, to have dinner and to talk about the local Ranch Handlers' Ball, the most important event in the Lubbock social calendar. Cliff is president of the Social Committee, so he's responsible for the success of the evening.

11pm, **Athens, Georgia.** Shirlee Lewis finishes dinner, washes up and watches the TV news. Her five children are asleep, so she tries to be very quiet.

2

GENERAL COMMENTS

Many students start reading or listening to a passage and then stop because they don't understand a particular word. Many reading and listening activity sequences in *Reward* begin with a task to encourage the students to extract the main ideas. This is designed not only to help them gain access to the heart of the passage but also to help them overlook any difficult words. If you explain every word, you may help extend the students' vocabulary, but you will not help them to become more effective readers. If you do need to explain words, limit them to five or six and ask the class to choose them carefully. Explain also that the vocabulary in the boxes is the vocabulary which is considered suitable for this level. Other words are less important. Finally, point out that it is difficult to learn more than ten or twelve words a lesson, and simply noting down more will not make the learning process more efficient.

Top photo: possible questions
What is a school bus? Do special buses take children to school in your country? Did you go to school this way?

Bottom photo: possible questions
Where do these two men come from?
Do they look like typical Americans?
What other clothes do Americans typically wear?
How else can you tell if someone is American?
Do you know any typical Americans?

READING AND VOCABULARY

1 Aim: to practise reading for main ideas.
● Explain that the passage the students are going to read is about a typical day in the life of some Americans. Before they begin, write these words on the board: *breakfast, station, train, school, sandwich, apple, home, meeting, artist, engineer, writer, lunch, go, shopping, woman, swim, telephonist, bank, bar, president, TV, children*

● Pre-teach these words which come from the passage by asking the students to say which ones they can use to describe what they can see in the photos.

● Ask the students to read the passage and decide who they can see in the photos.

> **Answers**
> Top: Annie Laurence, bottom: Cliff Renton

2 Aim: to practise saying times of the day.
● Write the actual time on the board and ask your students how many ways they can say it in their language. Explain that there are two ways of saying the time in English, and that it isn't common to use the twenty-four hour clock in English except for timetables. You can also mention that the afternoon begins after midday, the evening after six pm and the night at about ten or eleven pm.

● Ask them to write the times in the passage in two different ways.

● Explain that you can use the present continuous to describe actions which are going on at or around a particular time, and that the present continuous will be covered in Lesson 5.

● Ask the students to work in pairs and ask and say what they do at the different times mentioned.

3 Aim: to compare a typical day in the life of the USA with a typical day in the students' countries.
● Ask the students to read the passage again and find out similarities and differences.

● Ask the students to report back to rest of the class, giving some examples.

4 Aim: to present the verbs and nouns in the boxes and to focus on collocations.
● Explain that certain verbs are often followed by certain nouns. There may be many possible combinations but here are the most common routine activities.

> **Possible answers**
> **start:** breakfast, dinner, lunch, school, work
> **finish:** breakfast, dinner, lunch, school, work
> **come:** home
> **get:** home, breakfast, dinner, lunch
> **leave:** home, school, work
> **have:** breakfast, dinner, lunch
> **stop:** work
> **watch:** television

5 Aim: to present the vocabulary items in the box; to practise saying the time.
● Present the items of vocabulary in the box by saying when you do the routine activities. Write *When do you ...?* on the board. Ask questions and elicit answers from two or three students.

● Ask the students to work in pairs and to ask and answer questions about the routine activities.

GRAMMAR

1 Aim: to focus on the form of the third person singular in the present simple.
● Explain that the focus of this lesson is on the third person singular (*he/she/it* form) of the present simple. Ask the students to think of an activity they do as a routine every week, month or year.

● Ask the students to write down the third person singular of the verbs. Then ask them to check their answers by re-reading the passage.

● Ask the students to read the explanation about the present simple: third person singular in the grammar box.

2 Aim: to focus on the three endings of the third person singular in the present simple.
- Explain that there are three endings for the third person singular of the present simple. Draw three columns on the board and ask the students to suggest where each verb should go.

Answers
-s: gets, comes, joins, leaves, meets, says, stops, dresses, takes, wakes, walks, works, lives, makes
-es: goes, finishes, teaches, washes, watches, does, dresses
-ies: hurries, tries, flies, carries

3 Aim: to present the expressions of time in the grammar box.
- Explain that the expressions of time can be used when you want to be less specific about the time.

- Ask the students to work in pairs and ask and say what time of day they do the things in *Reading and vocabulary* activities 4 and 5.

LISTENING

1 Aim: to practise listening for main ideas.
- Explain to the students that they are going to listen to two people from *A day in the life of the USA* talking some more about their typical days. Remind them that it isn't necessary to understand every word. Ask them to listen and decide who is speaking. Play the cassette.

Answers
First person: Jo-Ann Rosenthal
Second person: George Markopoulos

2 Aim: to practise listening for specific information.
- Ask the students to listen again and find out what they are doing at the specific times mentioned. Play the tape.

Answers
Jo-Ann: 8am: leaves home and goes to work; 1pm: goes for a walk in the park; 6.30pm: takes the subway home; 11.30pm: goes to bed.
George: 8.15am: has breakfast; 12.30pm: has lunch; 5.30pm: meets friends and has a drink at the golf club; 7pm: goes home for dinner.

3 Aim: to practise saying the time and using verbs in the third person singular.
- Ask the students to check their answers to activity 2. Make sure that everyone is using and pronouncing the endings correctly.

SOUNDS

1 Aim: to show ways of pronouncing the letter *s* as an ending for the third person singular present simple; to practise pronouncing /s/, /z/ and /ɪz/.
- Write /s/, /z/ and /ɪz/ on the board and say each sound. Point to each sound in turn and ask the class to say the sound. You may like to draw the students' attention to the Pronunciation guide at the back of the book.

- Write 1, 2 and 3 by the three sounds. Ask the students to listen and write 1, 2 or 3 according to the sound they hear.

- Play the tape.

- Ask the students to write the words in the correct columns.

2 Aim: to check answers to activity 1; to practise pronouncing /s/, /z/ and /ɪz/.
- Play the tape.

Answers
/s/: takes sits asks talks
/z/: goes sings arrives offers has serves does
/ɪz/: finishes refuses washes watches

- Ask the students to say the words aloud as a group, and then in pairs. If necessary, play the tape and stop after every word so they can repeat it.

SPEAKING AND WRITING

1 Aim: to practise asking and answering questions about routine activities.
- Ask the students to find out when their partner does the five routine activities mentioned. Check that everyone is using the auxiliary *do*.

- Ask the students to make notes about their partner's answers.

2 Aim: to provide controlled practice of writing sentences from notes.
- Write the notes given as examples in the Student's Book *Speaking and writing* activity 1 on the board. Then add words to make full sentences.

- Ask the students to use their notes to write sentences.

3 Aim: to provide controlled practice of writing paragraphs from sentences.
- Add *and* and *then* to the sentences you wrote in activity 2. Point out that you join two sentences with *and*, and you start a sentence describing a subsequent activity with *Then*.

- Ask the students to link their sentences with *and* and *then*.

GRAMMAR

> **Present simple (2) for routines: third person singular**
> **You use the present simple to talk about routines.**
> *He **gets up** at 6.30.   She **works** in Seattle.*
> **You form the third person singular (*he/she/it*) of most verbs in the present simple by adding -*s*.**
> *He get**s** up at 6.30.   She work**s** in Seattle.*
> **You add -*es* to *do*, *go* and verbs which end in -*ch*, -*ss*, -*sh* and -*x*.**
> *He wash**es**.   She go**es** to school.*
> **Verbs which end in a consonant + -*y* change to -*ies*.**
> *carr**ies**   fl**ies***
> **The third person singular of *be* is *is*. You often use the contracted form '*s*.**
> *It**'s** a long day for Annie.*
> **The third person singular of *have* is *has*.**
> *She **has** lunch at school.*
>
> **Expressions of time**
> ***in** the morning   **in** the afternoon   **in** the evening*
> ***at** night*
> ***before** lunch   **after** dinner   **at about** seven o'clock*

1 Write down the third person singular of these verbs. (You can find them in the passage.)

come join finish get go hurry leave meet say stop take teach try wake walk wash watch work

2 Put the verbs in three columns.

-*s*	-*es*	-*ies*
gets	*goes*	*hurries*

Now add these verbs to the correct column.

do dress fly live make carry

3 Say what time of day you do the things in *Reading and vocabulary* activity 5. Use the expressions of time in the box above.

LISTENING

1 🔲 Listen and decide which people in *A day in the life of the USA* are speaking.

2 🔲 Listen again and find out what they do at these times:

Speaker 1: 8.00am   1.00pm   6.30pm   11.30pm
Speaker 2: 8.15am   12.30pm   5.30pm   7.00pm

3 Work in pairs and check your answers.
At 8.00am ___ leaves home and goes to work.

SOUNDS

1 There are three different ways of pronouncing the final -*s* in the third person singular present simple.

🔲 Listen to these verbs. Is the final sound /s/ , /z/ or /ɪz/? Put them in three columns.

takes goes finishes sits sings arrives refuses offers has asks talks serves washes watches does

2 Now say the words aloud.

🔲 Listen and check.

SPEAKING AND WRITING

1 Work in pairs and find out about your morning routines. What time does your partner do these things?

– wake up   – get up   – get dressed
– have breakfast   – go to work

Cecile, what time do you wake up?
I wake up at seven o'clock.

What time do you get up?
At a quarter-past seven.

Make notes about your partner's routine.
Cecile – wake up: 7am, get up: 7.15am.

2 Write sentences about your partner's morning routine.
Cecile wakes up at seven o'clock.
She gets up at seven fifteen.

3 Now write a paragraph about your partner's morning routine. Link the sentences you wrote in 2 with *and* and *then*.
Cecile wakes up at seven o'clock and gets up at seven fifteen. Then she…

Home rules

What does the word 'home' mean to you? How do you say the word in French? In Spanish? In your language? Although people usually know what the word means, it often has no exact translation. It's not surprising really, because the idea of home varies from country to country, and from person to person. A home is more than a roof and four walls. It's the cooking, eating, talking, playing and family living that go on inside which are important as well. And at home you usually feel safe and relaxed.

But it's not just that homes look different in different countries, they also contain different things and reveal different attitudes and needs. For example, in cold northern Europe, there's a fire in the living room or kitchen and all the chairs face it. In the south, where the sun shines a lot and it's more important to keep the heat out, there are small windows, cool stone floors and often no carpets. We asked some people about their homes.

What's the main room in your home?
'The kitchen, because its warm and we have breakfast, lunch and dinner there seven days a week.' **Jackie, Cork , Ireland**

Do you have a television? If so, where?
'In the bedroom. We like to watch it in bed.'
Maurice, Bruges, Belgium

Do you lock your door when you go out?
'In cities we do. Although when I was a child in the Tatra mountains, we left the door open with bread and dishes of food and something to drink, such as a glass of milk, on a table inside, so that visitors and travellers could stop and refresh themselves.' **Grazyna, Katowice, Poland**

How often do people move home in your country?
'In the USA many people move every ten years or more.'
Cheryl, Boston, USA

If you live in a town, do you stay there at weekends?
'Well, we live in the town, but only because I'm an architect and I work there. I really wouldn't call it home – that's what I call our house in the country where we go every weekend.'
Elizabeth, Sao Paulo, Brazil

What are typical features of homes in your country?
'In Britain, even in the town there's always a garden and sometimes a cellar. We have separate bedrooms and living rooms. But we don't often have balconies or terraces. The weather isn't warm enough!' **Pat, Exeter, England**

So *home* means different things to different people. What does it mean to you?

3

GENERAL COMMENTS

Articles
The explanation in the grammar box about the use of articles is brief for reasons of space. Your students will probably have already been exposed to how articles are used in English, and the explanation should be seen as consolidation. You may need to remind them that *an* is used before words beginning with a vowel. However, they may need more practice, especially if the English article is used in different ways to their own language.

Plurals
The general rule for forming plurals in English by adding *-s* is fairly straightforward, and the exceptions are few. It may sometimes be difficult to decide if a word is plural or singular. For example, *hair, police, news* and *money* are uncountable nouns (see Lesson 15) and therefore take singular verb forms. This aspect of the word needs to be taught at the time the item is first presented.

Optional extra material
It would be helpful for the early stages of *Vocabulary* activity 1 if you used magazine photos of items of furniture, household equipment, or rooms of the house.

Photo: possible questions
The topic of the lesson is houses, rooms and furniture. Student A, open your book at Lesson 3 and look at the photo. Student B, keep your book closed. Try to guess what room is in the photo, and what furniture there is. Student A, you can only answer yes or no.

VOCABULARY

1 Aim: to present the vocabulary items in the box and to prepare the students for the reading passage.
● Tell the students that the subject of the lesson is homes. Revise present simple questions by asking them
Where do you live?
Do you live in a flat or a house?
Do you live in the town or the country?

● Point to some items of furniture and equipment in the classroom and ask the students: *What's this?* Find out how many words to do with houses they know. Use the photo to present the vocabulary items.

● Write *types of housing* and *rooms* on the board. Ask the students to look at the vocabulary box and find two types of housing and six rooms.

Answers
types of housing: flat, house
rooms: bathroom, bedroom, dining room, kitchen, living room, toilet

You may like to explain that *apartment* is the American word for *flat*, and that *bathroom* is the American word for *toilet*.

2 Aim: to continue the presentation of the vocabulary items in the box and the process of categorisation.
● Ask the students to put the words for furniture and equipment with the rooms. You may be able to do this very effectively with extra magazine photos. Some items may go in more than one room.

Possible answers
bathroom: basin, bath, shower
bedroom: bed, carpet, chair, cupboard, curtains, lamp, table
dining room: carpet, chair, curtains, lamp, table
living room: carpet, chair, cupboard, curtains, lamp, sofa, table, video
kitchen: cooker, cupboard, dishwasher, fridge, sink, table, washing machine
toilet: basin. Explain that you can use *toilet* for both the room and the equipment in it.

● Write *garden* on the board and explain that it does not fit in the categories. *Door* and *window* fit all the rooms. Check that everyone knows what the words mean by using photos, drawings or pointing.

3 Aim: to practise using the vocabulary in the box with *Is there ...?*
● Write *Yes, there is* and *No, there isn't* on the board. Ask one or two students a few questions: *Is there a living room in your home? Is there a dining room?* Point to the replies on the board and ask them to reply. Make sure they reply using the full sentence.

● Ask the students to work in pairs and to continue asking and answering questions about their homes.

● Ask one or two students about what furniture and equipment there is in the rooms of their home. The answers may be different from the answers in activity 2. Make sure that if they use a plural noun in reply, they also say *Yes, there are two/three chairs.*

● Ask the students to work in pairs and note down what furniture and equipment there is in the different rooms. Encourage them to expand their lists with vocabulary of their own choice at this stage.

READING

1 Aim: to practise reading for specific information; to react to the text.
● It is sometimes difficult for students to distinguish between what they think, and what a passage actually says. This task directs them to look for the answer to a specific question and asks them if they agree. Ask them to read the first two paragraphs and then discuss what *home* means to the writer. Encourage them to distinguish between the general points and the specific examples the writer uses to illustrate these points.

● You may find that students have different attitudes to their homes. In *Speaking* activity 3, they will have the opportunity of brainstorming the words they associate with home.

2 Aim: to continue reading for specific information; to react to the text.

● Ask them to read the rest of *Home rules* and think about their answers to the questions. Suggest that they make notes at this stage, which they will use in *Speaking* activity 2.

GRAMMAR

1 Aim: to focus on the use of the article in English.

● Ask the students what the articles in English are, and if they have articles in their own language.

● Ask the students to read the explanation about articles in the grammar box. While they are doing this, write the nine uses of the article from the grammar box on the board and number them.

● Ask the students to look at the nine underlined words or expressions and decide which use of the article each example shows. They should put a number by it according to the use of the article. Make it clear that there is an example of every use of the article, definite, indefinite and zero, in the passage.

● The explanation in the grammar box covers the rules which the students need to learn at their present level.

2 Aim: to focus on the use of the article.

● Ask the students to do this activity and to think about which uses each sentence shows.

3 Aim: to focus on the formation of plurals.

● Ask the students if they know how to form plurals in English. Write a few simple words on the board and ask the students to tell you the plural form.

● Ask the students to read the explanation about plurals in the grammar box.

SPEAKING

1 Aim: to give practice in using the vocabulary and grammar presented so far in this lesson.

● Ask the students what room they think the photo shows, and if it looks like a room in their country. If it doesn't, can they explain why?

2 Aim: to check the students have understood the passage and have reacted to it.

● Ask one or two students to talk about their answers to the questions in the passage. In a mono-cultural group, it is likely everyone will have similar answers. Ask them if they find anything surprising or strange about the answers of the people quoted in the passage. In a multi-cultural group, encourage the students to discuss any similarities and differences.

3 Aim: to give further speaking practice and to examine attitudes towards homes.

● Write words which you associate with *home* on the board. Here are some ideas: *tea, dinner, family, sleep*.

● Ask the students to write down words which they associate with *home*. Explain that they should not write the first words which come to mind but should choose their words carefully. Give them two or three minutes for this stage of the activity.

● Ask the students to share their words and phrases with the rest of the class. Once again, are there any words or phrases which people find surprising or strange?

VOCABULARY

1 Look at the words in the box and find *two types of housing* and *six rooms*.

> basin bath bathroom bed
> bedroom carpet chair cooker
> cupboard curtains dining room
> dishwasher door flat fridge
> garden house kitchen lamp
> living room shower sink sofa
> table toilet video
> washing machine window

2 Work in pairs. In which rooms do you find the furniture and equipment in the box above? Which word is left?

3 Work in pairs. Ask and say what rooms there are in your homes.
Is there a living room?
Yes, there is.
Is there a dining room?
No, there isn't.

Now ask and say what furniture and household equipment there is in your home and where it is. Add words to the lists if you can.
There's a table and some chairs in the kitchen.
Is there a video?   No, there isn't.

READING

1 Read the first two paragraphs of *Home rules*. What does *home* mean to the writer? Do you agree?

2 Read the rest of the passage and think about answers to the questions for your country.

GRAMMAR

> ### Articles
> **You use the indefinite article *a/an*:**
> – **to talk about something for the first time:** *There's **a** kitchen and **a** dining room.*
> – **with jobs:** *I'm **a** teacher.   She's **an** engineer.*
> – **with certain expressions of quantity:** ***a** little food, **a** few beds, **a** couple of friends*
>
> **You use the definite article *the*:**
> – **to talk about something again:** *In **the** kitchen there's a table, and on **the** table there's a cat.*
> – **with certain places and place names:** ***The** Alps, **The** West, **The** USA*
> – **when there is only one:** ***the** president, **the** government, **the** weather*
>
> **You don't use an article:**
> – **with plural and uncountable nouns when you talk about things in general:** *It's got carpets and curtains.   There's lots of food.*
> – **with certain expressions:** *at home, at work, in bed, by car*
> – **with meals, languages, most countries and most towns:** *Let's have lunch.   Speak English.   We live in France.   I lived in Paris.*
>
> ### Plurals
> **You form the plurals of most nouns with *-s*:** *chair – chair**s**   cupboard – cupboard**s***
> **You add *-ies* to nouns of two or more syllables which end in *-y*:** *balcony – balcon**ies***
> **You add *-es* to nouns which end in *-ch*, *-ss*, *-sh*, and *-x*:** *church – church**es***
> **There are some irregular plurals:** *man – men   woman – women   child – children*

1 Look at the nine articles and phrases underlined in the passage. Which of the rules about articles in the grammar box do they illustrate?

2 Complete the sentences with *a/an, the* or put – if there's no article.
1 Last year we moved to ___ London.
2 ___ kitchen is ___ door on your left.
3 ___ weather is very hot in August.
4 There isn't ___ table in ___ kitchen.
5 Would you like ___ drink?
6 I'm sorry, he's still at ___ work.

3 Write the plural of these nouns.
parent house city family dish party bush country table fax feature

SPEAKING

1 Work in pairs and look at the photo. What room do you think it is? Does it look like a room in a house in your country?

2 Work in pairs and talk about your answers to the questions in *Home rules*.

3 What does the word 'home' mean to you? Write five words or phrases which you associate with the idea. Find out what other students in your class wrote.

Verb patterns (1): *-ing* form verbs; talking about likes and dislikes

First impressions

The British and the Americans speak the same language. But life in the two nations can be very different. We asked some Americans what they like or don't like about Britain...

'The police. They're very friendly and they don't carry guns.' Claude, Trenton

'The weather is awful. You don't seem to get any summer here. It's winter all year round.' Toni, San Francisco

'The tourists! The streets are so crowded. I think you should do something about them. And I can't stand the litter everywhere. It's a very dirty place.' José, Washington

'Walking and sitting on the grass in the parks, especially on a hot summer's day. Oh, and the green countryside. But why is the beer warm?' Max, Houston

'Well, they certainly seem rather unfriendly. Nobody ever talks on the buses. But maybe we haven't met any real English people yet.' Eva, Niagara Falls

'Feeling safe when you walk the streets. Oh, and the polite drivers who stop at a street crossing if they see someone waiting there.' Moon, Los Angeles

'Driving on the left. It's very confusing. I keep looking the wrong way.' Paula, San Diego

So then we asked some British people what they like or don't like about America...

'Arriving at the airport. Immigration is so slow, it takes hours to get through!' Geoff, London

'The waste of electricity. I just can't understand why their homes are extremely hot in winter and very cold in summer.' Louise, Southampton

'The people, they're so generous. If they invite you home, you're sure of a big welcome.' Amin, Bath

'Going shopping. I love it. It's so cheap everywhere – food, clothes, hotels, petrol.' Paul, Oxford

'I hate the insects. They're so big. In Texas the mosquitoes are enormous. But I suppose in Texas they would be!' Maria, Glasgow

'Lying on the beach in the sunshine. In California the sun shines all day, every day. It's great.' Rose, Cardiff

'Driving on the right. It's very confusing. I keep looking the wrong way.' Paula, St Albans

4

GENERAL COMMENTS

Verb patterns (1): -ing form verbs

In *Reward,* common constructions with verbs are referred to as *verb patterns.* In pre-intermediate *Reward,* there are two verb patterns, in this lesson and in Lesson 11. When there are two verbs in a sequence, the second verb in the sequence is either an infinitive or an *-ing* form. This lesson focuses on the *-ing* form. Some students may already know that with *love, like* and *hate,* the three main verbs presented in this lesson, you can put either *-ing* or an infinitive. If the question arises, explain that there is not much difference between the two patterns with these verbs, although if you are talking about a particular occasion, you usually use an infinitive. The most common expression is with *would like to: Would you like to go to the cinema this evening?*

Talking about likes and dislikes

A very common *-ing* form verb pattern is talking about likes and dislikes. The replies presented in this lesson are graded from strong to weak likes or dislikes, so the student has a suitable range of language to reply honestly to the questions. The lesson also focuses on short answers when two people have the same or different likes and dislikes.

Cross-cultural awareness

The passage is a collection of impressions about Britain and America. Obviously, they can be used at face value to stimulate a discussion about the two countries and how they appear to people from other cultures. But they should also be used as an opportunity to compare Britain and America with the students' own cultures and countries, and to raise their awareness of how people see their own and other countries. The key to cross-cultural awareness is to develop a sensitivity towards one's own culture.

Photos: Hyde Park in London; a truck outside a motel in the USA.

Possible questions

The lesson is about impressions of Britain and America. Write down words that you associate with Britain and America. Look at the photos and decide which shows a scene in Britain and which shows America. How can you tell? Imagine and describe two or three more photos showing typical scenes from Britain and America.

READING AND VOCABULARY

1 Aim: to practise reading and inferring; to encourage the students to react to the passage.

● When we read, the meaning is not always transparently clear, and this activity encourages the students to read between the lines. The impressions that the people have about Britain and America are in the form of noun objects or *-ing* form verbs. *Claude likes the police, Moon likes feeling safe in the streets.* Ask the students what impressions they have of Britain and America. It doesn't matter if they have never been there.

● Write a few impressions on the board. If someone has given an impression without ever having been to the country, ask them where they think the impression comes from. Mention that our impressions of other countries are not simply formed by direct experience but by different sources, such as films, books and television. Our impressions of another country may be fashioned by one or two influential sources.

● Ask the students to read the passage and decide if the impressions are positive or negative. Ask them also to compare these impressions with their own.

> **Answers**
> **Britain**
>   **positive:** the police, walking and sitting on the grass, feeling safe, polite drivers
>   **negative:** the weather, the tourists, English people, driving on the left
> **America**
>   **positive:** the people, going shopping, lying on the beach
>   **negative:** arriving at the airport, the waste of electricity, the insects, driving on the right

● Although it is helpful to draw attention to similarities between cultures, it is less remarkable than focusing on the differences. Anything which is surprising, strange, amusing, interesting or even shocking about a culture is generally a difference. For example, like Paula, some students may find it strange that the British drive on the left-hand side of the road. Encourage the students to respond with their own reactions to the passage.

2 Aim: to present the words in the vocabulary box and to show that the meaning of words may change according to the context.

● It may be useful to help the students distinguish between parts of speech at this stage of the course, as this skill becomes increasingly useful as the course continues. If necessary, they can use a dictionary.

● Start by asking the students to suggest which are the adjectives, and write them on the board.

● Ask the students if the adjectives are positive or negative according to the speakers and ask them to write the words in two columns.

> **Answers**
>
> | positive | negative |
> |----------|----------|
> | cheap | awful |
> | friendly | cold |
> | generous | confusing |
> | great | crowded |
> | polite | dirty |
> | | hot |
> | | slow |
> | | unfriendly |

- Point out that the meaning of the adjectives can change according to the context. For example, *They sell cheap clothes for children* is positive, but not *She wears cheap clothes.*

- Ask the students to remember which nouns the adjectives go with in the passage. Developing an awareness of collocation, of which words can go together, is important for two reasons. It helps the process of acquiring the vocabulary items, and it helps develop the students' vocabulary in general. Do this orally, without letting them look back at the passage.

- Ask the students to look at the passage to check which nouns the adjectives went with.

Answers
awful weather, cheap shopping, cold homes, confusing driving, crowded streets, dirty place, friendly police, generous people, great sunshine (colloquial use), hot homes, polite drivers, slow immigration, unfriendly people

3 Aim: to continue the presentation of the words in the vocabulary box.
- You may like to draw attention to the words which end in *-ing* and the nouns. The *-ing* form verbs are in fact gerunds, which are verbs that act like nouns.

- Ask the students to tick the things they like and put a cross by the things they dislike. In theory, they cannot answer the questions yet, as they have not read the grammar and functions box. However, it may be that the students are in fact revising this structure, so you may like to ask questions like *Do you like the beach? Do you like the countryside?* and elicit a suitable short answer *Yes, I do* or *No I don't.*

GRAMMAR AND FUNCTIONS

1 Aim: to practise saying if you have the same likes and dislikes as other people.
- Ask the students to read the information about verb patterns and talking about likes and dislikes in the grammar and functions box.

- Ask the students to respond truthfully to these statements. You may like to do this activity orally.

2 Aim: to check the students' answers to *Reading and vocabulary* activity 1 and to practise saying the full sentences with *like* and *dislike*.
- It may be easier to do this activity orally. Remind the students that the ✔ sign they wrote by each impression means they must use *likes* and the ✗ sign means they must use *dislikes*. Make sure the students pronounce the third person singular ending correctly. (See Lesson 2 for more information.) Encourage them also to use full sentences with *because*.

SOUNDS

Aim: to focus on the use of intonation to show strong likes and dislikes.
- Explain that intonation can be used to show strong likes or dislikes. All the sentences have a strong intonation. Usually, the key word is heavily stressed, and this has an effect on the intonation pattern which follows. Ask the students if they think their own language has similar intonation patterns for strong likes and dislikes.

- ▭ Play the tape.

- Play the tape again and stop after each sentence. Ask the students to say the sentences aloud. Make sure they use strong intonation.

- Ask the students for other likes and dislikes using strong intonation.

SPEAKING

1 Aim: to encourage the students to think about what they like and dislike about their town or country or the town where they are now.
- Ask the students to work alone and write down things they like and dislike about their town or country or the town where they are now. This is a good opportunity to prepare for the discussion of cross-cultural comparisons mentioned in the *general comments* at the beginning of these notes. You may need to help them at this stage with vocabulary.

2 Aim: to practise talking about likes and dislikes.
- Think of something you like, such as going shopping and go round asking the students *Do you like going shopping?* Ask the students to do the same. When they have found someone who likes the same things, ask them to find things they both dislike. If your classroom doesn't allow much room for students to move around, they can do this activity with their close neighbours.

3 Aim: to continue talking about likes and dislikes.
- Ask the students about their likes and dislikes, and write them on the board. Some people may like things which other people dislike. Use this as an opportunity to point out that even within our own culture there are different likes and dislikes, and that this should help the students be more aware of differences in other cultures.

- Try to get the class to rank the likes and dislikes in two *Top five* lists. This kind of ranking activity is very effective for stimulating discussion.

- You may like to ask the students to find out about the likes and dislikes of their family and friends and to write a few sentences in English using the passage in this lesson as a model.

READING AND VOCABULARY

1 Read *First impressions,* which is about people's impressions of Britain and America. Put a tick (✓) by the positive impressions and a cross (✗) by the negative impressions. Which is the most surprising impression?

2 Underline the adjectives in the box.

> awful beach cheap cold confusing countryside crowded dirty driving
> food friendly generous grass great gun hot insect park police polite
> shopping slow summer sunshine tourist unfriendly walking weather winter

Now write them in two columns, *positive* and *negative*, according to their meaning in the passage.

Can you remember which nouns they went with in the passage?

3 Look at the other words in the box. Put a tick (✓) by anything you particularly like and a cross (✗) by anything you dislike.

Think of two or three things that people like and dislike about your country.

GRAMMAR AND FUNCTIONS

> ### Verb patterns (1): *-ing* form verbs
> **You can put an *-ing* form verb after certain verbs.**
> *I **love** walking.   She **likes** swimming.   They **hate** lying on a beach.*
>
> ### Talking about likes and dislikes
> **Questions** **Short answers**
> ***Do* you *like* the weather?** *Yes, I **do**.*
> ***Does* he *like* the police?** *Yes, he **does**.*
> ***Do* you *like* walking in the park?** ***No, I** don't.*
> ***Does* she *like* the weather?** ***No, she** doesn't.*
>
> **Negatives**
> *I **don't** like the weather.   He **doesn't** like arriving at the airport.*
> *We **don't** like insects.   She **doesn't** like driving on the right.*
>
> **Expressing likes** **Expressing dislikes**
> *I love it.* *I hate it.   I can't stand it.*
> *A lot.* *Not at all.   I don't like it at all.*
> *A little.* *Not very much.*
>
> **Expressing neutrality**
> *It's all right.   I don't mind.*
>
> **Expressing the same likes and dislikes**
> *I like rock music.   So do I.*
> *I don't like jazz.   Neither/Nor do I.*
>
> **Expressing different likes and dislikes**
> *I like rock music.   I don't.*
> *I don't like jazz.   I do.*

1 Do you like or dislike these things? Respond to the statements.

1 I like going to parties.
2 I don't like cooking.
3 They like walking.
4 I don't like insects.
5 He doesn't like tennis.
6 She likes shopping.

1 I don't.

2 Look back at the passage and say what the people like or dislike and why.
Claude likes the British police because they're friendly and don't carry guns.

SOUNDS

Listen to the way these people use a strong intonation to express strong likes or dislikes.

1 I hate the insects.
2 I can't stand the litter.
3 He loves shopping.
4 I like walking very much.
5 She doesn't like the weather at all.
6 He hates the warm beer.

Now say the sentences aloud.

SPEAKING

1 Write down four or five things you like about your town or country, or the town where you are now.

2 Find people in the class who like the same things. Now talk about things you don't like.

3 Tell the rest of the class about your likes and dislikes. Make two class lists: the *Top five likes* and the *Top five dislikes* about your town or country, or the town where you are now.

5 | *Take a closer look*

Present simple and present continuous

SPEAKING

1 Work in pairs. Take a closer look at your partner! How much do you know about each other? Guess the answers to these questions.

Does he or she...

– play a musical instrument?
– paint or draw?
– travel by bus often?
– smoke?
– work hard?
– speak any foreign languages?
– listen to music?
– earn a lot of money?

I think she plays the piano.

2 Now ask and answer the questions about each other. Did you guess correctly in 1?
Do you play a musical instrument?   No, I don't.

3 Look at the *yes* answers. Is your partner doing these things at the moment?
Is he working hard?   Yes, he is.
Is she smoking?   No, she isn't.

4 Look at the photos. Choose one person in each photo and take a closer look. Imagine what their life is like and guess the answers to the questions in 1.
I think he plays a musical instrument.

5

GENERAL COMMENTS

Present continuous

This tense is also called the *present progressive* because it can be used to describe activities which are in progress at the time of speaking as well as repeated or continuous actions. It can also be used to talk about a future situation: *What are you doing this weekend? I'm staying at home.* There are two possible points of confusion. Firstly, many languages do not have a progressive aspect to their tense system and speakers of these languages may often have some trouble distinguishing it from the present simple. Secondly, the present participle has the same form as the gerund, although it has a very different function. It may be useful to point out that, for example, the word *smoking* in *I'm smoking a cigarette* does not have the same function as the same word in *smoking is bad for you.* Your students may also have problems with word order in questions, for example *Why is standing the woman there?* Most students will already have come across the difference between the present simple and continuous tenses, so this lesson should be to a large extent revision.

Photos

The photos in this lesson were taken by the French photographer Robert Doisneau, whose most famous photograph is *The Kiss.* It is common to use the present continuous to describe what is happening in a photograph, picture or scene, maybe to someone who cannot see it.

Photos: possible questions

Write on the board *1950, 1955, 1960, 1965, 1970, 1975, 1980, 1985, 1990.*
In what year were the photos taken?
In what town or country are the people in the photos?

SPEAKING

1 Aim: to revise the present simple and to create an opportunity to compare it with the present continuous later in the lesson.

● This activity will work best if your students know each other a little, but not well. You may want to separate those who know each other well, and ask them to work with people they know less well. If this is not possible, the questions are sufficiently varied to generate a meaningful question and answer exchange in activity 2. (If a student knows that his/her partner plays a musical instrument, it isn't meaningful to ask the question!)

● Say and write on the board what you think one or two students in the class do: *I think Xavier plays the piano. I think Maria works hard.* Ask the students *What about me?* and get them to think about you the teacher, and how you might answer the questions. Ask the students to write their guesses on the board.

● Ask them to work in pairs with people they don't know well and to guess the answers to the questions.

2 Aim: to practise asking and answering questions using the present simple.

● Ask one or two students the questions: *Xavier, do you play the piano? Maria, do you work hard?* and get them to say *yes* or *no.* Then get the students to ask their questions about you and give suitable answers.

● Ask the students to continue the activity, asking and answering the questions in activity 1 about themselves. Remind the students that they are using the present simple to ask and talk about habits, personal characteristics and general truths.

● Find out who guessed correctly in activity 1.

3 Aim: to present the main difference between the present simple and the present continuous.

● Write *at the moment?* on the board and ask the questions, using the present continuous but pointing to and stressing *at the moment.*

4 Aim: to focus on the present simple as preparation for its comparison with the present continuous.

● Ask the students to think about the people in the photos. Ask these questions.
Where are they? What town are they in? When was the photo taken?

● Ask the students to answer the questions in *Speaking* activity 1 about the people in the photos.

GRAMMAR

1 Aim: to present or revise present continuous questions and affirmative statements.

● Ask the students to read about the present simple and present continuous in the grammar box.

● Ask the students to work in pairs. Ask *What's the man in front doing?* and find someone who gives a suitable answer using the present continuous. Ask several more questions like this and obtain suitable answers, using vocabulary from the list in the Student's Book. Write *What is doing the man in front?* on the board, cross it out and rewrite it correctly.

● Ask the students questions with *why. Why is he playing the accordion? Why is he drawing?* and elicit suitable answers. Go round the class asking students to form questions with the correct word order.

● Ask the students to work in pairs and to ask and say what the people in the photos are doing, and why.

2 Aim: to focus on the formation of present participles.

● Orally, this activity does not pose great problems. It is only when you start to write the present participles that the issue of correct spelling arises. This activity shows the students the present participle, asks them to write down the information they know already, which is the infinitive form, and then reflect on the possible rules for forming the present continuous endings.

● It may be a suitable moment to explain that this is a very general rule, and that more complete information can be found in the Grammar review.

3 Aim: to focus on the difference between the present simple and present continuous in sentences using *and* and *but*.
● Ask the students to complete the sentences.

SOUNDS

Aim: to practice the sounds /n/ and /ŋ/.
● Read each phrase aloud so the students can hear the difference.

● ▭ Play the tape and repeat each phrase.

● Play the tape again, stopping after each phrase. Ask the students to say the phrases aloud.

LISTENING AND VOCABULARY

1 Aim: to prepare for the listening activity in 2.
● Ask the students to match the things with the places. Remind them to use the present simple tense for general truths, for example *You buy train tickets at the railway station.*

2 Aim: to expose the students to some natural language in transactional situations, and to encourage them to listen for main ideas; to give further practice in using the present continuous.

● ▭ The level of language in this activity is quite high, but the task is very simple and is designed to distract the students from trying to understand every single word. Explain that they need to be exposed to real-life English in the classroom context in order to prepare themselves for listening in a real-life context. They won't understand every word, but then nor would they in real life either. Explain that the task is to listen and decide where the people are.

● Check the students' answers to the activity by asking them where people are and what they are doing.

● If there were difficulties over choosing the correct preposition in the answers to this activity, explain that more work will be done on this in Lesson 16.

3 Aim: to present the words in the vocabulary box and to discuss their parts of speech.
● Ask the students to say if the words in the box are nouns or verbs, or both. You may like to take this opportunity of showing them the dictionary entries for some of these words. This may be the first time they have used dictionaries, so take plenty of time for this activity.

4 Aim: to continue the presentation of the vocabulary items in the box and to develop an awareness of words which go together.
● Ask the students to group any words which go together. This type of activity encourages the students to group the new vocabulary in word fields. It can be very effective to get them to choose their own word fields, although with concrete words like those in the vocabulary box, the fields are likely to be fairly predictable.

● It may be a suitable moment to draw the students' attention to the multi-part verbs *look at* and *wait for*, and put them in a special section of the *Wordbank*, which is at the back of their Practice Books. A full explanation of how multi-word verbs work is given in Progress check 31–35.

GRAMMAR

> **Present simple**
> **You use the present simple to talk about:**
> – a habit   *He smokes twenty cigarettes a day.*
> – a personal characteristic   *She plays the piano.*
> – a general truth   *You change money in a bank.*
> **There is an idea that the action or state is permanent.**
>
> **Present continuous**
> **You use the present continuous to say what is happening now or around now.**
> *It's raining. He's drawing a picture. I'm learning English*
> **There is an idea that the action or state is temporary.**
>
> **You form the present continuous with *is/are* + present participle (verb + *-ing*).**
> *I'm looking at the photos. She's waiting for a bus.*
>
> | **Questions** | **Short answers** | **Negatives** |
> |---|---|---|
> | *Is he drawing?* | *Yes, he is. No, he isn't.* | *He isn't drawing.* |
> | *Are you going home?* | *Yes, I am. No, I'm not.* | *I'm not going home.* |
>
> **You don't usually use these verbs in the continuous tenses:**
>
> *believe feel hear know like see smell sound taste think understand want*

1 Work in pairs and point at the people in the photos. Ask and say what they're doing at the moment and why.
What's the man in front doing?   He's playing the accordion.
Why is he playing the accordion?   Maybe he doesn't have any money.

Use these words and phrases to help you.

cross the street   draw   do the shopping   go home   hold an umbrella
listen to music   paint a picture   play the accordion   rain
shelter his instrument   sit on a suitcase   stand by the road   talk to someone
wait for a bus

2 Look at these verbs in the present continuous and write their infinitives.

drawing   getting   having   making   playing   shopping   putting   staying
draw   get …

What happens to infinitives ending in *-e, -t, -p* and *-y* when you form their present participle?

3 Complete these sentences with *and* or *but*.
1 I often go shopping at the supermarket ___ I'm going there now.
2 They usually eat at home ___ today they're having dinner in a restaurant.
3 She walks to work ___ this week she's taking the bus.
4 He smokes ten cigarettes a day ___ he's smoking a cigar at the moment.

SOUNDS

Listen and tick the phrase you hear. Is the underlined sound /n/ or /ŋ/?

1 carry i<u>n</u>/carry<u>ing</u> an umbrella
2 sitti<u>ng</u>/sit <u>in</u> there
3 sing <u>in</u>/sing<u>ing</u> tune
4 arrive <u>in</u>/arriv<u>ing</u> time
5 take <u>in</u>/tak<u>ing</u> money
6 stand <u>in</u>/standi<u>ng</u> there

Now say the phrases aloud.

LISTENING AND VOCABULARY

1 Say where you do these things.

buy train tickets   have dinner
change money
get some medicine   buy food

Choose from these places.

bank   chemist   post office
supermarket   railway station
restaurant

2 Listen to four conversations and decide where the people are. Choose from the places in 1. Now work in pairs. Say where the people are and what they're doing.

3 Look at the words in the box. Are they nouns, verbs or both?

> bank bus buy change chemist
> close cross draw food get
> hold look at medicine money
> paint play post office put
> queue railway station rain road
> shelter shop sit stand stay
> street suitcase supermarket take
> think ticket town umbrella
> wait for walk

4 Group any words which go together.
bank, money…

Progress check 1-5

VOCABULARY

1 Look at this crossword.

Work in pairs. Choose words in the vocabulary boxes from lessons 1 – 5 and put them in a crossword. How many words can you find?

2 Some words can be more than one part of speech. For example:

*cook: A **cook** (noun) is someone who **cooks** (verb) food.*
*orange: An **orange** (noun) is an **orange** (adjective) fruit.*

Use your dictionary to find out what parts of speech these words can be.

talk   head   drink   flat   start
rent   slice   heat

Write sentences showing their different parts of speech.
He talks all the time.
There's a talk on insects tonight.

3 Not every new word is useful to you. Look at the vocabulary boxes in lessons 1 – 5 again and choose ten words which are useful to you.

Start a *Wordbank* in your Practice Book for useful words and phrases. Write the ten words in your *Wordbank*.

GRAMMAR

1 You meet Tanya, from Russia, at a party in London. Here is some information about her. What are the questions?

1 I live in Moscow.
2 Yes, I am and we have three children.
3 I'm a scientist.
4 I start work at eight in the morning.
5 I finish at six in the evening.
6 In the evenings we have dinner.
7 At weekends we go to the country.
8 We usually go to a beach on the Black Sea.

1 Where do you live?

2 Put an adverb of frequency in each sentence so that it is true for you or your country.

always   (not) usually   (not) often   sometimes   never

1 We take off our shoes before we go into a house.
2 We offer guests something to eat.
3 We talk about politics and our families.
4 We have dinner at seven o'clock.
5 We give the hosts some wine or flowers.
6 We sit on the floor.
7 We wear smart clothes.
8 We arrive ten or fifteen minutes late.

3 Answer the questions with one of these expressions:

Yes, I do.   No, I don't.   I love it.   I hate it.
Not very much.   It's all right.   Not at all.

1 Do you like living in a town?
2 Do you like shopping?
3 Do you like walking?
4 Do you like hot weather?
5 Do you like warm beer?
6 Do you like fast food?

4 Write the *-ing* form of these verbs.

close   visit   fly   wear   like   sing   get   throw   cross   go   stay   cut

Progress check 1–5

GENERAL COMMENTS

You can work through this Progress check in the order shown, or concentrate on areas which may have caused difficulty in Lessons 1 to 5. You can also let the students choose the activities they would like to or feel the need to do.

VOCABULARY

1 Aim: to revise the vocabulary in Lessons 1 to 5 with a crossword.

● You may want to limit the time spent on this activity to ten minutes. It will give the students a good opportunity of looking very carefully at the vocabulary boxes and choosing words to fit their crosswords.

2 Aim: to focus on parts of speech.

● The students have already done some activities to do with parts of speech in Lessons 1 to 5.

> **Answers**
> talk *n* and *v*　head *n* and *v*　drink *n* and *v*
> flat *n* and *adj*　start *n* and *v*　rent *n* and *v*
> slice *n* and *v*　heat *n* and *v*

3 Aim: to explain the purpose of the *Wordbank*.

● It is important that the students realise that language acquisition in general and vocabulary acquisition in particular is only effective if they take an active part in the process. This means choosing words which are personally useful and recording them. Most students will only be able to retain about eight or nine words a lesson, so while the twenty or so words presented in the box are the most important ones to learn, the students must make the final selection. The *Wordbank* is designed to help the students organise their vocabulary records in a systematic way.

GRAMMAR

1 Aim: to revise asking questions in the present simple.

> **Answers**
> 1 Where do you live?
> 2 Are you married?
> 3 What do you do?
> 4 When do you start work? *or*
> 　 What time do you start work?
> 5 When do you finish work? *or*
> 　 What time do you finish work?
> 6 What do you do in the evenings?
> 7 What do you do at weekends?
> 8 Where do you spend your holidays?

2 Aim: to revise the position of adverbs of frequency in sentences.

● Make sure the students make true sentences for their own countries.

3 Aim: to revise short answers when talking about likes and dislikes.

● Make sure the students answer the questions according to their own likes and dislikes.

4 Aim: to revise the formation of present participles.

> **Answers**
> closing, visiting, flying, wearing, liking, singing, getting, throwing, crossing, going, staying, cutting

5 Aim: to revise the distinction between the present simple and the present continuous.

> **Answers**
> 1 Tanya **comes** from Russia.
> 2 She **is visiting** London.
> 3 She **is having** a holiday.
> 4 She **speaks** English quite well.
> 5 She **is staying** with friends in London.
> 6 She **is enjoying** her visit.
> 7 She **goes** shopping most days.
> 8 She **says** she wants to come back soon.

6 Aim: to revise the use of articles.

> **Answers**
> 1 There's **a** radio in **the** living room.
> 2 Would you like **a** cup of – tea?
> 3 They've got **a** large house in **the** centre of town.
> 4 We've got **a** son and – two daughters.
> 5 **The** children are outside in **the** garden.
> 6 What's **the** main room in your flat?
> 7 She spoke – very good French at – home.
> 8 I flew to – Lyon and spent – two weeks in **the** Alps.

SOUNDS

1 Aim: to practise /s/, /z/ and /ɪz/.

> **Answers**
>
> | /s/ | /z/ | /ɪz/ |
> | --- | --- | --- |
> | gets | does | finishes |
> | looks | goes | refuses |
> | smokes | leaves | |
> | wants | sings | |

2 Aim: to practise /ɪ/ and /iː/.
● 🔊 Play the tape and stop after each word. Ask the students to say the words aloud.

> **Answers**
>
> | /ɪ/ | /iː/ |
> | --- | --- |
> | it | eat |
> | fifty | fifteen |
> | live | leave |
> | sit | seat |

3 Aim: to focus on the use of intonation to show interest.
● 🔊 Play the tape and stop after each sentence. Ask the students to say the sentences aloud.

> **Answers**
> 1 What's your name? ✔
> 2 How old are you?
> 3 Where do you live? ✔
> 4 Do you live with your parents? ✔
> 5 Are you married?
> 6 Is that your brother?
> 7 Is that your husband? ✔
> 8 Do you have a sister?

SPEAKING AND WRITING

1 Aim: to present some very basic rules of punctuation in English.
● You may like to ask the students to read the rules of punctuation and simply decide if they are the same or different to the rules in their own language.

> **Answers**
> 1 We don't usually visit people without an invitation.
> 2 When we meet people for the first time we say, 'How do you do?'
> 3 When do you use first names in your country?
> 4 Your friend is called James Smith. Do you call him James or Mr Smith?
> 5 It's usual to use first names with people when you get to know them.

2 Aim: to practise using the present tenses and to revise the vocabulary areas of customs and habits.
● Ask the students to work in small groups and to discuss these questions. The main headings refer to some of the issues discussed in Lesson 1, but the questions may be different to ones answered in that lesson. Even if you are teaching students who all come from the same culture, there may be small differences in behaviour and attitudes. If so, you can use this to draw attention to the possibility of more important differences with people from other cultures.

3 Aim: to practise writing.
● The students can use this opportunity to write down the main points of their discussion in 2. If there isn't time, this can be done for homework.

5 Choose the correct verb form.

1 Tanya *comes/is coming* from Russia.
2 She *visits/is visiting* London.
3 She *has/is having* a holiday.
4 She *speaks/is speaking* English quite well.
5 She *stays/is staying* with friends in London.
6 She *enjoys/is enjoying* her visit.
7 She *goes/is going* shopping most days.
8 She *says/is saying* she wants to come back soon.

6 Complete these sentences with *a/an, the*, or put – if there's no article.

1 There's ___ radio in ___ living room.
2 Would you like ___ cup of ___ tea?
3 They've got ___ large house in ___ centre of town.
4 We've got ___ son and ___ two daughters.
5 ___ children are outside in ___ garden.
6 What's ___ main room in your flat?
7 She spoke ___ very good French at ___ home.
8 I flew to ___ Lyon and spent ___ two weeks in ___ Alps.

SOUNDS

1 Say these words aloud.

does finishes gets goes leaves looks refuses
sings smokes wants

Do they end in /s/ , /z/ or /ɪz/? Put them in three columns.

🔲 Now listen and check.

2 🔲 Listen and say these words aloud.

<u>ea</u>t <u>i</u>t fif<u>ee</u>n fi<u>f</u>ty l<u>i</u>ve l<u>ea</u>ve s<u>i</u>t s<u>ea</u>t

Is the underlined sound /ɪ/ or /iː/? Put the words in two columns.

3 🔲 Listen to these questions. Put a tick (✓) if you think the speaker sounds interested.

1 What's your name?
2 How old are you?
3 Where do you live?
4 Do you live with your parents?
5 Are you married?
6 Is that your brother?
7 Is that your husband?
8 Do you have a sister?

Now say the sentences aloud. Try to sound interested.

SPEAKING AND WRITING

1 Look at some rules for punctuation in English.

1 You use a capital letter:
 – at the beginning of a sentence: *I like cakes.*
 – for names: *Sue Jim Fiona*
 – for the first person singular pronoun: *I*
 – for nationalities: *She's British   the French*
2 You put a full stop at the end of a sentence:
 I don't speak Japanese.
3 You put a question mark at the end of a question:
 Do you speak English?
4 You put inverted commas and a comma around someone's actual words: *'I'm French,' he said.*
5 You put an apostrophe for contractions:
 I'm English.   He doesn't live here.

Now punctuate these sentences.

1 we dont usually visit people without an invitation
2 when we meet people for the first time we say how do you do
3 when do you use first names in your country
4 your friend is called james smith do you call him james or mr smith
5 its usual to use first names with people when you get to know them

2 Work in groups of three or four. Talk about customs and traditions of hospitality in your country. Talk about the following:

Invitations	Do you ever visit people without an invitation?
Greetings	How do you greet people when you meet them for the first time?
Names	When do you use first names, family names or titles?
Punctuality	When do you arrive for an appointment?
Clothing	What do you wear for a lunch or a dinner engagement?
Food and drink	What food or drink do you expect?
Conversation	What do you talk about? What don't you talk about?
Compliments	Do you make compliments about the food, the host's house or personal objects?
Leaving	When do you leave a dinner party?

3 Write some advice for visitors to your country about customs and traditions of hospitality and entertainment. Write about the points in 2 and say what people do or don't do.

Surprising behaviour

Past simple (1): regular and irregular verbs

VOCABULARY AND READING

1 Look at the photo, which shows a scene from the passage you are going to read. Where do you think the passage takes place?

2 Look at the words in the box. Decide which are nouns and which are adjectives. Look up words you don't know in the dictionary.

> bag smile broken cloud
> bunch coat people flat flower
> heavy low scarf shoes big
> sky warm

3 Group the words which can go together.

heavy bag, heavy cloud...

4 The passage is by Paul Theroux, an American travel writer. Read it and find out what country he is describing.

5 Are these statements about the passage true or false?

1 The woman had a friend with her.
2 She didn't have a dog.
3 The weather wasn't very good.
4 He didn't know her name.
5 He expected her to greet him.
6 She didn't say 'Good morning'.

Which statement describes what surprised him?

*A*s soon as I left Deal, I saw a low flat cloud, iron-grey and then blue across the Channel. The closer I got to Dover, the more clearly it was defined. I walked on and saw it was a series of headlands. It was France.

Ahead on the path was a person, down a hill four hundred yards away; but whether it was a man or a woman I could not tell. Some minutes later I saw her scarf and her skirt, and for more minutes on those long slopes we walked toward each other under the big sky. We were the only people visible in the landscape – there was no one behind either of us. She was a real walker – arms swinging, flat shoes, no dog, no map. It was lovely, too: blue sky above, the sun in the southeast, and a cloudburst hanging like a broken bag in the west. I watched this woman, this fairly old woman, in her warm scarf and heavy coat, a bunch of flowers in her hand – I watched her come on, and thought I am not going to say hello until she does.

She did not look at me. She drew level and didn't notice me. There was no other human-being in sight on the coast, only a fishing boat. Hetta Poumphrey – I imagined that was the woman's name – walked past me, and still stony-faced.

'Morning!' I said.

'Oh.' She turned her head to me. 'Good morning!'

She gave me a good smile, because I had spoken first. But if I hadn't, we would have passed each other, Hetta and I, in that clifftop meadow – not another soul around – five feet apart without a word.

Adapted from *The Kingdom by the Sea*, by Paul Theroux

6

GENERAL COMMENTS

Past simple tense

It is expected that most students will have already come across this tense before they start pre-intermediate *Reward*. For this reason, both regular and irregular verbs are presented in this lesson. This may be a suitable time to draw the students' attention to the list of irregular verbs on page 112 of the Student's Book. The two main areas of possible difficulty or error will be the pronunciation of the past simple endings, and the use of the past simple after *did* instead of the infinitive in questions. Specific coverage of the pronunciation point is given in the *Sounds* section of this lesson, and questions are presented in Lesson 7.

Level of difficulty

The passage has not been adapted for use by language students, but the tasks are suitable for pre-intermediate students. It is important to expose them to as much real-life English as possible, although it is equally important not to discourage them by giving them too much material at too high a level. In pre-intermediate *Reward*, authentic and un-graded material are accompanied by graded tasks.

Dictionaries

Using a monolingual or bilingual dictionary in class is a good way of preparing the students to take responsibility for their learning. At first you may want to do some directed activities with dictionaries, but after a while, most students use dictionaries if and when they need to. However, it is equally important to ensure that they don't overuse the dictionary and that they develop the necessary techniques for dealing with unfamiliar words. It may be necessary to impose a rule that they do not look up more than, say, six words in a passage. This will encourage the students to choose these words carefully, and not look up the words they can guess. It is also important not to treat a reading passage as an opportunity for vocabulary building. There are many words in this passage which are not essential at this level, and the words which are suitable for acquisition are included in the vocabulary box.

Cross-cultural awareness

The passage was chosen because it describes an incident when a person from one culture expected certain behaviour from a person from another culture and was surprised when it was not forthcoming. This situation can arise when, for example, someone wishes an English person *Bon appetit* and receives no answer, or when an English person expects someone to accompany a request with *please* and is offended when the politeness formula used in Britain is missing. In his book *The Kingdom by the Sea*, Paul Theroux is quite critical of the British for their behaviour, attitudes, customs and traditions. In a way, his intolerance is held up for scrutiny because he is subconsciously criticising the British for not behaving as he would. It is this kind of intolerance that the training in cross-cultural awareness is intended to examine and confront.

Photo: possible questions

What is the most famous landmark in your country?
What do visitors take photos of most often?
Think of the countryside in your country? Do any other countries have similar countryside?

VOCABULARY AND READING

1 Aim: to prepare for reading by predicting something of the background to the passage.

● The language of describing countryside is presented in Lesson 33, but you may want to use the photo to present a few items of vocabulary. Write the following words on the board: *cliff, sea, river, city, beach, desert, field, forest, hill, island, lake, village*. Ask the students which of these things they can see in the photo.

> **Answer**
> The scene is of the White Cliffs of Dover, England.

2 Aim: to present the words in the box and to prepare the students for the passage.

● Ask the students to underline the adjectives.

> **Answers**
> bag *n* smile *n* broken *adj* cloud *n* bunch *n*
> coat *n* people *n* flat *adj* and *n* flower *n* heavy *n*
> low *adj* scarf *n* shoes *n* big *adj* sky *n* warm *adj*

3 Aim: to develop an awareness of words which go together and to help the process of vocabulary acquisition.

● This activity involves work on collocation, an important skill to develop. Ask the students to group the words which go together. Some adjectives can go with more than one noun.

● Continue the activity by asking the students to think of other nouns which can go with the adjectives and other adjectives which can go with nouns.

> **Possible answers**
> broken bag, heavy bag, big bag, warm smile, big smile, broken cloud, broken shoes, flat cloud, heavy cloud, low cloud, big cloud, big bunch, heavy coat, big coat, warm coat, big people, warm people, flat shoes, big flower, bunch of flowers, heavy shoes, heavy sky, warm scarf, warm shoes, big scarf, big shoes

4 Aim: to practise reading for main ideas.

● The task is designed to ensure that the students read the passage for its main idea and don't get stuck on difficult vocabulary. Explain that the passage is at a slightly higher level of difficulty than the students' present level, but that they need to be exposed to examples of authentic English in class, in order to prepare them for real-life conditions of language use.

> **Answer**
> The country is England.

5 Aim: to present past simple affirmative and negative statements; to practice inferring.
- This activity involves simple statements which encourage the students to read between the lines and infer what the writer really says. Ask them to read the passage again on their own, and then to check their answers in pairs.

GRAMMAR

1 Aim: to focus on regular and irregular verbs.
- Ask the students to read the information about the past simple in the grammar box.

- This activity is designed to use the verbs in the passage to introduce, or remind the students about, the difference between regular and irregular verbs. The students should write the verbs down and then check them against the list on page 112. There are some verbs in the passage in the past perfect and third conditional; if anyone asks, say that you will deal with these tenses later in the course.

Answers
Regular: imagined, walked, watched
Irregular: was/were, got, gave, left, saw, thought

2 Aim: to focus on the formation of regular verbs.
- This activity allows the students to look at regular past simple endings and to guess the infinitive form.

Answers
carry, close, continue, dance, decide, like, live, stop, travel, try

- The students may not be able to work out this rule with complete accuracy, but the process of doing so will be generative enough to provide the basic principles. Suggest that they turn to the Grammar review to check their answers.

3 Aim: to focus on irregular verbs.
- This activity allows the students to work from given information about irregular verbs to produce their infinitive forms, which they should already know.

Answers
become, come, choose, cost, cut, do, hit, have, hear, know, make, meet, put, run, read, say, shut, take, tell, understand, go, write

4 Aim: to focus on patterns in irregular verbs.
- Despite the fact that the verbs in the list are irregular, there are nevertheless some patterns which make some of them similar.

Answers
Verbs which have the same form in the present and past simple: cost, cut, hit, put, read, shut
Verb which has the same form but sounds different: read (present simple /riːd/, past simple /red/)

SOUNDS

1 Aim: to focus on the sound of past simple endings.
- Explain that the -(e)d ending may be pronounced in three ways. Write the three phonemes on the board.
- Ask them to predict which column the words go in.
- 🎧 Play the tape and pause after each word. Ask a student to write the word in the correct column on the board.

Answers
/t/: finished walked danced
/d/: continued enjoyed called
/ɪd/: started wanted expected

2 Aim: to practise the pronunciation of past simple endings.
- 🎧 Explain that after a voiced sound the ending is pronounced /t/, after an unvoiced sound it is /d/, and after t or d it is pronounced /ɪd/.

SPEAKING

1 Aim: to give practice in using regular and irregular past verbs; to discuss reactions to the text.
- The writer was surprised that the woman didn't say hello first, because in his country it would be common to do so. Of course, his own behaviour is also surprising; if he expected to exchange greetings, he could have said hello first. The woman may have had a lot on her mind, or may have been nervous about speaking to a strange man in a lonely place. The passage describes what the writer considers is unsuitable behaviour, while at the same time revealing his own unsuitable behaviour.

- Find out if the students would expect to exchange greetings in this situation. In what other situations do strangers greet each other?

2 Aim: to give practice in using regular and irregular past verbs; to discuss behaviour in cross-cultural situations.
- It doesn't matter if your students have little or no experience of living within foreign cultures. Ask them to think about foreign visitors to their country, or even surprising behaviour of people from their own culture. Use this opportunity to remind the students that people from different cultures do not necessarily have the same customs and traditions as they do, and that their own behaviour and attitudes are not necessarily transferable to other cultural contexts.

- The students may like to write about their anecdotes for homework.

GRAMMAR

Past simple (1): regular and irregular verbs

You use the past simple to talk about a past action or event that is finished.
*We **walked** toward each other.*
*She **turned** her head to me.*

You form the past simple of most regular verbs by adding -ed.
*walk – walk**ed*** *I walked on.*
*watch – watch**ed*** *I watched this woman.*

Many verbs have an irregular past simple form.
leave – left see – saw be – was/were

For a full list of irregular verbs, see Grammar review.

Negatives
*She **did not** look at me. She **didn't** notice me.*

---

1 Look at the passage again. Find the past simple of these verbs.

be get give imagine leave see think walk
watch

Which verbs are regular? Which ones are irregular?

2 Look at the past simple form of these regular verbs and write their infinitive form.

carried closed continued danced decided liked
lived stopped travelled tried

What's the rule for forming the past simple of regular verbs ending in *-e, -y, -p, -l?*

3 Look at the past simple form of these irregular verbs and write their infinitive form.

became came chose cost cut did hit had
heard knew made met put ran read said
shut took told understood went wrote

4 Which verbs in 3 have the same form in the present and the past simple? Which verb has the same form but sounds different?

SOUNDS

1 🔊 Listen to the pronunciation of the past-tense endings of these verbs.

/t/	/d/	/ɪd/
liked	*lived*	*decided*
washed	*stayed*	*visited*

Put these verbs in the correct column.

continued finished enjoyed started walked
danced called wanted expected

2 🔊 Listen and check. What's the rule? Now say the verbs aloud.

SPEAKING

1 Work in pairs. Why didn't the woman in the passage say hello to the writer first? Why did he find this behaviour surprising?
Perhaps she was... Maybe she didn't...

2 Talk about a situation at home or in a foreign country when you found someone's behaviour surprising.
In England no one talks on the buses and underground. In my country we talk all the time!

Past simple (2): questions and short answers

When was the last time you had a holiday? And did you organise the trip or did you take a package tour? These days, most people choose a package tour, especially when they go abroad on holiday. They pay for their travel and accommodation in their own country, and they take traveller's cheques which they exchange for local money when they arrive in the foreign country. But in the past it was very different. In fact, before the middle of the nineteenth century, travelling for pleasure was rare and very expensive, and only a few rich people travelled abroad. The man who changed all this and brought in the age of mass tourism was Thomas Cook.

Thomas Cook was a printer in Leicester, England and the secretary of a local church organisation. In 1841 it was his job to arrange rail travel for members of his church to a meeting in Loughborough, a round trip of twenty-two miles. This was the world's first package trip. After this first success, he organised many more for his church. Then in 1845 he advertised a package tour to Liverpool for the general public, and before it took place he went to Liverpool to meet the hotel staff, and check the accommodation and restaurants.

He then started to organise trips all over Britain, including the Great Exhibition in London. In 1851 he published the world's first travel magazine which had details of trips, advice to travellers and articles and reports about the places to visit.

In 1854 he gave up his job as a printer. In 1855 he took his first group of tourists to Paris and later that year led a tour of Belgium, Germany and France. In 1863 he went to Switzerland and in 1864 to Italy. By then he had a million clients. The following year he opened an office in London, which his son John Mason managed. They introduced a circular ticket which gave the traveller a single ticket to cover one journey instead of a number of tickets from all the railway companies involved, and they organised a system of coupons which people bought at home and exchanged in the foreign country for a hotel room and meals.

In 1866 the first group of European tourists visited New York and the Civil War battlefields of Virginia. In 1868 the Cooks went to the Holy Land with tents because there were no hotels there at that time. After the Suez canal opened in 1869, Cook created his own fleet of luxury boats to travel up the river Nile.

It was dangerous to carry large amounts of cash, so in 1874 Cook introduced an early form of traveller's cheque, which travellers could cash at a number of hotels and banks around the world.

Thomas Cook died in 1892 at the age of 84, and his son John Mason seven years later. But the age of the package tour and mass tourism was born.

7

GENERAL COMMENTS

Past simple questions and short answers

Common mistakes are:
- the use of the past simple and not the infinitive after *did* in questions: *What did you saw there?*
- the inversion of the noun or pronoun and the infinitive: *What did see you there?*
- the omission of the auxiliary *did*: *What saw you there?*
As soon as anyone makes one of these mistakes, write the sentence on the board and cross it out. Then each time a student makes a similar error, you can point to the sentence and draw attention to the general pattern.

Thomas Cook

Some students may not have heard of Thomas Cook, but they may have heard of his name in connection with traveller's cheques.

Left-hand illustration: possible questions

Look at the title and the picture.
What do you think the theme of the lesson is?
Do you know who Thomas Cook was?
What nationality was he?
Where does this picture come from?
Is it an illustration for a book, an advertisement, or a painting in a gallery? How do you know?
Is it effective as an advert?
Does it make you want to travel somewhere by ship?
Do you like ships?
When was your last sea journey?
Where did you go? Did you enjoy it?

Right-hand photo: possible questions

Who do you think this man is?
When did he live?

READING

1 **Aim: to pre-teach some important words from the reading passage and to provide an opportunity for reading for specific information.**

● The words are not necessarily important vocabulary items, but they are necessary in order to understand the passage. Write the words on the board and find out if anyone knows what they mean. You may also like to give the students the definitions below and ask them to match them with the words.

Suggested explanations

mass tourism: in the past only the rich travelled for pleasure and education. Most people didn't have the money or the desire to visit foreign countries. But recently, travel has become less expensive and people have more money to spend. These days, more and more people go abroad on holiday, so we talk about the age of mass tourism.

package tour: a visit, usually to a foreign country, in which the travel arrangements and the accommodation are all organised for the tourist by a travel agency. They are usually paid for in the tourist's home country.

traveller's cheque: a form of money which is often bought in the tourist's home country and exchanged in a bank for foreign currency. It is also a secure way of carrying large amounts of money because the tourist signs it when he or she buys it and signs it again in front of the bank clerk, who compares the signatures before giving the tourist cash.

coupon: a voucher or piece of paper which shows the person has already paid for something such as hotel accommodation.

cash: money in the form of coins and notes, as opposed to cheques.

● Ask the students to read the passage and find out what the words have to do with Thomas Cook.

Answer
The five words represent concepts which Thomas Cook introduced to the world.

2 **Aim: to practise reading for specific information; to prepare for the presentation of negatives and questions in the simple past.**

● You may need to deal with some vocabulary issues before proceeding. In order to help the students develop the skill of dealing with unfamiliar words, ask them to choose up to six words which you will explain.

● Ask the students to read the passage and to mark the sentences true or false according to the passage.

Answers
1 True.
2 False. He was a printer.
3 False. The first public trip was in 1841.
4 False. The Great Exhibition was in London.
5 True.
6 False. He opened an office in 1865, a year after he went to Italy in 1864.
7 True.
8 False. There were no hotels there at the time.
9 True.
10 False. He died in 1892.

GRAMMAR

1 **Aim: to present past simple *yes/no* questions and short answers.**
- Ask the students to read the information about the past simple in the grammar box.

- Ask one or two students
Was travel very expensive before the middle of the nineteenth century?
Was Thomas Cook a travel agent?
Elicit the answers *Yes, it was* and *No, he wasn't.* You may want to do the whole exercise orally, like this.

- Ask the students to work in pairs and check their answers to *Reading* activity 2. Make sure they form the questions correctly and use short answers. Explain that only questions with *be* do not use *did/didn't* in answers.

2 **Aim: to practise writing *Wh-* questions.**
- Make sure the students have read and understood the information in the grammar box about the use of the auxiliary in subject and object questions. Explain that they must re-read the passage carefully in order to formulate the questions. Check they put a question mark at the end of the sentence.

Possible answers
1 Who travelled abroad before the middle of the nineteenth century?
2 What did Thomas Cook do?
3 Who did he meet in Liverpool?
4 What did he publish in 1851?
5 When did he give up his job as a printer?
6 What/where did the first group of European tourists visit in 1866?
7 Why did he introduce an early form of traveller's cheque?
8 When did John Mason die?

- You may want to continue this activity by writing on the board a few things which happened in the past, such as dates or important events and ask the students to write the questions. They can also write a list of dates and events themselves, and give them to other students to write questions.

VOCABULARY AND SPEAKING

1 **Aim: to present the words in the vocabulary box; to focus on parts of speech; to practise using the past simple.**
- These words are the most useful and important ones from the passage. Ask the students to decide which ones are verbs and what their past simple form is.

Answers
advertise – advertised, allow – allowed,
bring – brought, buy – bought, carry – carried,
change – changed, check – checked,
choose – chose, create – created, die – died,
exchange – exchanged, give up – gave up,
introduce – introduced, lead – lead (/liːd/ – /led/),
leave – left, meet – met, open – opened,
organise – organised, pay – paid,
publish – published, travel – travelled,
want – wanted

- Some students may know that *cash* can also be used as a verb.

2 **Aim: to develop an awareness of words that go together.**
- This activity is to encourage the students to look again at the passage and notice which words go together. It may be useful to remind them that an awareness of collocation will increase their vocabulary and their fluency in English.

3 **Aim: to practise using the past simple tense and to encourage the students to react to the passage.**
- Ask the students to reflect on the passage and to decide which were the most important events in Thomas Cook's career.

- You may like to list the different suggestions on the board and take a class vote.

4 **Aim: to practise using the past simple tense; to provide some further speaking practice.**
- Ask the students to talk about their last holidays or their experiences as tourists. Ask them to think about their best and worst experiences as tourists. This will prepare them for the topic of Lesson 8.

- Ask them to talk about how they feel about tourism. They may like to prepare a list of advantages and disadvantages of mass tourism for homework.

READING

1 What do these words and phrases mean?

mass tourism   package tour   traveller's cheque

Now read *The world's first package tours* and find out what the words and phrases have to do with Thomas Cook.

2 Read these statements about the passage. Are they true or false?

1 Travel was very expensive before the middle of the nineteenth century.
2 Thomas Cook was a travel agent in 1841.
3 The first public trip was in 1845.
4 The Great Exhibition was in Liverpool.
5 The first trip outside Britain went to France.
6 In 1866 he opened an office in London.
7 Some tourists wanted to visit the Civil War battlefields.
8 The Cooks stayed in hotels in the Holy Land.
9 The traveller's cheque allowed people to travel without large amounts of cash.
10 Thomas Cook died in 1891.

GRAMMAR

> **Past simple (2): questions and short answers**
>
> | **Questions** | **Short answers** |
> |---|---|
> | *Was the first package trip in 1841?* | *Yes, it **was**.   No, it **wasn't**.* |
> | *Did Cook live in Leicester?* | *Yes, he **did**.   No, he **didn't**.* |
>
> You can use *who, what* or *which* to ask about the subject of the sentence. You don't use *did*.
> *Who organised the first package trip?   Thomas Cook.*
>
> You can use *who, what* or *which* and other question words to ask about the object of the sentence. You use *did*.
> *Who did he take on the first package trip?   500 workers.*
>
> **Compare:**
> **Subject**   *Who **introduced** traveller's cheques?   Thomas Cook.*
> **Object**   *What **did** Thomas Cook introduce?   Traveller's cheques.*

1 Work in pairs and check your answers to *Reading* activity 2.

1 Was travel very expensive before the middle of the nineteenth century? Yes, it was.

2 Here are some answers about the passage. Write suitable questions.

1 A few rich people.
2 He was a printer.
3 Hotel staff.
4 The world's first travel magazine.
5 In 1854.
6 New York, and the battlefields of Virginia.
7 Because it was dangerous to carry lots of cash.
8 In 1899.

1 Who travelled abroad before the middle of the nineteenth century?

VOCABULARY AND SPEAKING

1 Underline the verbs from this lesson in the box.

> abroad advertise allow boat
> bring buy carry cash change
> check choose country create
> die exchange foreign give up
> holiday hotel introduce lead
> leave luxury meet open
> organise pay publish travel
> traveller's cheque want

What is the past simple tense of the verbs?

2 Look at the other words in the box. Try to remember which nouns, verbs or adjectives they go with in the passage.

Now look back and check.

3 Work in pairs. Choose the two most important events in Thomas Cook's career.
I think one was when he gave up his job as a printer.

4 Work in pairs. When was the last time you were a tourist? Where did you go? What did you do? Did you take a package tour?

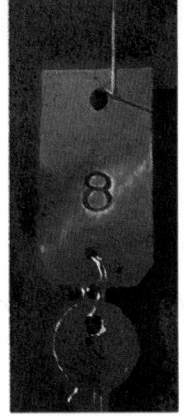

8 | *Something went wrong*

Expressions of past time; *so, because*

VOCABULARY

1 Work in pairs. Put the words and phrases in the vocabulary box with the following travel situations. Some words can go with more than one situation.

a train journey   a boat journey   a plane flight
hotel accommodation

> airport   bed and breakfast   boarding pass   book
> business class   cabin   check in   check out   connection
> delay   departure   double room   fare   ferry   harbour
> land   lift   luggage   passenger   platform   reservation
> return   single   single room   terminal   ticket   take off
> timetable

2 Think of two or three more words which go with each situation.

LISTENING AND SPEAKING

1 You are going to hear Ann, an English woman, telling a story about a travel situation when things went wrong. Look at how the story begins.

> A few years ago, I was going from
> London to Paris to join my husband
> and children. I checked in early and...

What do you think the travel situation is? Which words in the vocabulary box do you expect to hear?

2 🔲 Listen to Ann's story. As you listen, look at the vocabulary box again and tick (✓) the words you hear.

3 Work in pairs. Did you guess correctly in 1? Describe what happened.

🔲 Listen again and check.

4 Look at these phrases taken from another story about a situation where something went wrong. Decide which travel situation the speaker, Bob, is describing. What do you think happened?

sat down on the tiny single bed	☐
knocked on the door	☐
slept in the car that night	☐
didn't have a reservation	☐
wanted to check out	☐
asked if he had a room	☐
left my suitcase in my car	☐
picked up a key from behind the desk	☐
was frightened by the man downstairs	☐
showed me a very dusty room	☐

5 🔲 Listen to Bob's story. Did you guess correctly in 4?

6 Number the phrases in 4 in the order you heard them.

🔲 Now listen to Bob's story again and check.

7 Have you ever been in a situation where something went wrong? Tell the class about it.

8

GENERAL COMMENTS

Listening

There are several problems with using recorded material in class. Students are deprived of the visual and information clues which they would have if they were personally involved in the conversation. They are rarely confident of their listening ability because they are not in control of the speed at which the information needs to be processed and understood. They try to understand every word, and may panic if there is something difficult. The strategy for listening practice in *Reward* is to give students the chance to hear authentic or typical spoken English and to do tasks which focus more on the general sense than on specific words. The comprehension checks in *Reward* involve ticking and crossing, numbering, lettering or making some gesture of the pen, rather than producing language. Another major problem is simply getting the students to listen, so the preparation phase is particularly important in creating a pedagogical motivation. Finally, even if this motivation to listen is created, it has to be maintained by a task to complete while the students are listening. You may decide you want them to listen and follow the transcript at the back of the book, but this should supplement and not replace the tasks suggested. Using tapescripts generally involves focusing on every word, which is not helpful towards developing listening competence.

Main illustration: possible questions

Look at the picture. Where is it? Who is the man?
Would you like to stay in a hotel like this?
What adjectives describe the mood of the picture?
The picture illustrates one of the stories you're going to hear. What do you think it's going to be about?

VOCABULARY

1 Aim: to present the words in the vocabulary box, and to prepare for the listening passages.
● Before the students begin, write the four travel situations on the board and ask students to suggest words which go with them.

● Ask the students to match the words in the box to the four situations.

> **Possible answers**
> **a train journey:** book, connection, delay, departure, fare, luggage, passenger, platform, reservation, return, single, ticket, timetable
> **a boat journey:** cabin, delay, departure, fare, ferry, harbour, land, luggage, passenger, reservation, return, single, terminal, ticket, timetable
> **a plane flight:** airport, boarding pass, book, business class, cabin, check in, connection, delay, departure, fare, land, luggage, passenger, reservation, return, single, terminal, ticket, take off, timetable
> **hotel accommodation:** bed and breakfast, book, check in, check out, double room, lift, luggage, reservation, single room

● You may like to explain that in Britain some people offer a spare room in their home as *a bed and breakfast*; you pay much less than you would for a hotel, and get the opportunity to see inside a British home.

2 Aim: to expand vocabulary to do with travel situations; to help the students organise their vocabulary learning.
● Ask the students to add the words they suggested at the start of activity 1 to their lists.

LISTENING AND SPEAKING

1 Aim: to prepare for the first listening passage.
● It is important to prepare the students for listening. In real life they would probably know what they were going to listen to. Read the sentence aloud and ask them what the travel situation is likely to be.

> **Answer**
> The phrase *checked in* suggests the situation is going to be a plane flight.

● Ask the students to predict which of the words in the box they are likely to hear.

2 Aim: to practise listening to a passage of continuous English.
● ▭ Play the first ten or fifteen seconds of the first story and ask the students if they have understood. Make sure everyone understands the opening of the first passage. Then rewind and play the story again from the start. Remind the students that they have to tick the words in the box that they hear.

> **Answers**
> checked in, ticket, business class, boarding pass, delay, business class, passengers, take off, cabin, land, airport

3 Aim: to check that the students have understood the passage.
● Explain that you're going to ask a student to start recounting the story. The others must listen and ask questions. If the person recounting answers the question correctly, he/she continues. If he/she can't answer, the person who asked the question takes over. With a large class, the students could do this in groups.

● ▭ Play the tape a second time. The students may like to follow the tapescript as they listen to the story.

4 Aim: to prepare the students to listen to the second passage.
● Ask the students to look at the phrases taken from the second story and to predict the travel situation.

> **Answer**
> Hotel accommodation.

● Ask the students to think about the vocabulary they are likely to hear.

5 Aim: to practise listening to a passage of continuous English.
- ▣ Play the tape and ask if they guessed correctly in activity 4.

6 Aim: to check the students have understood the main ideas.
- Ask the students to work in pairs and to number the events in the order they heard them.

Answers	
sat down on the tiny single bed	[6]
knocked on the door	[2]
slept in the car that night	[10]
didn't have a reservation	[1]
wanted to check out	[7]
asked if he had a room	[3]
left my suitcase in my car	[9]
picked up a key from behind the desk	[4]
was frightened by the man downstairs	[8]
showed me a very dusty room	[5]

- ▣ Play the tape again. The students may like to follow the tapescript as they listen.

7 Aim: to practise using the past simple.
- Ask if anyone has been in a situation where something went wrong. It doesn't have to be a travel situation. Ask them to tell the rest of the class about it.

GRAMMAR

1 Aim: to practise using the past simple and to present expressions of past time.
- Ask the students to read the information about expressions of past time in the grammar box.
- Ask the students to do the activity.

Answers to 1 and 2
1 I **watched** television <u>last night</u>.
2 She **listened** to the news on the radio <u>yesterday</u>.
3 He **played** football <u>the day before yesterday</u>.
4 We **visited** my parents <u>last Sunday</u>.
5 I **stayed** at home <u>last weekend</u>.

2 Aim: to focus on expressions of past time.
- Ask the students to underline the expressions of past time in the sentences in 1.

3 Aim: to practise using the past simple and expressions of past time.
- Write on the board a sentence describing what you did *last night*, and one describing what you did *yesterday*. Ask one or two students what they did *the day before yesterday, last Sunday* and *last weekend*. These are the expressions they underlined in activity 2.
- Ask the students to use the expressions of past time in sentences about themselves.

4 Aim: to practise using the past simple and expressions of past time.
- Ask the students to think about important or memorable events in their lives.

5 Aim: to practise using the past simple and expressions of past time.
- Ask the students to work in pairs and to show each other their expressions of past time. They should ask and answer questions about what happened. Go round and check everyone is using the past simple correctly.

6 Aim: to focus on *because*.
- Ask the students to read the information about *so* and *because* in the grammar box.
- Ask them to join the two parts of the sentences. There's no need to write the sentences in full.

Answers
1 a 2 d 3 c 4 b

7 Aim: to focus on *so*.
- The students will have to write the sentences in full.

Answers
1 There wasn't much air traffic so the flight was on time.
2 They thought there was a bomb on the plane so they had to make an emergency landing.
3 He didn't expect to stay there long so he didn't have a hotel reservation.
4 The room was dirty and unpleasant so he decided to leave.

WRITING

1 Aim: to practise writing a story.
- Ask the students to form groups of five or six and if possible to sit in a circle. Then ask them to write an opening sentence on their own on a spare, detachable piece of paper (ie not fixed in a notebook).

2 Aim: to practise writing a story.
- Ask them to pass the piece of paper with their sentence to the person on their right, and receive a sentence from the person on the left. They continue the story from the point left off by the sentence on the new piece of paper and not the story they began writing in 1.
- They continue writing sentences and passing the paper on until the story is finished. They then all have a piece of paper with sentences written by other people.

3 Aim: to rewrite the story using *so* and *because*.
- Ask them to rewrite the story using *so* and *because*.
- Ask the students to read each other's stories.

4 Aim: to give further writing practice.
- If there isn't time, you can ask the students to do this for homework.

GRAMMAR

> **Expressions of past time**
> **You can use these expressions of past time to say when something happened.**
>
> *last night last Sunday last week last month last year*
> *yesterday yesterday morning yesterday afternoon*
> *the day before yesterday*
> *two days ago three weeks ago years ago ages ago*
> *in 1985 from 1987 to 1993*
>
> **So, because**
> **You can join two sentences with *so* to describe a consequence.**
>
> *She often took the plane, **so** she didn't look at the safety instructions.*
>
> **You can join the same two sentences with *because* to describe a reason.**
>
> *She didn't look at the safety instructions **because** she often took the plane.*

1 Complete the sentences with these verbs in the past simple. There are some extra verbs, and there may be more than one possible answer.

ask enjoy finish listen look open play start
stay walk visit wash watch

1 I _____ television last night.
2 She _____ to the news on the radio yesterday.
3 He _____ football the day before yesterday.
4 We _____ my parents last Sunday.
5 I _____ at home last weekend.

2 Look at the sentences in 1 and underline the expressions of past time.

3 Write sentences saying what you did, using the expressions of past time you underlined in 2.

4 Think of four or five important or memorable events in your life. Write down when they happened. Don't write down what happened.

five years ago last year three months ago...

5 Work in pairs. Show each other what you wrote in 4. Ask and say what happened.

What happened five years ago? I went to America.

6 Join the two parts of the sentence with *because*.

1 The flight was on time
2 They had to make an emergency landing
3 He didn't have a hotel reservation
4 He decided to leave

a there wasn't much air traffic.
b the room was dirty and unpleasant.
c he didn't expect to stay there long.
d they thought there was a bomb on the plane.

1 *The flight was on time because there wasn't much air traffic.*

7 Rewrite the sentences in 6 using *so*.

1 *There wasn't much air traffic so the flight was on time.*

WRITING

1 Write an opening sentence about what happened to Bob.

Bob arrived in a town at ten o'clock.

2 When you are ready, give your sentence to another student and you will receive an opening sentence from someone else. Read it, then write another sentence to continue the story.

Bob arrived in a town at ten o'clock.
He looked for a hotel.
He didn't have a reservation.
The main hotel was full.
He went down the road to a small guest house.
The lights weren't on.
He walked past it.

Continue writing and receiving sentences until the story is finished.

3 Now write the story in full by joining the sentences with *and, but, so* and *because*.

Bob arrived in a town at ten o'clock and looked for a hotel. But he didn't have a reservation and the main hotel was full. So he went down the road to a small guest house. Because the lights weren't on, he walked past it...

4 Write Ann's story in full.

9 | Family life

Possessive 's; possessive adjectives

VOCABULARY AND LISTENING

1 Look at the words in the vocabulary box. Put the words in pairs. Two words have no pairs. Which ones are they?

> aunt boy boyfriend brother child cousin daughter father friend girl girlfriend grandfather grandmother husband man mother nephew niece parent sister son uncle wife woman

aunt – uncle…

2 🔲 Listen to Corinne, who is French, talking about her family. Draw a line to show each person's relationship to her.

> Jacqueline Raymond Chantal Christine Tony Georges Vincent Marie
>
> mother sister father grandmother brother aunt husband uncle

GRAMMAR

> **Possessive 's**
> **You add 's to singular nouns to show possession.**
> *Corinne's father = her father*   *Vincent's wife = his wife*
> **You add s' to regular plural nouns.**
> *the parents' house = their house*   *the boys' mother = their mother*
> **You add 's to irregular plural nouns.**
> *the children's aunt = their aunt*   *the men's room = their room*
>
> **Possessive adjectives**
> **I – my   you – your   he – his   she – her   it – its   we – our   they – their**
> *Vincent is **my** husband.*

1 Work in pairs. Use the information in *Vocabulary and listening* activity 2 and say who these people are. Use the possessive 's.

1 Corinne (Raymond)   4 Georges (Marie)
2 Raymond (Chantal)   5 Marie (Corinne)
3 Chantal (George)   6 Corinne (Vincent)

1 Corinne is Raymond's sister.

2 Rewrite the sentences in 1 using possessive adjectives.

1 Corinne is his sister.

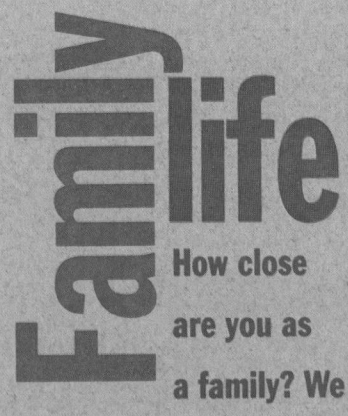

How close are you as a family? We talked to Corinne Mathieu, from Montpellier, France about her family life.

1 'We usually see each other at least once a month, maybe more often. We have lunch together on Sunday if we haven't got anything special to do. We live in Montpellier, which is about an hour and a half away, but we always come to Marseilles where my mother and father live. It's not so far. Usually my grandmother and my uncle and aunt are there too – we're quite a large family! Sometimes my brother and his girlfriend come over – they live nearby. The meal takes about four hours, we spend a lot of time chatting and there's always lots to eat.

2 'There's no one we call the head of the family, although my father's advice and opinion are very important in any decisions we take. My uncle Tony is in fact older than my father, so I suppose he's the real head of the family. When my grandfather was alive he liked to think that the whole family organised itself around him, but these days it's different. But we all try to discuss things together when we meet.

9

GENERAL COMMENTS

Possessive adjectives

Most students will already have come across possessive adjectives in a previous course, so this lesson will be revision for them. The main difficulty for speakers of languages in which nouns have a gender is that the possessive adjective refers to the person or thing it replaces rather than to the noun that follows. For example, *Stephen's wife = his wife* not *her wife, Caroline's husband = her husband* not *his husband*. Another common error, even among native speakers is to write *it's* for the possessive adjective. This is, in fact, the contracted form of *it is*.

Possessive 's

This is also known as the *Saxon genitive*. The exact position of the apostrophe depends on whether the noun is singular or plural, regular or irregular. Once again, it is common even among native speakers to place the apostrophe in the wrong position. Among purists, the seriousness of the error is compounded by the simplicity of this rule. Singular names ending in *-s* usually have *'s* , eg *Charles's sons, James's house*, but there are exceptions. You don't use an article when the first word is a proper name, eg *Vincent's wife* not *the Vincent's wife*. We often use the genitive *'s* with shops, eg *the chemist's, the baker's*. Sometimes the genitive is dropped when the noun can operate as an adjective, eg *the car door, the table leg,* but this is more advanced than can usefully be dealt with at this level.

Photo: possible questions

Where do you think this is?
What's happening?
Who are the people you can see?
Do you think they know each other well?
What time of year do you think it is?
Could this be a scene in your country?
Have you got any photos of a family group like this? Can you describe it?

Accents

The speaker in the listening passage is a French woman and speaks with the characteristic accent of the French speaking English. Students are more likely to hear English spoken by non-native speakers of the language than by native speakers, and certainly by speakers of British received pronunciation. It is important, therefore, that students are exposed to as many different accents as possible. However, please note that the accent in this lesson is not proposed as a model for pronunciation.

VOCABULARY AND LISTENING

1 Aim: to present the words in the box and to pre-teach new vocabulary in the listening passage.

- Write *aunt - uncle, boy - girl* on the board. Then ask the students to pair the other words. It's probably best to do this activity orally with the whole class.

> **Answers**
> **Pairs:** aunt - uncle, boy - girl, boyfriend - girlfriend, brother - sister, child - parent, daughter - son, father - mother, grandfather - grandmother, husband - wife, man - woman, nephew - niece
> **No pairs:** friend, cousin

- You may need to point out that the child - parent pair is the only one not based on gender. Ask the students to write down the words in their vocabulary books.

2 Aim: to practise listening for specific information.

- Explain that the students are going to listen to a French woman, Corinne, talking about her family. Some of the members of her family are in the photo. Ask the students to predict which words she is likely to use.

- 🔊 Play the tape and ask the students to draw a line to show the relationship between Corinne and the people she mentions.

> **Answers**
> Jacqueline-grandmother   Tony-uncle
> Raymond-brother   Georges-father
> Chantal-sister   Vincent-husband
> Christine-aunt   Marie-mother

GRAMMAR

1 Aim: to focus on the possessive 's.

- Ask the students to read the information about possessive adjectives and the possessive *'s* in the grammar box.

- Ask them to work in pairs and to do the exercise.

> **Answers**
> 1 Corinne is Raymond's sister.
> 2 Raymond is Chantal's brother.
> 3 Chantal is Georges's daughter.
> 4 Georges is Marie's husband.
> 5 Marie is Corinne's mother.
> 6 Corinne is Vincent's wife.

2 Aim: to focus on the possessive adjectives.

- Go round and point to objects which belong to people, eg *Mario's book, Hitomi's pen.* Then say *his book, her pen.*

- Ask the students to do the exercise.

> **Answers**
> 1 Corinne is his sister.
> 2 Raymond is her brother.
> 3 Chantal is his daughter.
> 4 Georges is her husband.
> 5 Marie is her mother.
> 6 Corinne is his wife.

SOUNDS

1 Aim: to focus on the /ə/ sound.

● The /ə/ sound or *schwa* is one of the most common sounds in English. It represents an unstressed syllable. The large number of unstressed syllables in English is what gives the impression that English is spoken rather quickly or gabbled.

● 🔲 Play the tape and ask the students to underline the *schwa*.

> **Answers**
> broth<u>er</u> daught<u>er</u> fath<u>er</u> husb<u>a</u>nd moth<u>er</u>
> par<u>e</u>nt sist<u>er</u> wom<u>a</u>n

● Ask the students to say the words aloud. In words ending in -er, pronouncing the /r/ sound will immediately make the accent sound more American.

2 Aim: to focus on contrastive stress.

● Remind the students that the words the speakers consider to be important are the words they stress. In this activity, wrong information is corrected and stressed, to underline the contrast between the correct and wrong information.

● 🔲 Play the tape and ask a student to correct the information, changing the stressed word each time. You may need to stop the tape at each pause.

> **Answers**
> 1 No, Vincent is Corinne's **husband**.
> 2 No, **Vincent** is Corinne's husband.
> 3 No, Vincent is **Corinne's** husband.
> 4 No, Vincent is Corinne's **husband**.
> 5 No, Vincent is **Corinne's** husband.
> 6 No, **Vincent** is Corinne's husband.

● 🔲 Play the tape again and ask the class to do the activity together.

READING AND SPEAKING

1 Aim: to practise reading for main ideas.

● The activity is designed to encourage the students not to get distracted with unfamiliar vocabulary. Try to resist answering vocabulary questions at this stage. Ask the students to match the questions with each paragraph.

> **Answers**
> a paragraph 2
> b paragraph 1
> d paragraph 4
> e paragraph 3
> The extra question is c.

● You may want to answer a few questions about vocabulary now. Try to limit it to six or seven words in total.

2 Aim: to practise separating general from specific information.

● It is important to be able to see when a writer is making a generalisation and when he/she is making a specific point. Ask the students to re-read the passage and to find out which paragraphs contain generalisations and which contain specific information.

> **Answers**
> **Specific information:** paragraphs 1, 2 and 4
> **Some general information:** paragraphs 3 and 4

3 Aim: to talk about reactions to the passage.

● In many cultures, it is acceptable to talk about your family to people you don't know very well. If this is the case with your students, ask them to answer the questions in activity 1 and talk about their families. If your students come from a culture where it is less acceptable to talk about personal matters, ask them to answer the questions in a very general way. Ask the students to work in groups of three or four and make a few points in answer to the questions.

● Ask the students to report back to the rest of the class and tell each other about family life in their country or countries. It doesn't matter if your students all come from the same cultural background. There may still be varying reactions that will make the students more sensitive to cultural differences.

3 'In most families, it's a small family group who live in the same house, mother, father and the children before they get married. But if one of the grandparents dies, the other usually sells their home and goes to live with their children. So it's quite common to have one grandparent living with you, but not more.

4 'In France most children leave home when they get married, and not before. I lived in Marseilles with my mother and father until I got married. But there are some people who want to lead independent lives and they find a flat as soon as they start their first job, even before they get married. Of course, the main problem is that flats are so expensive to rent here, and we simply have to live with our parents.'

SOUNDS

1 🔊 Listen and underline the /ə/ sound.

brother   daughter   father   husband   mother
parent   sister   woman

Now say the words aloud.

2 Look at this true sentence.
Vincent is Corinne's husband.

🔊 Listen and correct the statements below with the true sentence. Change the stressed word each time.

1 Vincent is Corinne's uncle.
2 Tony is Corinne's husband.
3 Vincent is Marie's husband.
4 Vincent is Corinne's brother.
5 Vincent is Chantal's husband.
6 Georges is Corinne's husband.

*1 No, Vincent is Corinne's **husband.***

READING AND SPEAKING

1 Read *Family life* and match the questions with each paragraph. There is one extra question.

a Who's the head of the family?
b How often does the family get together?
c How many people live in your house?
d How long do people live with their parents?
e How many people live in the same house?

2 Which paragraphs give specific information about Corinne's family? Which paragraphs give more general information?

3 Work in pairs. In your country, do you talk about your family to people you don't know? If so, answer the questions above with specific information about your family. If not, answer the questions above with general information about family life in your country.

10 | *The town where I live*

Have got

VOCABULARY AND LISTENING

1 Look at the words in the box. Tick (✓) any features or facilities your town has got.

> art gallery beach cathedral cafe sports stadium museum nightclub park swimming pool university port cinema theatre factory restaurant

2 📼 Listen to Catherine, an English woman who lives in Mestre near Venice. Tick (✓) the features or facilities in the box which she mentions.

3 Write down what Catherine considers to be the advantages of living in Mestre.

> 📼 Listen again and check.

4 Work in pairs. Underline the adjectives in the box.

> architecture bad beautiful boring busy cheap climate cold crowded dangerous dirty entertainment excellent expensive food good interesting large medium-sized modern old safe safety shops size small traffic

Which adjectives can you use to describe the features and facilities of your town? Can you think of other adjectives to describe them?

5 📼 Listen to six people talking about their home towns. Each one is talking about a different aspect of their town. Put the number of the speaker by the aspect they are talking about.

architecture ☐	climate ☐	cost of living ☐
entertainment ☐	food ☐	public transport ☐
safety ☐	traffic ☐	size ☐

GRAMMAR

> *Have got*
>
> **You use *have* or *have got* when you talk about facilities, possession or relationships. You can use the contracted form with *have got*, but this is not common with *have*.**
>
> *I **have** a new car.*　　　　or　　　*I've got/have got a new car.*
> *My town **has** a museum.*　　or　　*My town's got/has got a museum.*
>
> **In spoken English it is much more common to use the contracted form if possible.**
>
> **Negative**
> *I **don't have** any cigarettes.*　　or　　*I **haven't got** any cigarettes.*
> **You don't usually use *have got* in the past.**
> **When you talk about facilities, you can also use *there is/are*.**
> *It's got an art gallery.*　　　or　　　*There's an art gallery.*

10

GENERAL COMMENTS

Has/have got

This is a particularly common structure in English, but is likely to be revision for most students. The focus of the lesson is, therefore, more on the vocabulary and the listening activities.

Main photo: the Grand Canal in Venice and the church of Santa Maria della Salute; in the distance is the industrial town on the mainland, Mestre.

Possible questions

Where is the photo?
What can you see in the foreground?
What words can you use to describe Venice?
What can you see in the distance?

Right-hand photo: possible questions

What can you see?
What's special about getting around in Venice?
What advantages or disadvantages are there in having no cars or lorries?

Vocabulary

The vocabulary area covered in this lesson, town facilities and related adjectives, is a very large one. It may be advisable to restrict active acquisition to items which are personally relevant to the student and the town where he/she lives.

VOCABULARY AND LISTENING

1 Aim: to present the words in the vocabulary box and to pre-teach important items to be heard in the listening passage.

● Before the students begin this activity, write the title of the lesson on the board and ask them to suggest any words they are likely to see in the lesson.

● Ask the students to look at the words in the box and tick the ones which they can use to describe their town. These are the words which they should make an effort to learn. They should also list other words which they have come across which can be used to describe their town. Some of the vocabulary in Lessons 4 and 5 will be useful, so suggest that they should look again at the relevant vocabulary boxes and notes to see which ones they can add to their lists.

2 Aim: to practise listening for specific information.

● 🔲 Explain that the students will hear some of the words in the box, and that they should tick them as they hear them.

Tapescript and answers

INTERVIEWER What's Mestre like?

CATHERINE   Oh, it's not bad. I mean, people don't really visit Mestre at all, because they're on their way through to Venice. But in a way, it's easier to live in Mestre, rather than Venice.

INTERVIEWER Why's that?
CATHERINE   Well, it's cheaper, for one and it's a lot more convenient being on the mainland.
INTERVIEWER Oh, right.
CATHERINE   But it's quite industrial, really, it's got a **port** and quite a few **factories**. I mean, it's not a pretty place to look at.
INTERVIEWER What's the rest of the town like?
CATHERINE   Well, as far as culture goes, it hasn't got a **cathedral** or an **art gallery**. They're all in Venice. But the shops are very good, I think, and it's got a nice **park** as well.
INTERVIEWER Any sports facilities?
CATHERINE   Unfortunately, nothing very big, no.
INTERVIEWER Would you say the night life is good?
CATHERINE   No, not really, I mean, it's got **restaurants** – the food can be very good – and a **cinema**, which we go to quite often. But if you want **theatre** or discos you have to go to Venice.

● The other facilities Catherine mentions are shops and discos. Note that the words the students tick are the words she mentions, not necessarily the facilities in Mestre. Play the tape again and ask the students to identify the facilities Mestre has got.

3 Aim: to check comprehension; to infer information.

● Ask the students to think about the passage and to select the advantages and disadvantages of living in Mestre. Write a + sign and a - sign on the board and ask one or two students to suggest advantages and disadvantages. Sometimes the advantage or disadvantage is signalled simply by the preceding discourse marker, eg *but, unfortunately,* etc.

● Ask the students to work in pairs and to write their own lists of advantages and disadvantages.

Answers

+: easier to live there than in Venice, cheaper, more convenient, good shops, nice park, restaurants, cinema
–: quite industrial, not a pretty place, no culture, no sports facilities, no theatre or discos

● You may like to ask the students to discuss whether they agree with what Catherine considers to be advantages and disadvantages. For example, some people may think it is an advantage to have no discos.

4 Aim: to present the words in the vocabulary box; to distinguish between parts of speech.

● Ask the students to underline the adjectives in the box.

Answers

bad, beautiful, boring, busy, cheap, cold, crowded, dangerous, dirty, excellent, expensive, food, good, interesting, large, medium-sized, modern, old, safe, small

- Point out that in fact *public* is part of the phrase *public transport*. The other words are nouns and are *aspects of a town*. Ask them to suggest words to describe aspects of their town and write them on the board.

- Ask the students to suggest adjectives which can go with the aspects of their town. Write the adjectives on the board under the aspects.

5 Aim: to practise listening for main ideas.

- 🔲 Explain to the students that they are going to listen to six people talking about different aspects of their town. The students should listen for the main idea and not worry about difficult vocabulary.

> **Answers**
> architecture   [3]
> climate   [5]
> cost of living   [1]
> safety   [6]
> traffic   [2]
> size   [4]

- You may want to play the tape again when you have checked the answers.

GRAMMAR

1 Aim: to focus on contractions in *has/have got*.

- Ask the students to read the information in the grammar box.

- Ask the students to do the activity orally, saying the sentences with contractions.

- Ask the students to write the sentences with contractions.

> **Answers**
> 1 The university's got a park.
> 2 It's got a modern tram system.
> 3 He's got a swimming pool.
> 4 I've got tickets for the theatre.
> 5 Rio's got some beautiful beaches.
> 6 She's got a good view from the window.

2 Aim: to practise using *have got*.

- Ask the students to say what their town has or hasn't got. You can ask them questions using the words in *Vocabulary and listening* activity 1 to help them, eg *Has it got an art gallery? Has it got a beach?* and elicit a short answer *Yes, it has or No, it hasn't.*

- Ask the students to talk about their towns in pairs. Go round and check they are using *have got* properly.

WRITING

1 Aim: to read for specific information; to read a passage to be used as a model for guided writing practice.

- The main purpose of this passage is to provide a model which the students will use in activity 6 to write a similar passage about a town of their choice. This task is simply designed to make the students read the model passage with some purpose in mind.

> **Answers**
> **Aspects mentioned:** size, public transport, cost of living, shopping, entertainment

2 Aim: to read and infer advantages and disadvantages.

- Ask the students to make a list of the advantages and disadvantages which Anna mentions.

> **Answers**
> **Advantages:** small city, full of history, no cars or lorries, good public transport, not far to walk
> **Disadvantages:** lots of tourists, expensive, not much to do

3 Aim: to link two parts of a sentence with *and* and *but*.

- All of the advantages can be linked together using *and*. Ask the students to join two advantages or two disadvantages with *and*.

- Not all of the advantages and disadvantages can be linked using *but*, so make sure that the students choose the pairs carefully.

4 Aim: to write notes about the advantages and disadvantages of towns.

- If the students all come from the same city, you can do this orally with the whole class. Ask the students to suggest advantages and disadvantages and write them on the board. They can include aspects which have not been mentioned so far.

5 Aim: to write linked sentences about the advantages and disadvantages of towns.

- Ask the students to join together the notes they made in activity 4 with *and* and *but*, and write full sentences.

6 Aim: to write a short description of a town.

- Ask the students to use the sentences they wrote in activity 5 in a short description of their town. The reading passage about Venice can be used as a model. Suggest that they include some more personalised information in their descriptions.

- This activity can be done for homework.

1 Rewrite these sentences with contractions.

1 The university has got a park.
2 It has got a modern tram system.
3 He has got a swimming pool.
4 I have got tickets for the theatre.
5 Rio has got some beautiful beaches.
6 She has got a good view from the window.

2 Work in pairs. Say what features and facilities your town has and hasn't got.

It's got a cathedral, but it hasn't got a theatre.

WRITING

1 Read this description of Venice by Anna, a student. Which aspects in *Listening* activity 5 does she mention?

2 Read the passage again and write down all the advantages and disadvantages of living in Venice.

Advantages: small city, full of history...
Disadvantages: lots of tourists, expensive...

3 Join two advantages or two disadvantages together using *and.*

It's quite a small city and it's full of history.

Then join an advantage and a disadvantage together with *but.* Make sure they are all features which can go together.

*There are no cars and lorries **but** there are lots of tourists.*

4 Make a list of the advantages and disadvantages of living in your town.

Advantages	**Disadvantages**
good public transport	too much traffic
excellent shops	very expensive

5 Write sentences joining advantages or disadvantages using *and.* Then write sentences joining an advantage with a disadvantage using *but.*

It's got good public transport and the shops are excellent.
There's too much traffic but the public transport is good.

6 Write a short description of your town. Use Anna's description and the sentences you wrote in 5 to help you.

The town where I live

I love living in Venice, it's full of history. I like it because it's quite a small city. I think it's got a population of about 200,000 people, but there are lots of tourists, especially in the summer, and there's not enough room for them all. Of course, there are no cars or lorries, but there are water buses on the main canals all day and night, and it's not far to walk anywhere in the city. But you need lots of money to live here. Flats are very expensive and everything comes from the mainland, so the prices of everyday supermarket items are rather high. And apart from cinemas and theatres, there's not much to do in the evenings. But I still love it here.

Progress check **6-10**

VOCABULARY

1 Work in pairs and make a *word chain*. Each word must be associated with the word immediately before it. Start like this:

hotel   reception   reservation   double room
bed   breakfast

2 Some nouns go together to make a new word. Sometimes you write them as two words: *double room,* and sometimes as one word: *timetable.*

Put a noun from list A with a noun from list B and make at least 12 new words.

A package   travel   round   return   train   first
booking   boarding   air   suit

B journey   pass   terminal   class   case   tour   trip
agent   ticket   office

3 It is useful to write down new vocabulary under headings, such as *travel, towns, family* etc. You can also use more personal or impressionistic categories, such as words that sound nice, or words that look like words in your own language. Look at the vocabulary boxes in lessons 6 – 10 again. Choose words which are useful to you and group them under headings of your choice in your *Wordbank*.

GRAMMAR

1 Write these regular verbs in the past simple tense.

ask   carry   change   continue   decide   enjoy   finish
happen   like   listen   live   look   open   play   start
stay   stop   talk   test   try   travel   visit   walk   watch

2 Write these irregular verbs in the past simple tense.

be   become   do   go   get   have   know   leave
make   put   run   see   sit   sleep   take   tell

3 Complete this conversation with *did, didn't, was, wasn't, were* or *weren't.*

1 '___ you enjoy your holiday?' 'Yes, I ___ .'
2 '___ the weather good?' 'Yes, it ___ .'
3 '___ you go to the local museums?' 'Yes, we ___ . They ___ very interesting.'
4 '___ you send any postcards?' 'No, we ___ .'
5 '___ you buy anything?' 'No, we ___ have any money.'
6 '___ you pleased with hotel?' 'Yes, we ___ .'

4 Write short answers to these questions about Thomas Cook. Give the correct information.

1 Did he live in Nottingham?
2 Was his first trip to Loughborough?
3 Did he take people to Manchester in 1845?
4 Did he open an office in London?
5 Were tourists interested in the Civil War battlefields?
6 Was it dangerous to carry large amounts of cash?

5 Write questions about Thomas Cook using the prompts below.

1 what/do?
2 what/be son's name?
3 where/go on first trip abroad?
4 when/publish the first travel magazine?
5 what/take to the Holy Land?
6 when/die?

6 Look back at the passage on page 16 and write answers to the questions in 5.
He was a printer.

7 Complete these sentences with *I, my, you, your, he, his, she, her, we, our, they* or *their.*

1 I know ___ face. Is she famous?
2 We know them quite well. ___ children go to the same school as ours.
3 Philip, this is ___ friend, Mary.
4 What do you do, Peter? What's ___ job?
5 I'll call you later. What's ___ telephone number?
6 That belongs to me. It's got ___ name on it.

Progress check 6–10

GENERAL COMMENTS

You can work through this Progress check in the order shown, or concentrate on areas which may have caused difficulty in Lessons 6 to 10. You can also let the students choose the activities which they would like to or feel the need to do.

VOCABULARY

1 Aim: to revise vocabulary and to remember words which can be grouped together.
- This activity is not meant to be a serious word-association test, so give the students between five to ten seconds to think of a suitable word. Play one round of the game with the class as a whole.

- Ask the students to work in pairs and continue playing the game.

2 Aim: to focus on compound nouns.
- In some languages it is not possible to combine two nouns to make a compound noun. In English the first noun operates like an adjective, but the compound usually represents a single concept. The rules for writing them as one, two or hyphenated words are complex, so suggest to the students that they have to take note of this information at the same time as they learn the new compound noun.

> **Answers**
> package tour, travel agent, round trip, return journey, return trip, return ticket, train journey, train trip, train ticket, first class, booking agent, booking office, boarding pass, air terminal, air ticket, suitcase

3 Aim: to help the students organise their vocabulary learning.
- Categorising is a useful way of organising new vocabulary and making the learning process more effective. It is often even more effective if the students choose these categories themselves. In *Reward* there are often categories which are suggested in the activity and then categories which the students are invited to create for themselves. This activity will give the students an opportunity to go back over lessons 6 to 10, to remind themselves of the words presented in the vocabulary boxes, and to create new categories which consolidate the process of acquisition.

GRAMMAR

1 Aim: to revise the past simple of regular verbs.

> **Answers**
> asked, carried, changed, continued, decided, enjoyed, finished, happened, liked, listened, lived, looked, opened, played, started, stayed, stopped, talked, tested, tried, travelled, visited, walked, watched

2 Aim: to revise the past simple of irregular verbs.

> **Answers**
> was/were, became, did, went, got, had, knew, left, made, put, ran, saw, sat, slept, took, told

3 Aim: to revise the use of auxiliaries in questions and short answers.

> **Answers**
> 1 **Did** you enjoy your holiday? Yes, I **did**.
> 2 **Was** the weather good? Yes, it **was**.
> 3 **Did** you go to the local museums? Yes, we **did**. They **were** very interesting.
> 4 **Did** you send any postcards? No, we **didn't**.
> 5 **Did** you buy anything? No, we **didn't** have any money.
> 6 **Were** you pleased with the hotel? Yes, we **were**.

4 Aim: to practise using the past simple.

> **Answers**
> 1 No, he didn't. He lived in Leicester.
> 2 Yes, it was.
> 3 No, he didn't. He took people to Liverpool.
> 4 Yes, he did.
> 5 Yes, they were.
> 6 Yes, it was.

5 Aim: to practise writing past simple questions.

> **Answers**
> 1 What did he do?
> 2 What was his son's name?
> 3 Where did he go on his first trip abroad?
> 4 When did he publish the first travel magazine?
> 5 What did he take to the Holy Land?
> 6 When did he die?

6 Aim: to practise using the past simple.

> **Answers**
> 1 He was a printer.
> 2 His son's name was John Mason.
> 3 He went on his first trip abroad in 1855.
> 4 He published the world's first travel magazine in 1851.
> 5 He took tents to the Holy Land.
> 6 He died in 1892.

7 Aim: to practise possessive pronouns.

> **Answers**
> 1 I know **her** face. Is she famous?
> 2 We know them quite well. **Their** children go to the same school as ours.
> 3 Philip, this is **my** friend, Mary.
> 4 What do you do, Peter? What's **your** job?
> 5 I'll call you later. What's **your** telephone number?
> 6 That belongs to me. It's got **my** name on it.

SOUNDS

1 Aim: to practise /ð/ and /θ/.

> **Answers**
> /ð /: this they
> /θ/: theatre cathedral think thank you

2 Aim: to practise /ɒ/ and /əʊ/.

> **Answers**
> /ɒ/: not lots shop what opera
> /əʊ/: know boat photo video don't go disco opening

3 Aim: to focus on friendly intonation.

> **Answers**
> 1 Did you arrive late this morning? No, I didn't. ✔
> 2 She didn't say hello. Yes, she did.
> 3 Did you give me your passport? Yes, I did. ✔
> 4 Did they pay you for the tickets? No, they didn't. ✔

WRITING

1 Aim: to practise writing a story and to create a reading comprehension check.
- This activity is designed to show the students that they can predict a great deal about a passage by looking at the questions. Make quite sure they write full answers to the questions.
- You may like to do this orally and write the sentences on the board. Every time someone guesses information, ask them *Are you sure?* If they aren't sure, then simply leave a blank.

2 Aim: to practise reading for specific information.
- The sentences the students have written in activity 1 constitute the reading passage with blanks. They have in fact written their own reading comprehension check. Ask them to read and fill in the blanks in their versions.

SPEAKING

1 Aim: to practise speaking using the past simple.
- Ask one or two students the following questions:
  Did you leave home early this morning?
  Did you have a holiday three months ago?
  If they say *yes,* write their names on the board.

- Ask the students to go round the class with a pen and paper, asking and answering the questions and writing down the names of people who answer *yes*. Give them about four or five minutes to do this activity.

2 Aim: to practise talking about families.
- To illustrate the expression *family tree*, draw your family tree on the board.

- Encourage the students to be as inventive or imaginative as they like about their false relatives. They can invent as much information about them as they like.

3 Aim: to practise talking about families.
- Ask the students to show other people their family trees and to talk about their families. Explain that they must guess who the false relatives are.

- Ask the class to report back on the most unlikely false relatives they found.

- If there isn't time, ask them to write a short description of their family and false relatives. They can bring the descriptions to the next class and put them on the wall for the others to read at their leisure.

SOUNDS

1 📼 Listen and say these words aloud.

<u>th</u>is <u>th</u>eatre ca<u>th</u>edral <u>th</u>ink <u>th</u>ank you <u>th</u>ey

Is the underlined sound /ð/ or /θ/? Put the words in two columns.

2 📼 Listen and say these words aloud.

n<u>o</u>t kn<u>ow</u> l<u>o</u>ts b<u>oa</u>t ph<u>o</u>to vide<u>o</u> sh<u>o</u>p d<u>o</u>n't g<u>o</u>
wh<u>a</u>t disc<u>o</u> <u>o</u>pera <u>o</u>pening

Is the underlined sound /ɒ/ or /əʊ/? Put the words in two columns.

3 📼 Listen to the sentences. Put a tick (✓) if you think speaker *B* sounds friendly.

1 **A** Did you arrive late this morning?   **B** No, I didn't.
2 **A** She didn't say hello.   **B** Yes, she did.
3 **A** Did you give me your passport?   **B** Yes, I did.
4 **A** Did they pay you for the tickets?   **B** No, they didn't.

Now work in pairs and say the sentences aloud. Try to sound friendly.

WRITING

1 You are going to write a story called *The least successful annual conference*. Look at the questions below and try to guess the answers. Write full answers to the questions. Leave a blank when there is any information that you don't know.

> ### THE LEAST SUCCESSFUL ANNUAL CONFERENCE
> •
> **1 Where did the Association of British Travel Agents go for their annual conference in 1985?**
>
> **2 Why was the flight from Gatwick to Naples delayed?**
>
> **3 Why were many people ill?**
>
> **4 Who did the organisers ask to give a speech to the delegates in the forum at Pompeii?**
>
> **5 What did a local travel agent decide to do as a gesture of friendship?**
>
> **6 What happened as the Minister began his speech?**
>
> **7 Where did the flowers land?**
>
> **8 Why did no one hear the speech?**
>
> **9 How many times did the plane pass over Pompeii?**
>
> **10 Where did the roses land each time?**
>
> **11 What happened the last time it flew over the delegates?**

1 In 1985 the Association of British Travel Agents went to ___ for their annual conference.

2 Now turn to Communication activity 10 on page 99 and read the story. Fill in the blanks in your version.

SPEAKING

1 Find someone in your class who:

– left home early this morning
– travelled by plane last month
– had a holiday three months ago
– went on a bus yesterday
– walked to school/work ten days ago
– took a train last week
– stayed in a hotel last year
– went abroad in 1993

2 Draw your family tree, but include two 'false' relatives who don't exist. Think of some interesting or unusual information about the homes of each relative and invent information for the two relatives who don't exist.

3 Work in pairs. Ask and answer questions about your families. Try and guess who the 'false' relatives are.

How ambitious are you?

Verb patterns (2): *to* + infinitive; *going to* for intentions, *would like to* for ambitions

READING

1 How ambitious are you? Put a tick (✓) by the ambitions you have. Have you any other ambitions?

– stop work
– learn to fly
– learn to ski
– have children
– go back to university
– live in the country

– move house
– run a marathon
– work abroad
– write a novel
– earn a lot of money
– go to Disneyland

– change your job
– learn a foreign language
– be healthy and happy
– travel around the world
– become famous

2 Read *How ambitious are you?* and answer the questions.

3 Turn to Communication activity 9 on page 99 and find out how ambitious you are.

How ambitious are you?

1 Your neighbour buys a new Porsche. What do you say?
  a I don't need an expensive car.
  b One day I'm going to buy one, too.
  c I'd like to let the air out of its tyres.

2 Your boss leaves very suddenly. What do you think?
  a They're going to give me his/her job.
  b They're going to appoint someone else.
  c They're going to say it's my fault.

3 Ihich ambition do you have?
  a I'd like to be rich.
  b I'd like to be famous.
  c I'd like a drink.

4 You aren't happy with your job. What would you like?
  a More responsibility.
  b More money.
  c More weekend.

5 Your best friend is writing a novel. What do you say?
  a I'm going to write a best-selling novel.
  b I'm going to write a postcard.
  c What a coincidence! You're writing a novel and I'm reading one.

6 What does your future hold for you?
  a I'm going to be president.
  b I'm going to be happy.
  c I'm going to be late.

7 There's a marathon race on television. What do you say?
  a I'm going to do that next year.
  b I'd like to do that but I'm not very fit.
  c What's on the other channel?

8 Which of these statements do you agree with?
  a Every day, in every way, I'm getting better and better.
  b Tomorrow is the start of the rest of my life.
  c If you don't succeed, try again. Then give up.

11

GENERAL COMMENTS

Verb patterns (2): *to* + infinitive

In Lesson 4 the first verb pattern was presented. It was explained that when there are two verbs in a sequence the second verb is either an infinitive or an *-ing* form. This lessons focuses on *to* + infinitive.

Going to for intentions, *would like to* for ambitions

A common use of *to* + infinitive is to talk about plans, ambitions, hopes and preferences. This is the first of a series of lessons which presents ways of talking about the future.

Questionnaires

In British magazines, questionnaires are very common. Unlike other kinds of reading material, they address the reader directly and establish a form of interaction. This is the intention of many of the passages, especially the questionnaires, in *Reward*. The questionnaire in this lesson is not meant to be taken too seriously.

Illustration: possible questions

Look at the two people.
Where are they now?
What are they doing?
Are they pleased with what they're doing?
Where would they like to be?
What would they like to do?

READING

1 Aim: to prepare for reading *How ambitious are you?*; to create the framework for practising the structures in the lesson.

● Ask the students *Are you ambitious?* If they answer *yes*, ask them to describe what being ambitious means. Ask them if they have had the chance to achieve any of their ambitions yet?

● Ask the students to look at the ambitions in the list and to tick the ones they have. Strictly speaking, they do not yet have the language to talk about their ambitions and say *I'd like to...*, but they can at least say which ones they ticked. In *Grammar* activity 2, they will return to these ambitions.

2 Aim: to read and answer a questionnaire.

● The questions in the questionnaire demand a detailed understanding in order to be answered properly. Ask the students to write down the letter corresponding to their answers. Explain, if necessary, that the questionnaire is meant to be lighthearted.

3 Aim: to evaluate the students' answers to the questionnaire.

● Ask the students to turn to the Communication activity on page 99 and find out how ambitious they are. Ask them if they agree with the analysis.

● The students may like to share their reactions with other people. Ask them to find out how other people answered each question.

● Finish the activity by finding out how the class as a whole answered the questionnaire.

GRAMMAR

1 Aim: to focus on the difference between *would like to* and *going to*.

● Ask the students to read the information about verb patterns in the grammar box.

● Ask the students to do the exercise on their own.

● Check the answers to the exercise orally.

> **Answers**
> 1 He's got a place at Essex University. He is going to study there.
> 2 She's got her plane ticket and she's going to go to Canada.
> 3 He'd like to buy a new car, but it's too expensive.
> 4 I'd like to work in television but there aren't many jobs.
> 5 She enjoys her job. She isn't going to change it.
> 6 He's got a new job with a foreign company. He is going to work abroad.

2 Aim: to practise using *would like to* and *going to*.

● Ask the students to look back at the ambitions they ticked in *Reading* activity 1. Ask one or two people to say what they'd like to do.

● Ask the students to work in pairs and to talk about their ambitions.

● Invite the students to discuss at this stage if they have any other more personal ambitions. Which do they think is the most extraordinary ambition?

VOCABULARY AND WRITING

1 Aim: to present the words in the vocabulary box; to focus on words which go together.

● The verbs in the box collocate with the nouns. Ask the students to put the verbs and the nouns together.

> **Answers**
> work abroad, change jobs, earn money, move house, study a foreign language, learn a foreign language, run a marathon, read a novel, write a novel, stop work

2 Aim: to read a passage to be used as a model for guided writing practice.

● The passage is presented as a model for guided writing practice. The students will use the passage as a model when they reach activity 9. The four prompts can also be used to guide the students in the writing of their own passages.

> **Answers**
> **what she'd like to do:** learn to fly a light airplane
> **why she'd like to do it:** to know what it feels like to be in control
> **what she needs to do to achieve it:** take flying lessons
> **what she's going to do:** take her pilot's test

3 Aim: to write notes about ambitions.

● Ask the students to list their ambitions. Remind them of the ambitions they talked about in *Grammar* activity 2.

4 Aim: to expand notes about ambitions.

● Ask the students to make notes about why they'd like to achieve one of their ambitions.

5 Aim: to expand notes about ambitions.

● Ask the students to expand their notes by saying what they need to do to achieve this ambition.

6 Aim: to expand notes about ambitions.

● Ask the students to make notes about what they're going to do to achieve their ambition. They can be as inventive as they wish at this stage!

7 Aim: to find out about other students' ambitions and to make notes.

● Explain that the passage the students are going to write is about their partner's ambitions, so ask them to work in pairs and to ask and answer questions. They need information about *what he/she'd like to do, why he/she'd like to do it, what he/she needs to do to achieve it* and *what he/she's going to do*. It may be helpful to write these four points on the board.

8 Aim: to turn notes into sentences using *because* and *so*.

● Remind the students that they used *so* and *because* in Lesson 8. Check that they can still use these link words correctly by asking for a few notes from one or two students and write them on the board. Then rewrite them using the link words.

9 Aim: to turn sentences into a paragraph.

● By this stage, the notes should be tight enough to be turned into a paragraph. The passage the students read in activity 2 can be used as the model.

● If time is short, this activity sequence can be completed for homework.

GRAMMAR AND FUNCTIONS

> ### Verb patterns (2): *to* + infinitive
> **You can put *to* + infinitive after many verbs.**
> I **want to** leave now.   He **decided to** drive to work.
> She's **learning to** fly.   She **needs to** pass her exam.
>
> ### Going to, would like to
> **You can use *going to* + infinitive to talk about future intentions or plans which are fairly certain.**
> I'm studying medicine. I**'m going to** be a doctor.
> I**'m not going to** be an accountant.
>
> **You can use *would like to* + infinitive to talk about ambitions, hopes or preferences.**
> I**'d like to** speak English fluently.   I **wouldn't like to** run a marathon.
>
> **Remember that *like* + *-ing* means *enjoy*.**
> I like learning English. = I enjoy learning English.
>
> **For more information see Verb patterns (1) on page 9.**

1 Choose the correct verb pattern.

1 He's got a place at Essex University. *He would like to/He is going to* study there.
2 She's got her plane ticket and *she'd like to/she's going to* go to Canada.
3 *He'd like to/He is going to* buy a new car, but it's too expensive.
4 *I'd like to/I'm going to* work in television but there aren't many jobs.
5 She enjoys her job. *She wouldn't like to/She isn't going to* change it.
6 He's got a new job with a foreign company. *He'd like to/He is going to* work abroad.

2 Work in pairs and talk about the ambitions you ticked in *Reading* activity 1.
I'd like to run a marathon.

VOCABULARY AND WRITING

1 Match the verbs with the nouns in the box below.

| abroad | change | earn | jobs | house | a foreign language | learn | a marathon |
| money | move | a novel | read | stop | run | study | work | write |

Can you think of other nouns which can go with the verbs?

2 Look at this passage about Hanna's ambition. Find out:

– what she'd like to do   – what she needs to do to achieve it
– why she'd like to do it   – what she's going to do

> **H**anna is forty-five years old and is a technician in Ludwigshafen, Germany. She wants to learn to fly a light airplane because she loves flying. 'I'm always a passenger and I'd really like to know what it feels like to be in control,' she says. So she's taking flying lessons at her nearest private airport, and she's going to take her pilot's test in three weeks' time. 'It's not going to be easy, but I love it. It's going to change my life.'

3 Think about your ambitions. Make a list of what you'd like to do.
earn a lot of money, write a novel...

4 Choose one ambition, and make notes about why you'd like to do it.
earn a lot of money – buy a big house...

5 Now make notes about what you need to do to achieve it.
change my job, learn a foreign language...

6 Finally, make notes about what you're going to do to achieve your ambition.
look at job advertisements,
go to evening class...

7 Work in pairs. Ask questions about your partner's ambition and take notes. Answer questions about your ambition.

8 From your notes, join what your partner wants to do and why with *because*.
Imogen would like to change her job because she wants to work abroad.

Join what your partner needs to do and what he/she's going to do with *so*.
She needs to decide where she wants to live, so she's going to travel round Europe.

9 Write a paragraph about your partner's ambition. Use the prompts and passage in 2 to help you.

12 English in the future

Will for predictions

VOCABULARY

1 Look at the words in the box and put them under two headings: *jobs* and *subjects*.

> accountant actor arithmetic banker biology chemistry computer science
> dancer doctor economics engineer geography history journalist languages
> maths nurse physics physical education politician secretary

2 Work in pairs and look at the lists you made in 1. Which jobs do you need English for? Which subjects do you need English in order to study? Are there any other jobs and subjects you need English for?

LISTENING

1 Think about learning English in the future in your country. Which of these predictions do you agree with? Put a tick (✓) if you agree and a cross (✗) if you disagree.

	You	Lynne	Greg	Your partner
Children will learn English from the age of six.				
There will be few adults who don't speak English.				
More lessons at school will be in English.				
Everyone will need to learn about British and American culture.				
Everyone will need English for their jobs.				
Everyone will learn English at home by television and computers.				
It will be more important to speak English than your own language.				

2 🔲 Listen to two English teachers talking about the statements. Put a tick (✓) if the speaker agrees with the statements, a cross (✗) if he or she disagrees and put *?* if it's not clear.

3 Work in pairs. Can you remember what Lynne and Greg said?

🔲 Listen again and check.

12

GENERAL COMMENTS

Will and shall

This lessons focuses on the use of *will* to make predictions or express an opinion about the future. Some older grammar books suggest that *shall* is always used for the first person singular and plural, eg *I shall see you tomorrow. We shall all speak English in the future.* This is not true nowadays. *Shall* is often used when someone is offering to do something, eg *Shall I do that for you?* or as a question tag *I'll do that, shall I?* This is covered in Lesson 30. In any case, the contraction *'ll* is used in everyday speech for *shall* and *will*.

English in the future

The theme of this lesson has been chosen to stimulate some discussion about the role and uses of English as a medium of international communication. Some students, particularly those still in secondary education, may simply see English as another school subject to pass a test in, and may not realise its potential use beyond the classroom. Making students aware of how much English there is around them, even in towns away from international centres or tourist attractions, is an important student training and language awareness objective.

Illustration: possible questions

The people and objects in the illustrations all have something to do with English. What you can see and what are the people doing?
How is English useful to them?

VOCABULARY

1 Aim: to present the words in the vocabulary box and to pre-teach any items which may occur in the listening passage.

● Write the two headings: *jobs* and *subjects* on the board and ask one or two students to put the words under the headings. Say the word *accountant* and elicit the reply *job*. Do several words like this.

● Ask the students to write the words under the two headings.

2 Aim: to practise using the words in the vocabulary box and to start thinking about the theme of the listening material.

● The students may not know which jobs they might need English for, and this will vary from country to country, so no answer is given. It is, therefore, important that you check you can answer the question for your country. In general, however, English is often needed for the following jobs: banker, doctor, journalist, politician and secretary. The subjects you often need English in order to study are computer science, economics and perhaps science in general.

LISTENING

1 Aim: to prepare for listening.

● This activity is designed to encourage the students to prepare their opinions on the role of English. It is also intended to let them prepare for the listening passage. Read out one or two of the opinions and ask students if they agree or disagree with them.

● Then ask them to put a tick or a cross by all the other predictions in the chart.

2 Aim: to practise listening for main ideas.

● 🔲 This listening passage makes no particular concessions to the non-native speaker, and especially not the student at pre-intermediate level. It is not expected that the students will understand every word, but it is hoped that they will understand enough of the general sense to be able to complete the task. Play the tape and ask the students to put a tick if the speaker agrees and a cross if the speaker disagrees with the statements, and to put a question mark if it's not clear.

Answers	Lynne	Greg
Children will learn English from the age of six.	✔	✔
There will be few adults who don't speak English.	✔	✔
More lessons at school will be in English.	✔	✗
Everyone will need to learn about British and American culture.	✗	✔
Everyone will need English for their jobs.	✔	✗
Everyone will learn English at home by television and computers.	✗	✔
It will be more important to speak English than your own language.	✗	✗

3 Aim: to discuss the main ideas and details about what the speakers said.

● Ask the students to work in pairs and discuss what Lynne and Greg said. You could ask them to fill in the chart with as much detail as they can remember.

● Ask one student from each pair to go round other pairs and find out more details. Ask them to take this information back to their partner and add it to their charts.

● 🔲 Play the tape a second time and ask the students to check their answers.

GRAMMAR

1 Aim: to practise using *I will* and *I won't* for predictions.

● Ask the students to read the information about *will* in the grammar box. Then ask them to do the exercise.

● This activity is designed to make students think about how much progress they have made and are likely to make in English by the end of the course. You may like to do the activity orally and to discuss the answers with the rest of the class.

2 Aim: to practise asking questions with *Do you think + will?*

● Ask the students to write full questions with the prompts.

> **Answers**
> 1 Do you think computers will replace secretaries and accountants?
> 2 Do you think journalists will disappear because there won't be any newspapers.
> 3 Do you think economics will be the most important school subject?
> 4 Do you think teaching by television will be very common?
> 5 Do you think people won't need to study maths because computers can do the calculations?
> 6 Do you think that few people will need geography and history?

3 Aim: to practise asking and answering questions about the future.

● Ask the students to work in pairs and to ask and answer the questions they wrote in activity 2.

● Ask the students to go round the class asking and answering the questions they wrote in activity 2. Ask them to put a tick by the questions which people answer with yes and a cross by the questions people answer with *no*. They should finish the activity with a number of ticks and crosses by each question.

● Find out how many people answered each question with *yes* or *no*. If there is time, ask the students to write a brief paragraph describing the results.

SOUNDS

1 Aim: to focus on word stress.

● Copy the columns on the board and number them 1, 2 and 3. Point out that the numbers 1, 2 and 3 refer to the syllable that is stressed in the words which go in the columns. Say one or two of the words and ask *First, second or third syllable?* Elicit the right answer and write the word in the correct column.

● Ask the students to list the words according to their stressed syllable.

> **Answers**
>
> | First | Second | Third |
> |---|---|---|
> | actor | accountant | economics |
> | banker | arithmetic | education |
> | chemistry | biology | engineer |
> | secretary | computer | politician |
> | geography | | |

● ▭ Play the tape and allow the students to check their answers.

2 Aim: to focus on /e/ and /eɪ/.

● ▭ Ask the students to listen and repeat the words. Play the tape. Pause after each word. Make sure they distinguish between /e/ and /eɪ/.

> **Answers**
> /e/: lesson chemistry education secretary friend
> /eɪ/: age railway station holiday

● It is important to draw the students' attention to the different spellings represented by /eɪ/.

SPEAKING AND WRITING

1 Aim: to practise using *will*.

● Ask the students to check their answers to *Listening* activity 2, and to complete the chart. Ask them to make other predictions about English in the future.

2 Aim: to practise using *will*.

● Ask the students to go round asking and answering questions about the future of English based on the predictions they made in activity 1. As they go round, they should put a tick or a cross by the statements.

3 Aim: to write a report of the class's opinions about the future of English.

● Write the statements on the board. When the students have interviewed several people, read each statement aloud and ask them to raise their hand if they agree with it. Count the number of people who agree and write the number by the statement. Read the statement again and ask them to raise their hand if they disagree with it. Write the number by the statement. Do the same for each statement.

● The students will now be able to write a paragraph saying what people in the class think. Use the example to draw attention to the verb endings with *all, most, some* (plural) and *nobody* (singular). If there is no time, students can do this for homework.

GRAMMAR

> *Will*
>
> **You use *will* to make a prediction or express an opinion about the future.**
>
> *Children **will learn** English from the age of six.*
> *I think most people **will need** English for their jobs.*
> *I'm sure everyone **will speak** some English.*
>
> **You form the future simple with *will* + infinitive. You often use the contracted form *'ll*.**
>
> *At the end of the course I**'ll speak** English fluently.*
>
> **Negatives**
>
> *There **won't be** traditional language classes in school.*
> *There definitely **won't be** many teachers.*
>
> | **Questions** | **Short answers** |
> |---|---|
> | ***Will** we **use** English at work?* | *Yes, we **will**.* |
> | | *No, we **won't**.* |

1 Think about the end of your course. Make predictions about your level of English. Use *I will* or *I won't*.

1 speak English fluently
2 be able to read an English newspaper
3 be able to understand radio broadcasts
4 be able to write reports in English
5 speak with a perfect English accent
6 be able to understand English songs
7 know a lot of vocabulary
8 use English for my work

2 Look at these predictions about jobs and studying in the future. Do you agree or disagree? Write questions to ask another student.

1 computers – replace secretaries, accountants
2 journalists – disappear – because no newspapers
3 economics – the most important school subject
4 teaching by television – very common
5 people – no need to study maths – because computers can do calculations
6 few people – need geography, history

1 Do you think computers will replace secretaries and accountants?

3 Now work in pairs and find out what your partner thinks.

Do you think computers will replace secretaries and accountants? No, I don't.

SOUNDS

1 Say these words aloud. Which words are stressed on the first syllable? Which words are stressed on the second syllable? Which words are stressed on the third syllable? Put them in three columns.

accountant actor arithmetic banker biology
chemistry computer economics education
engineer geography politician secretary

□▪	▪ □ ▪	▪ ▪ □ ▪
actor	*accountant*	*economics*

🔊 Listen and check.

2 🔊 Listen and say these words.

a̲ge le̲sson che̲mistry e̲ducation se̲cretary ra̲ilway
sta̲tion fri̲end ho̲liday

Is the underlined sound /e/ or /eɪ/? Put the words in two columns.

SPEAKING AND WRITING

1 In pairs, check your answers to *Listening* activity 2. Find out what your partner thinks and complete the *Your partner* column in the chart.

2 Find out what other people in your class think about the future of English and make notes.

3 Write a paragraph saying what people in your class think.

All of us think that children will learn English from the age of six. Most of us think that we'll use English for our jobs. Some of us think that learning English will be more important than our own language. Nobody thinks that all lessons at school will be in English.

Foreign travels

Going to for plans and *will* for decisions;
expressions of future time

LISTENING

1 You're going to hear Duncan, an English student, talking to a friend, Cathy, about a visit he's going to make to South America. He's going to visit some of these places. Do you know which countries the places are in?

Places

The Amazon	☐	Lake Titicaca	☐	Rio	☐
Bel Horizonte	☐	Lima	☐	Santiago	☐
Buenos Aires	☐	Machu Picchu	☐	Sao Paulo	☐
Caracas	☐	Montevideo	☐	Valparaiso	☐
Cordoba	☐	Patagonia	☐		

2 Look at the map in Communication activity 21 on page 101 and check your answers. (No prizes if you live in South America!)

3 📼 Listen to the conversation and number the places in 1 as Duncan mentions them.

4 Work in pairs. Try to remember what Duncan plans to do in each place. Put the number of the place by what he's going to do there.

lie on the beach	☐	spend a week in the jungle	☐
visit the ruins	☐	go round the museum	☐
do some sightseeing	☐	take the cable car up the mountain	☐
stay in a hotel	☐	buy some souvenirs	☐
meet his girlfriend	☐	hire a car	☐
travel by coach	☐	learn to dance the samba	☐

📼 Now listen again and check.

5 📼 Listen to the rest of the conversation. Underline the verbs used.

CATHY Well, have a nice time! Have you got a good guide book?

DUNCAN No, I haven't. But *I'll get/I'm going to get* one. It's on my shopping list of things to buy before I go.

CATHY Well, they say the best one is the South American handbook.

DUNCAN Really? Well, *I'll get/I'm going to get* it when I go into town.

CATHY Look, *I'll go/I'm going* into town right now because I need to do some shopping. *I'll buy/I'm going to buy* it for you at the bookshop, if you like.

DUNCAN Really?

CATHY Yes, of course.

DUNCAN Well, *I'll give/I'm going to give* you the money for it right now.

CATHY OK, and *I'll bring/I'm going to bring* it round tonight. Who knows, perhaps *I'll borrow/I'm going to borrow* it from you one day.

DUNCAN OK. Thanks very much.

13

GENERAL COMMENTS

Going to and *will*

It is sometimes difficult for students to understand the difference between *going to* and *will*. In pre-intermediate *Reward* they have come across *going to* in Lesson 11 and *will* in Lesson 12, but this lesson may be the first occasion they have come across the two structures together. After they have studied the explanation in the grammar box, it may be necessary to give further practice in the distinction between the two structures using material from the Grammar review at the back of *Reward* Student's Book, the Practice Book or the Resource Pack. Reassure any student who finds the distinction difficult to absorb that it will become clearer with practice. It is a subtle distinction but the structures are very common, even at this level, and need to be acquired relatively early in the course.

Photo: The Sugar Loaf mountain in Rio de Janeiro, Brazil.

Possible questions

What can you see in the photo? Where is it?
What words come to your mind when you think of Rio?
Does it look like an interesting place?
Do you like Rio? or Would you like to go there? Why/Why not?
Are there any other famous landmarks in South America?

LISTENING

1 Aim: to prepare for listening and to show how the names of certain countries and places are pronounced in English.

● It is to be hoped that most students by now know how the name of their own country is pronounced in English, but they may not be acquainted with the English pronunciation of many other countries as well. It is not possible to do the listening activity without being able to recognise the names of the countries and places in the box, so write the following countries on the board *Argentina, Brazil, Chile, Peru, Uruguay, Venezuela*, and say the places in the box and the countries aloud. Then ask if anyone knows which country the places are in.

Answers
Argentina: Buenos Aires, Cordoba, Patagonia
Brazil: The Amazon, Bel Horizonte, Rio, Sao Paulo
Chile: Santiago, Valparaiso
Peru: Lake Titicaca, Lima, Machu Picchu
Uruguay: Montevideo
Venezuela: Caracas

2 Aim: to prepare for listening and to check the English pronunciation of places in South America.

● Ask students to check their answers with the map in Communication activity 21.

3 Aim: to practise listening for main ideas.

● 🔊 The main idea to be identified is the route Duncan is planning to take. Ask the students to listen and number the places in activity 1 in the order in which Duncan mentions them.

Answers
[1] Rio	[2] Santiago
[3] Valparaiso	[4] Lima
[5] Machu Picchu	[6] The Amazon

4 Aim: to listen for specific information.

● The specific information in this activity is what he is planning to do. Strictly speaking, the students do not yet have the structures necessary for saying what he's *going to* do, but they should be able to talk about what he *plans* to do. Ask them to try to remember what they heard in the conversation.

● 🔊 Play the tape again and ask the students to put the number of the place by what Duncan is going to do there.

Answers
lie on the beach	[3]
spend a week in the jungle	[6]
visit the ruins	[5]
do some sightseeing	[2]
take the cable car up the mountain	[1]
meet his girlfriend	[4]

5 Aim: to present the difference between *going to* and *will*.

● 🔊 In this activity the students listen and observe the verbs Duncan uses for decisions taken before the moment of speaking (*going to*) and decisions taken at the moment of speaking (*will*). Explain that the tapescript is a continuation of the dialogue the students have been listening to.

Answers
CATHY Well, have a nice time! Have you got a good guide book?
DUNCAN No, I haven't. but **I'm going to get** one. It's on my shopping list of things to buy before I go.
CATHY Well, they say the best one is the South American handbook.
DUNCAN Really? Well, **I'll get** it when I go into town.
CATHY Look, **I'm going** into town right now because I need to do some shopping. **I'll buy** it for you at the bookshop, if you like.
DUNCAN Really?
CATHY Yes, of course.
DUNCAN Well, **I'll give** you the money for it right now.
CATHY OK, and **I'll bring** it round tonight. Who knows, perhaps **I'll borrow** it from you one day.
DUNCAN OK. Thanks very much.

● You may like to point out that one use of *will* is to make an offer.

GRAMMAR

1 Aim: to ask the students to work out for themselves the difference between *will* and *going to*.

● Go through the dialogue line by line with the students and ask them to decide when the decision was taken.

> **Answer**
> You use *going to* for decisions taken before the moment of speaking and *will* for decisions taken at the moment of speaking.

● Ask the students to read the information about *going to* for plans in the grammar box.

2 Aim: to focus on things which are sure to happen and things which are not so sure.

● Suggest to the students that they use their answers to *Listening* activity 4 to do this activity. Make sure the students use *going to* for things which are sure to happen and *will probably* for things which are not so sure.

> **Answers**
> He's going to fly to Rio.
> He'll probably take the cable car up the mountain.
> He's going to fly to Santiago.
> He'll probably do some sightseeing.
> He's going to lie on the beach in Valparaiso.
> He's going to meet his girlfriend in Lima.
> They're going to visit the ruins in Machu Picchu.
> They'll probably go somewhere in the Amazon.
> They'll probably spend a week in the jungle.

3 Aim: to focus on the difference between *going to* and *will*.

> **Answers**
> 1 'I need a strong bag.' 'I **will** get you one.'
> 2 I bought a good map, because **I'm going** to South America.
> 3 'Where are you going to stay?' '**We'll** stay with friends, probably.'
> 4 I need some fresh air. Perhaps **I'll** have a walk in the park.

4 Aim: to focus on expressions of future time.

● Ask the students to write sentences about what they're sure they're going to do and what they'll probably do at some time in the future. Make sure they use the expressions of future time in the grammar box.

SPEAKING AND VOCABULARY

1 Aim: to develop speaking skills.

● Ask the students to work in groups of two or three and to think of a place to visit. Try to make sure each group chooses somewhere different, and make sure they don't tell the other groups the place they have chosen.

2 Aim: to focus on the words in the vocabulary box.

● Ask the groups to think about the words in the box and decide which things they need to take with them. Ask them to add three or four more things.

3 Aim: to practise using *will* for decisions taken at the moment of speaking.

● Ask the groups to decide who will get each thing. Make sure everyone uses *will* for a decision taken at the moment of speaking.

4 Aim: to practise using *going to* for decisions taken before the moment of speaking.

● Ask the students to check who is going to get which things. Make sure they use *going to* for decisions taken before the moment of speaking. Remind them, if necessary, that they took the decision in activity 3.

● You can finish this activity sequence by asking one group to write the list of things it needs for the trip on the board. The others must try to guess where it's going and what it's going to do there. The group can only answer *yes* or *no* to any questions:
Are you going somewhere in Europe? No.
Are you going somewhere in Asia? Yes.
Are you going to the beach?
When the class has guessed where the group is going, they can ask five more questions.

● Make sure each group has a turn at writing their list of things on the board.

GRAMMAR

Going to for plans

You use *going to* to talk about things which are arranged or sure to happen.
I'm going to visit South America.   *I'm going to* visit Buenos Aires.

You use *going to* for decisions you made before the moment of speaking.
I'm going to buy a guide book.

Will for decisions

You use *will* for decisions you make at the moment of speaking.
I'll give you the money right now.

You often use *will* for offers. (For more information about offering, see Lesson 30.)

You usually use the present continuous with *go* and *come*.
He's going to South America.	not	*He's going to go* to South America.
She's coming with us.	not	*She's going to come* with us.

Expressions of future time

You can use these expressions of future time to say when you are going to do things.

tomorrow	*morning afternoon evening*
next	*week month year*
in	*two days' time three months' time five years' time*

---

1 Does Duncan make his decisions before or at the moment of speaking to Cathy? Which verb form do you use to talk about decisions? Which one do you use to talk about plans?

2 Work in pairs and check your answers to *Listening* activity 4. Which things is Duncan sure he is going to do, and which things is he not sure will happen?
He's going to fly to Rio. He'll probably take the cable car up the mountain.

3 Choose the correct verb form.
1 'I need a strong bag.' '*I'm going to/I will* get you one.'
2 I bought a good map, because *I'm going to/I will* go to South America.
3 'Where *will you /are you going to* stay?' '*We'll /We're going to* stay with friends, probably.'
4 I need some fresh air. Perhaps *I'm going to/I'll* have a walk in the park.

4 Choose six expressions of future time and write three sentences about things you're sure you're going to do, and three things you're not sure will happen.
I'm going to have a holiday next month.
I'll probably move abroad in five years' time.

SPEAKING AND VOCABULARY

1 Work in groups of two or three. You're planning a trip somewhere. Decide where you'd like to go.

2 In your groups, look at the words in the box and decide which things you will need for the trip. Add three or four more things.

---

aspirin backpack map camera
currency food guide book
handbag medical kit passport
penknife razor scissors tent
sleeping bag suitcase wallet
toothbrush toothpaste watch
traveller's cheques walkman

---

3 Tell the others in your group which things you'll get.
I'll get the map.
OK, I'll buy some food.
Yes, and I'll bring my walkman.

4 Check who is going get which things.
So, Mario is going to get the map and Tanya is going to buy some food. Thanks for your help Paco, but...

14 | *In Dublin's fair city*

Prepositions of place; asking for and giving directions

VOCABULARY AND READING

1 Work in pairs. Look at the map of Dublin, the capital city of Ireland. Can you see any of these town features on the map? What are the places called?

bank bus station college
hospital law courts library
parliament pub railway station
river square town hall

The Bank of Ireland...

2 The article you are going to read is about the pubs in Dublin. Which of these words can you use to talk about pubs or bars in your country?

traditional conversation jokes
old-fashioned welcome private
charm modern beautiful local
famous thirsty lively literary
sophisticated elegant left-wing

3 Read *In Dublin's fair city* and decide which pub(s) you'd like to visit and why.

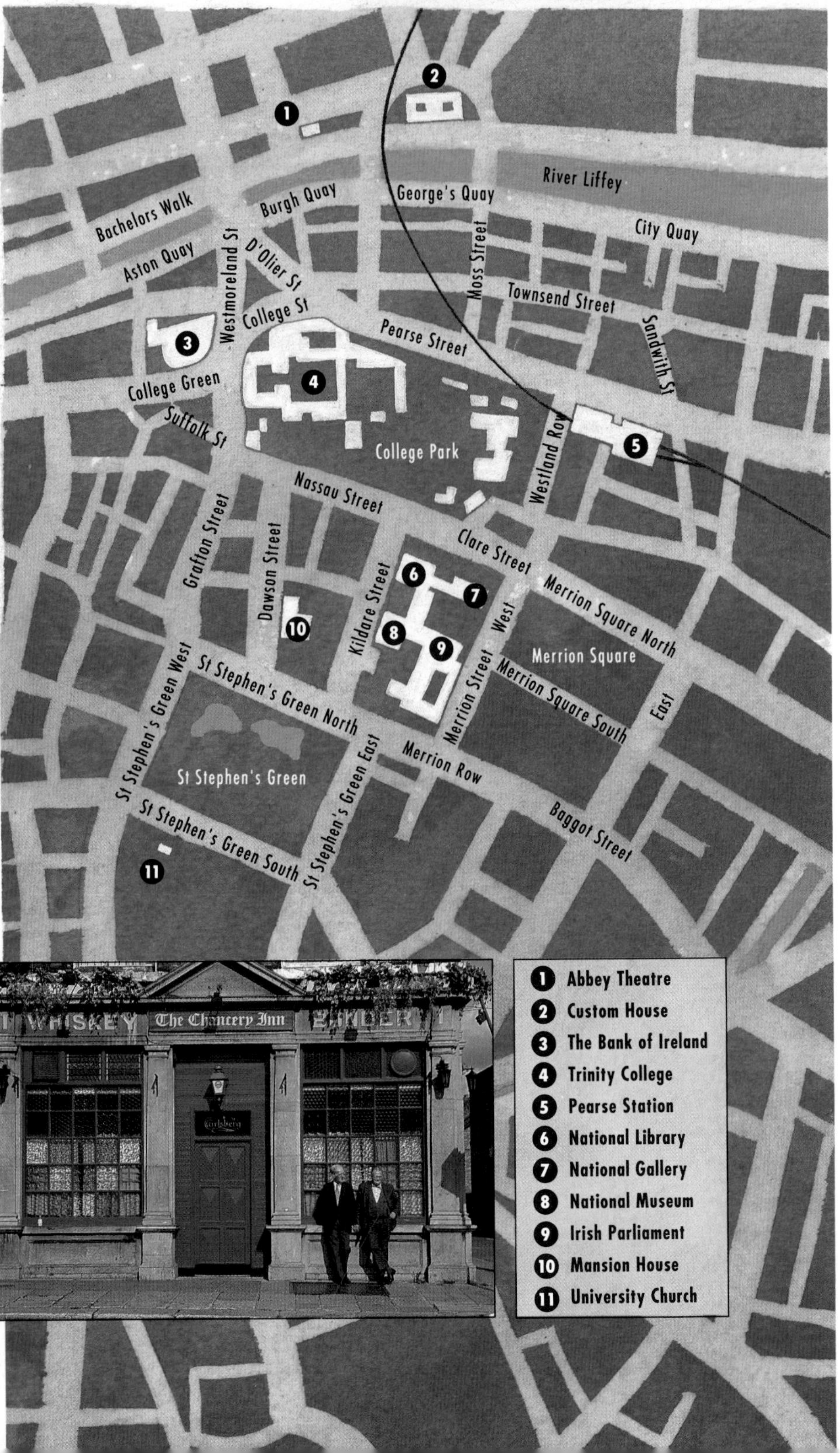

LISTENING

1 [cassette] Listen to a tour guide at the start of a tour. Where is he standing?

2 [cassette] Listen to the guide explaining the route of the tour. Draw the route on the map.

1 Abbey Theatre
2 Custom House
3 The Bank of Ireland
4 Trinity College
5 Pearse Station
6 National Library
7 National Gallery
8 National Museum
9 Irish Parliament
10 Mansion House
11 University Church

14

GENERAL COMMENTS

Prepositions; asking for and giving directions

In contrast to Lesson 13, the grammar and functions in this lesson will pose the students less of a problem. They will already have used prepositions of place in a number of lessons in *Reward* Student's Book, and may have studied asking for and giving directions in a previous course.

In Dublin's fair city

The title of the lesson is a line from an old folk song, *Cockles and Mussels* (two types of seafood):

> *In Dublin's fair city,*
> *Where the girls are so pretty,*
> *I first set my eyes on sweet Molly Malone*
> *As she wheeled her wheelbarrow*
> *Through streets broad and narrow*
> *Crying 'Cockles and mussels, alive, alive-o.'*

Dublin is famous for its pubs, some of which you can see in the photos, and the good conversation you can find in them. The word *pub* is short for *public house* – a house which is open to the public. This idea quite effectively conveys some of the conviviality and informality which characterises the Irish and British pub. The opening hours are quite strict; as the passage suggests, pubs in Ireland only open at 10.30 in the morning, and at 11am in Britain. In both countries they have to close at 11pm, although it is a common joke in Ireland that some rural pubs continue to serve drinks behind closed curtains after this time.

Photos: Various pubs in Dublin

Possible questions

What do you know about Dublin? Where is it? What is it famous for?
Do you like Dublin? or Would you like to go there? Why/Why not?
The photos show typical pubs in Dublin. Do they look like bars in your country?
How would you describe the atmosphere in bars in your country? Do you think it's similar in Dublin?

VOCABULARY AND READING

1 Aim: to present the words in the vocabulary box.

● Ask the students to look at the map and to try to find examples of the words in the box.

> **Answers**
> The Bank of Ireland, Trinity College, the National Library, Irish Parliament, Pearse Station, the River Liffey, Merrion Square

● You may like to ask students to say if the features mentioned in the box can be found in their own town, and if so, where.

2 Aim: to present the words in the vocabulary box and to pre-teach some vocabulary from the reading passage.

● For the British and the Irish, the main characteristic of a pub is its atmosphere. All of the words can be used to describe a British or Irish pub, but they may be surprising when used to describe a bar. Ask students which words they can use to describe a pub or bar they know.

3 Aim: to read and react to a text.

● You may need to give students some background information:
Guinness is a strong, dark beer with a white 'head', and is regarded by some as the national drink of Ireland. The people mentioned were or are all famous Irish men:
Jonathan Swift (1667-1745) was a clergyman, who wrote *Gulliver's Travels*.
Oliver Goldsmith (1728-74) was a poet and a playwright, who wrote *She Stoops to Conquer*.
James Joyce (1882-1941) wrote *Ulysses* and *Finnegan's Wake*.
WB Yeats (1865-1939) wrote poems, stories and plays and was active in Irish politics.

● Ask the students to read the passage and decide which pubs they'd like to visit and why.

● When the students have finished, ask them to talk about their reactions to the pubs mentioned.

● Ask the students to compare the different characteristics of pubs and bars. If they've never been to Britain or Ireland, ask the students if they know of an English or an Irish pub in their town.

LISTENING

1 Aim: to listen for main ideas.

● Explain to the students that they are going to hear a tour guide at the start of a tour around Dublin. Ask them to think about the kinds of background noise they might hear.

● 🔲 Mention that the speaker has a noticeable Irish accent. Ask them to listen and decide where on the map the guide is standing.

> **Answer**
> The guide is standing in St Stephen's Green on the corner of Kildare Street.

2 Aim: to listen for specific information and to follow the route of a guided tour.
● 📼 Make sure everyone understands that the tour begins in St Stephen's Green on the corner of Kildare Street. Play the guided tour and ask the students to follow the route on the map.

Answer

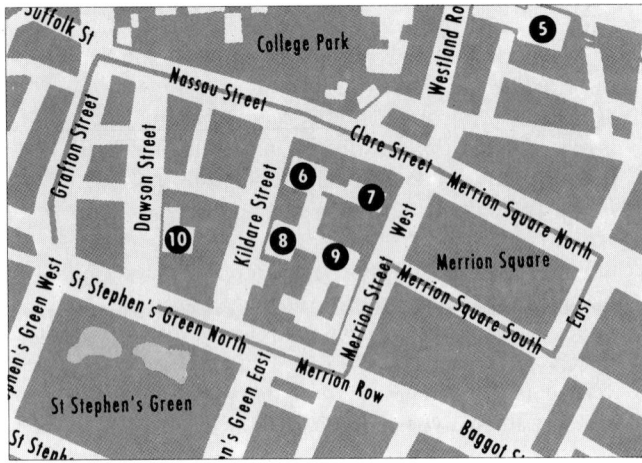

● You may want to play the tape a second time and pause after each sentence. Repeat each line carefully so the students can check their answers.

GRAMMAR AND FUNCTIONS

1 Aim: to focus on the use of prepositions of place.
● Ask the students to look at the information on prepositions of place and asking for and giving directions in the grammar box.

● Ask the students to do the exercise.

> **Possible answers**
> The Irish Parliament is in Merrion Street West, next to the National Museum.
> The Mansion House is between Dawson Street and Kildare Street.
> Merrion Square is next to the National Gallery.
> Kildare Street is between St Stephen's Green and Nassau Street.
> The National Gallery is behind the National Library.

2 Aim: to focus on giving directions.
● 📼 You may want to play *Listening* activity 2 again to help the students write the guided tour.

> **Answers**
> See tapescript, *Listening* activity 2.

SOUNDS

1 Aim: to practise /ɑː/, /æ/ and /ʌ/.
● 📼 These three phonemes can be confused, although they do not usually pose any problem for speaker or listener when pronounced in isolation.

> **Answers**
> /ɑː/: article charm far bar
> /æ/: bank map actor national
> /ʌ/: pub Dublin number others bus

2 Aim: to focus on word stress in questions.
● Ask the students to predict which words are likely to be stressed. Remind them, if necessary, that the stressed words in a sentence are the words the speaker considers to be important.

● 📼 Play the tape and ask the students to check their answers. Pause the tape so they can ask the questions aloud.

> **Answers**
> 1 <u>Excuse</u> me, how do I <u>get</u> to the <u>post</u> <u>office</u>?
> 2 <u>Excuse</u> me, is there a <u>bank</u> near <u>here</u>?
> 3 <u>Excuse</u> me, could you tell me the <u>way</u> to the <u>station</u>, please?
> 4 <u>Excuse</u> me, <u>where's</u> the <u>nearest pub</u>?

SPEAKING

1 Aim: to practise using prepositions of place.
● Ask the students to work in pairs and follow the pairwork instructions. Remind them to use some of the prepositions in the grammar and function box.

2 Aim: to practise giving directions.
● Ask the students to work in pairs and to turn to their instructions in the Communication activities section on pages 99 and 100. Make sure they realise they can use the map of Dublin.

● If both students in each pair come from a town they know well, they can continue this activity by describing the route from one well-known place to another, without saying where they are going. The student who is listening must follow the route in his/her head. They can prepare this activity for homework; ask the students to write a guided tour. Then at the beginning of the next lesson, spend five minutes asking the students to read other people's guided tours and to follow the route in their heads.

IN DUBLIN'S FAIR CITY

'I went into a pub one morning and asked for a Guinness. The barman said, "Sorry, we're closed. We open in fifteen minutes." Then he paused. "But would you like a drink while you're waiting?"'
– An English visitor to Dublin

Dublin means many things to many people. To some it is a city of writers, the city of Jonathan Swift, Oliver Goldsmith, James Joyce and W B Yeats. For others, it is the city of talkers, its pubs full of Guinness and jokes, and the source of the writers' inspiration.

You can still find many traditional pubs in Dublin and they're open all day from 10.30am on Mondays to Saturdays and most of the day on Sunday. You're sure of a warm welcome, especially if you offer to buy the locals a drink. Here are a few to visit in the centre.

Doheny and Nesbitt's in Baggott Street is a lively, old-fashioned pub, long and narrow, with a small private bar at each end. Government ministers, civil servants and journalists use it for meetings.

The Horsehoe Bar of the elegant Shelbourne Hotel on St Stephen's Green keeps its charm and intimacy in its modern, sophisticated surroundings.

Mulligan's in Poolbeg Street opened in 1872 and has a number of sections: students in the entrance on the left, journalists and TV people on the right, and left-wing politicians in the far bar.

Neary's, in Chatham Street, is a beautiful, old-style pub visited by actors from the Gaiety Theatre.

The Bailey, in Duke Street, is a very traditional pub, and famous for its literary connections. It was in James Joyce's Ulysses, where it is known as Burton's.

There are hundreds of other pubs as well, where visitors can enjoy something of the real Dublin. You don't need to be thirsty to enjoy the pub; just enjoy the conversation!

GRAMMAR AND FUNCTIONS

Prepositions of place
*The Bank of Ireland is **opposite** Trinity College.*
*Pearse Station is **next to** Trinity College.*
*College Park is **in** Trinity College.*
*Dawson Street is **between** St Stephen's Green and Nassau Street.*
*The National Library is **behind** the National Museum.*
*The National Museum is **on the corner of** College Green and Westmoreland Street.*
*The National Museum is **in front of** the National Library.*

Asking for and giving directions
How do I get to ...?	*Go straight ahead/on.*
Go down/along...	*Go to the end of...*
Cross over...	*Turn left/right into...*
It's on the left/right.	*It's at the crossroads.*
Take the first/second turning on the left/right.	

1 Look at the map and say where these places are. Use the prepositions in the grammar box.

The Irish Parliament   the Mansion House
Merrion Square   Kildare Street   The National Gallery

2 Write directions for the tour you heard in *Listening* activity 2. Use the prepositions in the grammar box, and the map to help you.

🔲 Listen again and check.

SOUNDS

1 Say these words aloud.

bank pub map article Dublin number others
charm far bar actor bus national

Is the underlined sound /ɑː/, /æ/ or /ʌ/? Put the words in three columns.

🔲 Listen and check.

2 Underline the stressed words in these questions.

1 Excuse me, how do I get to the post office?
2 Excuse me, is there a bank near here?
3 Excuse me, could you tell me the way to the station, please?
4 Excuse me, where's the nearest pub?

🔲 Listen and check. Ask the questions aloud.

SPEAKING

1 Work in pairs.

Student A: Think of a well-known place in the town where you are now. Imagine you are looking at it and describe where you are to Student B. Don't say what the place is.

Student B: Listen to Student A describing a well-known place in the town where you are now. Guess what the place is.

Change round when you're ready.

2 Work in pairs.

Student A: Turn to Communication activity 7 on page 99.
Student B: Turn to Communication activity 15 on page 100.

15 An apple a day

Expressions of quantity (1): countable and uncountable nouns, *some* and *any*, *much* and *many*

VOCABULARY

1 What sort of things do you eat, drink or use in your cooking? Look at the words in the box and put them in lists under these headings: *every day, twice a week, every week, on special occasions, never.*

apples bananas beef beer biscuits bread butter cabbage carrots cheese chicken coffee eggs fish fruit grapes ham juice lamb lettuce meat milk oil onions oranges pasta peaches peas pork potatoes rice salad strawberries tea tomatoes vegetables water wine

Can you think of two or three more things to add to each list?

2 Work in pairs and compare your lists. Which of the things can you see in the photo?

15

GENERAL COMMENTS

Countable and uncountable nouns

Countable (or count) nouns and uncountable (or mass) nouns often cause students some difficulties. It's not always obvious from the noun if it is countable or uncountable. Sometimes a word can have a countable meaning in one context and an uncountable one in another (country, cold, wine, chicken), and logic doesn't necessarily help the student decide. While this lesson was being piloted, one student was extremely confused to be told that *money* is uncountable! (Money is uncountable because you cannot have two monies, although you can have two pounds or dollars.) It is important to treat the matter as an issue of vocabulary, and to encourage students to annotate each new word and meaning as C or U.

Jigsaw listening

This is a technique in which a class is divided into three groups. Each group is given a cassette containing a third of a listening passage (each group having a different third), and a cassette player, and is asked to listen to their part of the passage. They then work together to check that they have all understood their part. Then members of each group form a new group and reconstitute the passage as a whole. The technique is effective in pedagogical terms in that it creates an information gap between members of the new group (you know something I don't, I know something you don't) which is central to communicative language teaching. But there are practical problems: a teacher may not have three cassette players, or may not have the space to allow three groups to work separately. In this lesson, there is a jigsaw listening activity with one tape recorder, in which students work in groups of three, all listening to the same passage, but for different information. The information gap is not as tight, but teachers who have tried the technique agree that it is sufficient to generate a highly motivated sharing of information.

Photo

A display of fruit and vegetables in the market in Venice.

Possible questions

Work in pairs. Student A, open your book and look at the photo. Student B, keep your book closed. Student A is looking at a display of food in a market. Ask what food there is, like this: Is there any cheese? Are there any apples? Student A can only answer yes or no.

VOCABULARY

1 Aim: to present the words in the vocabulary box.

● This activity is designed to present the new words to the students and to encourage them to put them into personalised categories. There may be too many items to learn, but it is to be expected that by the end of the *Vocabulary* activities, they will have been exposed to and have understood all of the words, and will probably retain the words they have placed in the *every day* and *twice a week* categories.

● Ask the students to categorise the words under the headings. Encourage them to think of two or three other items to add to the list, so that the vocabulary acquisition process is supported by the personal relevance of the words to be retained.

● Ask one or two students what they eat *every day* and *never*. Then ask other students to compare their lists.

2 Aim: to practise using the new words.

> **Answers**
> apples, cabbage, carrots, fruit, grapes, lettuce, onions.
> Other fruit and vegetables in the photo are: beans, cauliflower, peppers, pineapple.

3 Aim: to present some common words to describe quantity.

● Ask the students to look at the words in the box and the examples below. Ask the class *Can you buy a bottle of apples? Can you buy a can of bananas?* and find someone who says *no*.

● Write the words in the box on the board as headings, and ask students to come and write the food words under the suitable headings.

> **Possible answers**
> **bottle:** beer, juice, milk, oil, water, wine
> **cup:** coffee, tea
> **glass:** beer, juice, milk, water, wine
> **kilo:** butter, coffee, pasta, rice, tea, and all the fruit, vegetables and meat, but not bread and liquids
> **loaf:** bread
> **packet:** biscuits, butter, cheese, coffee, pasta, rice, tea
> **piece:** bread, cheese, fruit
> **slice:** beef, bread, cheese, chicken, fruit, ham, lamb, meat, onion, orange, peach, pork, tomato
> **tin:** carrots, fruit, peaches, peas, potatoes, strawberries, tomatoes, vegetables

● The answers depend on the item in question, whether it is liquid or solid and on matters of usage. For example, we don't talk about *a glass of coffee* because we don't serve coffee in a glass. *Tin* is not usually used for liquids; you usually say a *can* of beer. *Cup* is used as a measurement in American English, and so it can be used with all the words which you can use *kilo* with. There will be other combinations in certain contexts which are not covered by the answers above.

● Say a food item, eg *flour* and elicit the response *a packet of flour*. Try this with several food items.

● If you have time, use the activity for some cross-cultural awareness. Would the students say in their language *a glass of coffee, a packet of cheese, a slice of potato, a cup of wine, a kilo of bread*?

GRAMMAR

1 Aim: to focus on whether items of food and drink are countable or uncountable.

● Ask the students to read the information about expressions of quantity in the grammar box and then do the exercise.

● It was mentioned in the introduction that some words can be either countable or uncountable depending on the context. Generally speaking an uncountable noun becomes countable when you talk about a kind or variety of things, eg *Chilean wines are among the best in the world*. The answers below refer to the context of food and drink, and shopping.

> **Answers**
> apples C, bananas C, beef U, beer U, biscuits C, bread U, butter U, cabbage U, carrots C, cheese U, chicken U, coffee U, eggs C, fish U, fruit U, grapes C, ham U, juice U, lamb U, lettuce U, meat U, milk U, oil U, onions C, oranges C, pasta U, peaches C, peas C, pork U, potatoes C, rice U, salad U, strawberries C, tea U, tomatoes C, vegetables C, water U, wine U

2 Aim: to focus on *some, any, much* and *many*.

● ▣ Ask the students to complete the dialogue and then play the tape to check.

> **Answers**
> A We need **some** water. How **many** bottles do we need?
> B Two. And we haven't got **any** fruit. Shall we get **some** peaches?
> A OK. Have we got **any** tea?
> B No, how **much** do we need?
> A Just a packet.

LISTENING AND SPEAKING

1 Aim: to pre-teach some difficult words from the listening passage.

● Make sure the students understand the food items. They can use their dictionaries if they wish. They should be able to understand most of the other vocabulary in the listening passage.

2 Aim: to listen for specific information.

● ▣ Ask the students to work in three groups, A, B and C. The members in each group should look at their instructions in the Communication activities section. Each group has different instructions and different information to listen for. Ask them to listen and to write notes in answer to their specific questions. Point out to the students that Pat is a man, and is interviewed first.

● Ask the students to discuss their answers in groups, and check that everyone has got the same information.

3 Aim: to discuss what they have heard, to reconstitute the two listening passages and to complete the chart.

● Ask the students to form new groups of three people, with one person from group A, one person from group B and one person from group C. They should work together, and talk about the notes they made in activity 2. It may be that some students can remember information which they were not asked to write down. But the rule is that they can only talk about information which they were specifically asked to listen for and write down. Ask them to fill the chart in with as much detail as possible. They can copy the chart if there is not enough space in their books.

● You may like to check their answers to the chart by drawing it on the board and asking the students to fill it in orally.

> **Answers**
>
> | | Karen | Pat |
> |---|---|---|
> | Typical breakfast | cereal, toast orange juice tea | cereal, fruit, juice |
> | Typical lunch | sandwich, baked potato | pasta, vegetables, sandwich |
> | Typical dinner | Chinese, Japanese, Indonesian food, dumplings with meat and vegetables | seafood, vegetables, salad, pie |

4 Aim: to practise using the vocabulary and grammar presented in this lesson.

● Ask the students to talk about their typical meals and to complete the *You* column of the chart in activity 3.

5 Aim: to continue talking about eating habits; to practise using expressions of quantity.

● Explain that the expressions of quantity in the list are in a graded sequence from *a lot* to *not any*. You can do this activity with the class as a whole, simply asking them for their reactions to Karen's and Pat's eating habits, using some of the expressions.

6 Aim: to continue talking about eating habits.

● Ask the students to write a list of ten things they often eat and drink.

● Ask the students to go round asking people if they eat and drink the same things, like this: *Do you eat cheese? Yes, I do. I drink wine. So do I.* Even in a mono-cultural class there will be plenty of small differences.

● Ask the students to find out how much they eat, drink or use. Check they are using *how much* with uncountable nouns and *how many* with countable nouns.

● You may like to ask the students to write a brief report on the class's eating and drinking habits for homework.

3 Look at these words.

bottle cup glass kilo loaf packet piece slice tin

a **bottle** of beer   a **cup** of tea   a **glass** of water
a **kilo** of potatoes   a **loaf** of bread
a **packet** of biscuits   a **piece** of cheese
a **slice** of bread   a **tin** of peaches

Which other items in the vocabulary box can you use with these words?

GRAMMAR

Expressions of quantity (1): countable and uncountable nouns
Countable nouns have both a singular and a plural form.
an apple – two apples   a peach – two peaches
Uncountable nouns do not usually have a plural form.
bread, beef, butter, coffee, water

Some and *any*
You usually use *some* in affirmative sentences.
*I'd like an orange, two apples, **some** peaches and **some** water.*
You usually use *any* in negative sentences and questions.
*We haven't got **any** butter.   Are there **any** eggs?*

Much and *many*
You usually use *much* and *many* in negative sentences and questions. You use *many* with countable nouns.
*We haven't got **many** carrots.*
*How **many** eggs would you like?*
You use *much* with uncountable nouns.
*There isn't **much** cheese.   How **much** butter do you need?*

1 Look at the words in the vocabulary box again. Write *C* (countable) or *U* (uncountable) by them.

2 Complete the dialogue with *some, any, much* or *many*.

A We need ___ water. How ___ bottles do we need?
B Two. And we haven't got ___ fruit. Shall we get ___ peaches?
A OK. Have we got ___ tea?
B No, how ___ do we need?
A Just a packet.

🔊 Now listen and check.

LISTENING AND SPEAKING

1 Work in groups of three. You are going to hear Karen, who lives in Hong Kong, and Pat, who lives in San Francisco, talking about a typical breakfast, lunch and dinner. First, make sure you understand these food items they mention:

cereal   pie   steamed dumplings   toast
sandwich   chowder

2 *Student A:* Turn to Communication activity 1 on page 98.

Student B: Turn to Communication activity 12 on page 100.

Student C: Turn to Communication activity 20 on page 101.

3 Now work together and complete the columns *Karen* and *Pat*.

	Karen	**Pat**	**You**
Typical breakfast			
Typical lunch			
Typical dinner			

4 What do you have for a typical breakfast, lunch and dinner? Complete the *You* column in the chart.

5 Compare the speakers' typical meals with your typical meals. Use these expressions.

a lot of/lots of   quite a lot of   a few /a little
not much/many   hardly any   not any

Karen eats a lot of meat, and so do I. Pat eats hardly any vegetables, but I eat lots.

6 Find out what sort of things other people in your class eat, drink or use in their cooking.
Fadia, do you drink tea?   Yes, I do.

Then find out how much they eat, drink or use.
How much tea do you drink every day? How many cups a day do you drink?

Progress check **11-15**

VOCABULARY

1 Word maps are a good way of remembering and organising new vocabulary. Make a word map of your town or city.

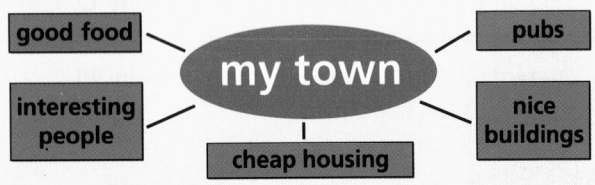

2 Look at these words for jobs. Which come from verbs? Write the verb.

actor banker dancer footballer journalist manager musician politician teacher writer

actor – to act

Which ones come from other nouns? Write the nouns.

banker – a bank

You can make other words using a suffix. Look at these suffixes.

*act**or** bank**er***

Underline all the suffixes in the words above. Words with these suffixes are often jobs.

3 Look at the vocabulary boxes in lessons 11 – 15 again. Choose words which are useful to you and group them under headings of your choice in your *Wordbank.*

GRAMMAR

1 Write questions about Jane with *going to.*

1 be an accountant	4 learn Italian
2 live in the USA	5 visit South America
3 start her own company	6 start a new life

1 Is she going to be an accountant?

2 Write answers to the questions you wrote in 1.

1 be a doctor	5 travel round Spain and
2 move to Spain	Portugal
3 work in a hospital	6 stay in contact with her old
4 learn Spanish	friends

1 Is Jane going to be an accountant?
No, she isn't. She's going to be a doctor.

3 You're going to start a new, exciting job tomorrow. Say what you are or aren't going to do.

– walk to work – arrive early – drink less coffee
– work hard – be friendly – stay late

4 Complete these sentences with *(be) going to* or *would like to.*

1 He's got his ticket and he _____ fly to Madrid.
2 I _____ buy a car, but I haven't got any money.
3 It _____ rain today.
4 We _____ go on holiday but we're too busy.
5 He's got his coat on and he _____ leave now.
6 We sold our flat last week and we _____ live abroad.

5 Complete the sentences with *will* or *(be) going to.*

1 It's very early. Maybe I ___ go back to bed.
2 He ___ fly to Bel Horizonte next week.
3 She ___ have a baby next July.
4 'I'm so tired.' 'I ___ take you home by car.'
5 He works in Lisbon, so he ___ move there.
6 Maybe we ___ have dinner in a restaurant.

6 Make predictions about the following.
Use *I think/perhaps/maybe + will.*

1 next weekend	4 your family
2 your job	5 next year
3 next holidays	6 your friends

7 Write sentences saying where these places are in your town. Use these prepositions:

at on opposite next to in between behind
on the corner of in front of

1 the post office	5 the swimming pool
2 the bank	6 the cinema
3 the library	7 the football stadium
4 the supermarket	8 the park

Progress check 11–15

GENERAL COMMENTS

You can work through this Progress check in the order shown, or concentrate on areas which may have caused difficulty in Lessons 11 to 15. You can also let the students choose the activities which they would like to or feel the need to do.

VOCABULARY

1 Aim: to present word maps as a way of remembering and organising new vocabulary.
- Word maps can be useful devices to consolidate the acquisition of new ideas and to organise vocabulary into vocabulary fields. The students should do this activity on their own.
- Ask someone to write their word map on the board.
- Then ask the students to compare their word maps.

2 Aim: to focus on noun suffixes.
- Explain that suffixes can often indicate the part of speech of a new word. For example, -er is often used for the names of people or things that do something. This activity focuses on the suffixes which often indicate jobs.

Answers
verbs: act, dance, manage, teach, write
nouns: bank, football, journal, music, politics

3 Aim: to help students organise their vocabulary learning.
- You may want to suggest that the students extend their *Wordbanks* beyond the space available in the Practice Book.

GRAMMAR

1 Aim: to revise questions with *going to*.

Answers
1 Is she going to be an accountant?
2 Is she going to live in the USA?
3 Is she going to start her own company?
4 Is she going to learn Italian?
5 Is she going to visit South America?
6 Is she going to start a new life?

2 Aim: to revise *going to*.

Answers
1 No, she isn't. She's going to be a doctor.
2 No, she isn't. She's going to move to Spain.
3 No, she isn't. She's going to work in a hospital.
4 No, she isn't. She's going to learn Spanish.
5 No, she isn't. She's going to travel round Spain and Portugal.

6 No, she isn't. She's going to stay in contact with her old friends.

3 Aim: to revise *going to*.
- The students should write their own answers to these questions.

4 Aim: to revise *going to* and *would like to*.

Answers
1 He's got his ticket and he**'s going to** fly to Madrid.
2 I**'d like to** buy a car, but I haven't got any money.
3 It**'s going to** rain today.
4 We**'d like to** go on holiday but we're too busy.
5 He's got his coat on and he**'s going to** leave now.
6 We sold our flat last week and we**'re going to** live abroad.

5 Aim: to revise *will* and *going to*.

Answers
1 It's very early. Maybe I**'ll** go back to bed.
2 He**'s going to** fly to Bel Horizonte next week.
3 She**'s going to** have a baby next July.
4 'I'm so tired.' 'I**'ll** take you home by car.'
5 He works in Lisbon, so he**'s going to** move there.
6 Maybe we**'ll** have dinner in a restaurant.

6 Aim: to revise making predictions.
- The students should write their own sentences.

7 Aim: to revise the use of prepositions.
- The students can write their own sentences.

8 Aim: to revise giving directions.
- The students should write their own directions.

9 Aim: to revise countable and uncountable nouns.

Answers
egg C, money U, orange juice U, apple C, sugar U, potato C, butter U, rice U, strawberry C, cheese U

10 Aim: to revise *How much* and *How many?*

Answers
How many eggs have you got?
How much money have you got?
How much orange juice have you got?
How many apples have you got?
How much sugar have you got?
How many potatoes have you got?
How much butter have you got?
How much rice have you got?
How many strawberries have you got?
How much cheese have you got?

- Check that the students form the irregular plurals properly, *potatoes* and *strawberries*.

11 Aim: to revise *some* and *any*.

Answers
1 Have you got **any** oranges?
2 I'd like **some** wine please.
3 I don't have **any** money with me.
4 Is there **any** water?
5 We've got **some** chicken but we haven't got **any** salad.
6 I'll get you **some** bread, if you like.

SOUNDS

1 Aim: to revise the sound /ə/.
- The students have already practised the /ə/ sound in Lesson 9. Ask them to say the words.
- 🔲 Play the tape and ask the students to listen and check.

Answers
pizza polite police theatre cinema opera performance dinner weather

2 Aim: to practise the sounds /tʃ/ and /ʃ/.
- Ask the students to say the words aloud.
- 🔲 Play the tape and ask the students to listen and check.

Answers
/tʃ/: charm chicken cheese peach
/ʃ/: politician traditional she old-fashioned fish

3 Aim: to practise contrastive stress.
- 🔲 Make sure the students understand that they are going to respond to the prompts on tape with the same sentence, but stressing a different word each time. Play the tape and stop it at each pause.

Answers
1 No, **Joe** is going to study maths at university.
2 No, Joe is going to study **maths** at university.
3 No, Joe is going to study maths at **university**.
4 No, Joe is **going** to study maths at university.

4 Aim: to focus on polite intonation.
- 🔲 Even in transactional situations, such as in a shop or at the railway station, British people try to sound polite. The intonation of a sentence can make a major contribution to how polite a speaker sounds; other factors would include the language the speaker uses.

Answers
How do I get to the station? ✔
How do I get to the hospital?
Could you tell me where the town hall is? ✔
Could you tell me where the bus station is? ✔
Where's the post office?
Where's the river? ✔

- Although it isn't the aim of cross-cultural training to make students acquire the behaviour of a foreign culture, it is nevertheless designed to help them be sensitive to people from other cultures. Tell your students that in Britain, it is important to sound as polite as possible most, if not all, of the time, even to strangers. Repeating these sentences is not designed to make the students acquire British behaviour, but to be aware of how intonation can change the way something may be perceived by someone.

SPEAKING

1 Aim: to provide speaking practice.
- Ask the students to plan the lunch they are going to arrange. They should decide the dish they are going to make and the place where they are going to have lunch. Go round the class and check that everyone is using *going to* and *will* correctly.

2 Aim: to practise writing a note with directions.
- Ask each student to choose someone from another group and to write to him or her saying where they're going to have lunch and how to get there.

3 Aim: to practise using *will* for decisions and the vocabulary of food.
- Tear some paper into small pieces and give five pieces to each pair of students. Ask them to write one of the main ingredients for the meal on each piece of paper.

4 Aim: to practise using the vocabulary of food.
- Collect all the pieces of paper with the ingredients, shuffle them and then give out five new pieces of paper to each pair of students. These pieces of paper will obviously not have the ingredients the students wrote in activity 3. The students' task is to find the ingredients they need for their meal, using the language shown in the example. The only rule is that if someone asks for an ingredient, the other person must hand it over even if he or she needs it.

- Each student should check with his or her partner now and then. When they have got the ingredients for their meal, they have won. (It doesn't matter if they have pieces of paper which they didn't originally write; only the ingredients matter.)

8 Write directions from where you are now to the following places.

1 the railway station
2 the bus station
3 a petrol station
4 the cinema

9 Are these things countable or uncountable? Write *C* or *U*.

egg money orange juice apple sugar potato butter rice strawberry cheese

10 Write sentences using *How much ... have you got?* or *How many ... have you got?* and the words in 9.
How many eggs have you got?
How much money have you got?

11 Complete these sentences with *some* or *any*.

1 Have you got ___ oranges?
2 I'd like ___ wine please.
3 I don't have ___ money with me.
4 Is there ___ water?
5 We've got ___ chicken but we haven't got ___ salad.
6 I'll get you ___ bread, if you like.

SOUNDS

1 Say these words aloud. Underline the /ə/ sound.

pizza polite police theatre cinema opera performance dinner weather

🔊 Now listen and check.

2 Say these words aloud.

charm chicken politician traditional cheese she old-fashioned peach fish

Is the underlined sound /tʃ/ or /ʃ/? Put the words in two columns.

🔊 Listen and check.

3 Look at this true sentence.
Joe is going to study maths at university.

🔊 Listen and answer the questions below with the true sentence. Change the stressed word each time.

1 Is Tim going to study maths at university?
2 Is Joe going to study physics at university?
3 Is Joe going to study maths at school?
4 Did Joe study maths at university?

*1 No, **Joe** is going to study maths at university.*

4 🔊 Listen to these questions. Put a tick (✓) if the speaker sounds polite.

1 How do I get to the station?
2 How do I get to the hospital?
3 Could you tell me where the town hall is?
4 Could you tell me where the bus station is?
5 Where's the post office?
6 Where's the river?

Now say the questions aloud. Try to sound polite.

SPEAKING

1 Work in pairs. One of the other pairs is coming for lunch. Decide:
– what dish you'll make
– where you'll have lunch.

2 Write a note to another pair. Say:
– where you're going to have lunch
– how to get there.

3 You need to buy the ingredients for the dish you chose in 1. Write each ingredient on a piece of paper. With your partner, discuss which ingredients you'll get.

A *I'll get the tomatoes.*
B *And I'll buy some onions.*
A *OK. You're going to buy some onions and I'm going to buy the tomatoes.*

4 Give the pieces of paper with ingredients to your teacher, and you will receive some different ingredients. Go round asking other people if they have got the ingredients that you're going to get, and saying what you've got.

Have you got any tomatoes?
Yes, I have./No, I haven't.

Give your ingredients to anyone who asks for them, even if you need them! The first pair to get all their ingredients is the winner.

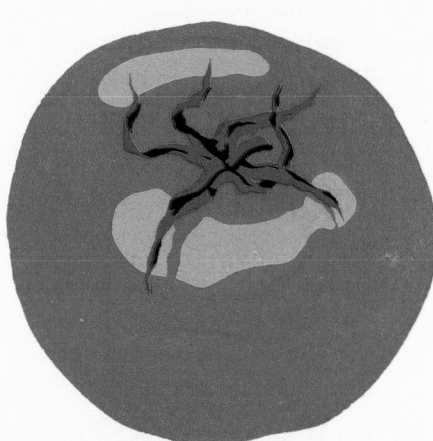

What's on?

Prepositions of time and place; making invitations and suggestions

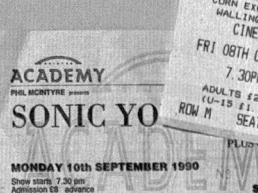

VOCABULARY

Look at the words in the box. Which words go under these headings: *What's on? Where?*

> ballet cinema closing time club concert disco
> exhibition film gallery interval match musical
> museum opening hours opera opera house painting
> performance play row sculpture seat stadium
> theatre ticket

What's on?	*Where?*
ballet	*theatre*
film	*cinema*

Now put the remaining words in a third column: *Related words.*

What's on?	*Where?*	*Related words*
ballet	*theatre*	*ticket, row, seat*

LISTENING AND SPEAKING

1 Work in groups of three. You are going to hear people talking about typical entertainment in Japan and Argentina. Ken talks about *karaoke* in Japan, and Philippa talks about *tango* in Argentina.

Student A: Turn to Communication activity 2 on page 98.
Student B: Turn to Communication activity 19 on page 101.
Student C: Turn to Communication activity 11 on page 100.

2 How much do you know about karaoke and tango now? Work together and complete the chart.

	karaoke	tango
Type of entertainment		
Place of entertainment		
Performers		
Type of music		
Reasons why people enjoy it		

3 Talk about a typical type of entertainment in your country. Use the chart to help you.

16

GENERAL COMMENTS

Prepositions of time and place
The students have already been using prepositions of time and place in pre-intermediate *Reward*. This lesson is designed to consolidate and organise their knowledge of this grammar point.

Cross-cultural training
The theme of this lesson is entertainment in different countries. It is not the aim of *Reward* to give extensive and specific information about different cultures, because that would involve much more time and space than is available. The aim is to present aspects of certain *sample* cultures with which the students can compare their culture.

Jigsaw listening
There is another example of jigsaw listening with one tape recorder. For details of the rationale behind this technique, see the general comments in Lesson 15.

Main illustration: possible questions
These tickets are all for different types of entertainment. Look at them carefully and find as many different types of entertainment as possible. Don't write anything down. Now close your books.
Write down as many different types of entertainment shown on the tickets as you can remember.

VOCABULARY

Aim: to present the words in the vocabulary box.
● Ask the students how they like to spend their spare time. Write their hobbies and interests on the board and put a tick by each one for each student who enjoys it. Which is the most popular hobby or interest in the class?

● On the board write *What's on?* and *Where?* and to illustrate their meaning, suggest a film or a play which is on in your town and then say where. Go round the class asking students to suggest what's on in their town and where it's taking place.

● Ask the students to put the words in the box under the headings. They can do this on the board and then copy the lists into their notebooks.

● Explain that not all of the words can go into the two categories, and should be put into a third category *Related words*. Tell the students that some related words can go with more than one category. Ask them to continue their work until they have categorised all the words in the vocabulary box. They may like to do this stage in pairs.

Answers		
What's on	Where	Related words
ballet	opera house, theatre	ticket, row, seat, performance, interval
concert	opera house, theatre	ticket, row, seat, performance, interval
film	cinema	ticket, row, seat
musical	theatre	ticket, row, seat, performance, interval
opera	opera house	ticket, row, seat, performance, interval
play	theatre	ticket, row, seat, performance, interval
match	stadium	ticket, row, seat, interval
exhibition	gallery, museum	painting, ticket, closing time, opening hours, sculpture
disco	club	ticket

● Explain that you talk about an *exhibition of sculpture or paintings.*

● Ask the students to decide which of the words they can use to talk about their hobbies and interests.

LISTENING AND SPEAKING

1 Aim: to listen for specific information.
● 🔊 Ask the students to work in three groups, A, B and C. The members in each group should look at their instructions in the Communication activities section. Each group has different instructions and different information to listen for. Ask them to listen and to write notes in answer to their specific questions.

● Ask the students in each group to discuss their answers, and check that everyone has got the same information.

2 Aim: to discuss what they have heard, to reconstitute the two listening passages and to complete the chart.
● Ask the students to form new groups of three people, with one person from group A, one person from group B and one person from group C. They should work together, and talk about the notes they made in activity 1. Remember (from Lesson 15) the rule is that they can only talk about information which they were specifically asked to listen for and write down. Ask them to fill the chart in with as much detail as possible. They can copy the chart if there is not enough space in their books.

● You may like to check their answers to the chart by drawing it on the board and asking the students to fill it in orally.

3 Aim: to practise speaking about types of entertainment.

● The students should now be ready to talk about a typical entertainment in their country. If they are all from the same country, ask them to suggest typical types of entertainment, and to talk about each one using the chart to help them.

● Alternatively, ask the students to work in small groups and to copy and complete the chart but leaving out the name of the entertainment. When they are ready, ask them to tell other groups about the place, performers, type of music and the reasons why people enjoy it, and ask them to guess what the type of entertainment is.

GRAMMAR AND FUNCTIONS

1 Aim: to focus on prepositions of time and place.

● Ask the students to read the information about prepositions of time and place and making invitations and suggestions in the grammar and functions box.

● This exercise is simply another way of organising the information in the grammar and functions box.

2 Aim: to focus on the prepositions you use to talk about time.

3 Aim: to practise using prepositions of time and place.

● In 2 it is possible to say *in the National Gallery,* which stresses that the exhibition is inside the building. *At* refers to the National Gallery as a location.

4 Aim: to practise making, accepting and refusing invitations.

● This activity involves some movement around the class. If this is not possible, the students can ask their nearest neighbours without leaving their seats. The students should begin the activity with a blank diary, and complete parts of it as they accept invitations.

● You may like to explain that if you refuse an invitation, a British person and an American would expect you to give an explanation. Ask the students if this is the case in their cultures.

WRITING

1 Aim: to practise writing an invitation.

● Ask the students to look at the invitation and find out what's on, where and when. You can write this information on the board.

● The students should then copy out the letter and complete it with the information on the board.

2 Aim: to practise accepting or refusing a written invitation.

● Ask the students to work in pairs and to exchange invitations. Ask them to write a reply to the invitation, using the layout of the letter as a model.

● Remind the students to include their address at the top of the letter, the date, to write *Dear* (name) and to finish it *Best wishes* and their name. Point out that this is an informal way of closing a letter.

● If there isn't much time left, the students can do this activity for homework.

Preparation for Lesson 17

● Ask the students to bring some magazine photos of some famous people for Lesson 17.

GRAMMAR AND FUNCTIONS

> **Prepositions of time and place**
>
> **at**   *the Dominion Theatre   8pm   the football match*
> **on**   *Monday   31st July*
> **in**   *June   1996   London   England*
> **to**   *go to work   go to the cinema   go to a party*
>
> **Making invitations and suggestions**
> **Would you like to** *come to the cinema?*
> **How about** com**ing** *to the cinema?*
> **Let's** *go to the cinema.*
>
> | **Accepting** | **Refusing** |
> | --- | --- |
> | *I'd love to.* | *I'm sorry, I can't. I'm busy.* |

1 Look at the prepositions of time and place in the grammar box. Write the phrases in two columns: *time* and *place*.

2 Complete the rules about when we use *at*, *on* or *in* to talk about time.

1 You use ___ to talk about months and years.
2 You use ___ to talk about days of the week and dates.
3 You use ___ to talk about a point of time in the day (eg *eight o'clock*).

3 Complete these sentences with *at, in, on* or *to*. Sometimes you can use more than one preposition. Is there a difference in meaning?

1 The match starts ___ 3pm ___ Saturday.
2 ___ the National Gallery ___ June and July there's a Van Gogh exhibition.
3 I'd like to go ___ the opera when we're ___ Moscow.
4 ___ Paris there's a World Cup football match ___ Saturday.
5 I'm going ___ a party ___ 20th June.
6 Would you like to come ___ the theatre ___ London ___ Sunday?

4 Make a list of what's on in your town at the moment. Go round the class inviting people to go out with you on different days. Accept or refuse their invitations, and try to fill up your diary.
Would you like to come to the match on Friday?
Yes, I'd love to!
How about coming to the cinema on Saturday?
I'm sorry, I'm busy. How about Sunday?

WRITING

1 Look at the formal invitation below.

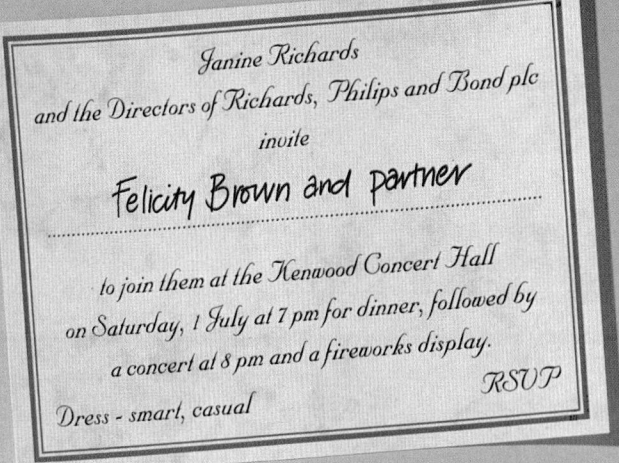

> *Janine Richards*
> *and the Directors of Richards, Philips and Bond plc*
> *invite*
> **Felicity Brown and partner**
> *to join them at the Kenwood Concert Hall*
> *on Saturday, 1 July at 7 pm for dinner, followed by*
> *a concert at 8 pm and a fireworks display.*   *RSVP*
> *Dress - smart, casual*

Work in pairs. Invite your partner to go with you. Complete the letter.

> *45 Thame Street*
> *Garsington, OX7 1LS*
> *3rd June*
>
> *Dear _____*
>
> *I've got an invitation from*
> *_____*
> *to join them on _____*
> *for _____ followed by a _____*
> *and a _____ . Would*
> *you like to come? It starts at _____ .*
>
> *Best wishes*
>
> *_____*

2 Give your partner your invitation. Now reply to his/her invitation. Use these phrases to help you.
Thank you for the invitation to ... I'd love to come.
Shall we meet at...   *I'm afraid I'm busy.*

17 | *Famous faces*

Describing people

VOCABULARY

1 Work in pairs and look at the words in the box. Which are adjectives and which are nouns?

> attractive bald beard beautiful black blonde brown curly
> dark face fair fat glasses good-looking hair head kind
> long man middle-aged moustache nice old pretty
> medium-height round short shy slim square straight tall
> teenager thin ugly woman young

Group any nouns and adjectives which often go together.
attractive face...

Which adjectives can you use to talk about the following?
– height – age – looks – build – character

2 Think of a famous person. Choose three or four words from the vocabulary box which you can use to describe his/her appearance.

Now tell your partner the name of your famous person. He/she must guess which words you chose.
My famous person is Mickey Mouse.
Did you choose short, dark, middle-aged?
Yes... and bald!

FUNCTIONS

> **Describing people**
> **Appearance**
> **You use *look like* to describe people's appearance.**
> ***What* does she *look like*? *She's tall and she's got fair hair.*
> *She **looks like** a banker.*
> ***Who* does he *look like*? *He **looks like** his father.*
>
> **Character**
> **You use *be like* to describe people's character.**
> ***What's* he *like*? *He's nice.*
> ***Who's* he *like*? *He's **like** his father.*
>
> *She's **quite** nice.   He's **about** twenty.   She's **about** one metre sixty.*
> *He's **very** tall.   He's **in** his **mid**-twenties.   She's **in** her **mid**-thirties.*
> *He's **really** handsome.   She's about thirty, **with** dark hair.*

17

GENERAL COMMENTS

Describing appearance and character
Some students may be shy or embarrassed about describing their or other people's appearance and character. If you think this is likely with your group, you may like to bring some magazine photos of people which can be used for description instead of other people in the class.

Paintings
portraits: Left hand side, top: *Jeanne Hebuterne* by Amadeo Modigliani (1884–1920, Italian); middle: *Self portrait* by Vincent van Gogh (1853–90, Dutch); bottom: *Self portrait* by Gwen John (1876–1939, English); right hand side: *Self portrait* by Stanley Spencer (1891–1959, English)

Possible questions
Do you recognise any of the people?
Do you recognise the painters?
Which one do you like best?
Do they look like people you'd like to know?

Optional extra material
Magazine photos of people. (See above.)

VOCABULARY

1 **Aim: to focus on the words in the vocabulary box and to distinguish between parts of speech.**
   ● Ask the students to look at the list of words and simply choose any which they could use to describe themselves.

   ● You can ask the students to tell you about their choice of words, or you can move onto the next stage.

   ● Ask the students to separate the nouns from the adjectives in two lists.

> **Answers**
> **adjectives:** attractive, bald, beautiful, black, blonde, brown, curly, dark, fair, fat, good-looking, kind, long, middle-aged, nice, old, pretty, medium-height, round, short, shy, slim, square, straight, tall, thin, ugly, young
> **nouns:** beard, face, glasses, hair, head, man, moustache, teenager, woman

   ● You may need to point out that to describe someone as *fat* or *ugly* is very direct and potentially rather rude in many cultures. People may also be sensitive about being *bald, old* and *short,* so the words are presented here mainly for receptive use. *Pretty* is usually used only for girls and women, and *good-looking* mostly for men.

   ● Ask the students to group words which often go together. This collocation activity will be very useful during the rest of the lesson.

> **Possible answers**
> **beard:** black, blonde, curly, dark, fair, long, short
> **face:** attractive, beautiful, fat, kind, long, nice, pretty, round, square, thin, ugly
> **glasses:** dark, round, square
> **hair:** attractive, beautiful, black, blonde, brown, curly, dark, fair, long, nice, pretty, short, straight, thin
> **head:** bald
> **man/teenager:** attractive, black, blonde, dark, fat, fair, good-looking, kind, middle-aged, old, nice, short, shy, slim, tall, thin, ugly, young
> **woman/teenager:** attractive, beautiful, black, blonde, dark, fat, fair, good-looking, kind, middle-aged, old, nice, pretty, short, shy, slim, tall, thin, ugly, young

   ● Only do the last stage of the activity if the students had difficulty with the collocation work.

> **Answers**
> **height:** short, tall, medium-height
> **age:** middle-aged, old, young
> **looks:** attractive, beautiful, good-looking, pretty, ugly
> **build:** fat, round, slim, thin
> **character:** kind, nice

   ● You may like to put the collocations and the final grouping on the board to refer to during the lesson.

2 **Aim: to practise using the words in the vocabulary box.**
   ● Ask the students to suggest the name of a famous person. Make sure everyone knows this person, then ask them to suggest adjectives chosen from the box to describe the person.

   ● Ask the students to think of a famous person, or to choose someone from the magazine photos, if you have brought some in. They should follow the instructions in the book.

FUNCTIONS

1 Aim: to focus on the use of *like* for appearance and character.

● Ask the students to read the information about describing appearance and character in the functions box.

● Ask the students to do the exercise.

2 Aim: to focus on questions and answers about personal appearance.

● Ask the students to do this on their own and then check their answers orally.

3 Aim: to practise describing appearance and character; to present *quite, very* and *really.*

● Ask the students to use the language in the box to describe members of their family and friends. You may need to explain that in this context *quite* means *fairly.* There is a chance that a student will have heard *quite* used to means *extremely,* eg *He's quite brilliant.* This use of *quite* is more advanced. The intonation is different and it's usually only used with extreme adjectives.

SPEAKING

1 Aim: to practise describing people's appearance.

● Ask the students to work in pairs. Choose someone in the paintings and describe him or her. Ask the students to point to the person you're describing. Go round and check that they are pointing to the right person.

● Ask the students to choose someone in the paintings and describe him or her to their partners. The other student must guess who is being described.

2 Aim: to practise describing people's character.

● Ask the students to imagine that they can meet the people in the paintings. Ask them to think about who they would like to meet and why. You can do this activity with the whole class.

● You can extend activities 1 and 2 with the magazine photos, if the students have brought any.

WRITING

1 Aim: to prepare for writing a letter describing your appearance.

● If your students chose adjectives to describe themselves at the beginning of *Vocabulary* activity 1, ask them to write them down. If they did not, ask them to choose adjectives now and write them down. If you have got a class with a sense of humour, you can ask them to choose adjectives that would describe how they would like to look!

2 Aim: to write sentences about your appearance.

● Ask the students to write full sentences describing their appearance.

3 Aim: to practise joining sentences.

● Ask the students to join the sentences using *and.* You may need to help them choose the best sentences to join. Usually, we only join a couple of sentences with *and,* and then perhaps add a phrase beginning *with,* as in *I'm in my mid-twenties and I'm quite tall with brown hair.*

4 Aim: to write a letter describing appearance.

● Ask the students to rewrite their sentences in the form of a letter, using the model provided. Make sure they put their address at the top, with the date, begin with *Dear...* and end *Yours sincerely.* Explain this is quite a formal letter to someone you don't know, so it is less usual to use contractions.

5 Aim: to practise describing appearance and character.

● If your students have not yet exaggerated the description of their appearance, now is their chance to do so! Ask them to follow the instructions. Encourage them to be as inventive as they like about the false pieces of information.

● If you haven't got much time, you may like to ask the students to do the preparation for this activity for homework.

WRITING

1 You are going to the station to meet someone who doesn't know you, and you're going to write a letter describing your appearance. First of all, make notes about these aspects of your appearance:
– age   – height   – looks   – hair   – build
age: mid-twenties   *height: quite tall*

2 Now write sentences describing your appearance.
I'm in my mid-twenties. I'm quite tall.

3 Then join the sentences using *and*.
I'm in my mid-twenties and I'm quite tall.

4 Write a letter describing what you look like.

(Write your address here)

(Write the date here)

Dear Mr. Freeman,

I am looking forward to meeting you at the station next Monday. I will be there at ten o'clock and will wait for you on the platform.

(Describe your appearance)

Yours sincerely,

(Write your full name here)

1 Complete the sentences with *like* if necessary.

1 What does she look ___?   She looks ___ very kind.
2 Who's she ___?   She's ___ her mother.
3 What's he ___?   He's ___ lovely!
4 Who does he look ___?   He looks ___ his brother.

2 Match the questions and the answers.

1 How old is he?   a One metre seventy-eight.
2 How tall is he?   b Blond.
3 What colour is his hair? c He's quite young, good-
                                looking and slim.
4 What does he look like? d Twenty-one.

3 Look at the sentences in the grammar box above. Write four sentences describing people you know with *quite, very* and *really*.
My father is very old and really intelligent.

SPEAKING

1 Work in pairs. Choose someone in the pictures and describe him or her to your partner. Can they guess who you're describing?

2 Talk about what you imagine the people in the pictures are like. You can use the words in *Vocabulary* activity 1 to describe them. Which person do you think you'd like to meet? Explain why.

5 Write five true pieces of information about your appearance and two false pieces. Show them to another student. Can he/she find the false information?

18 | *Average age*

**Making comparisons (1):
comparative and
superlative adjectives**

VOCABULARY

1 Work in pairs. Choose five words
to describe yourself. Use a
dictionary if necessary.

careful interesting clever cold
confident fit funny imaginative
intelligent kind lazy nervous
optimistic patient pessimistic
polite quiet calm rude sad
sensitive nice serious tidy
thoughtful

Think of other words you can
use.
honest, friendly...

Discuss your choice of words with
your partner.
*I think I'm usually optimistic.
And I'm always polite!*

Does he/she agree with you?

2 Think of three people you
admire very much. They can be
politicians, musicians, sports
personalities etc. or people you
know personally. Choose the
person you admire most and
think of three adjectives to
describe this person.

Then choose the second and
third person you admire and
think of three more adjectives
for each person to explain why.

Now turn to Communication
activity 13 on page 100.

READING

1 Read *Average age* and find things which are different from your experience
or from the experience of people you know.

A v e r a g e   a g e

10 Ten is the year of the closest friendships – though not with the opposite sex. It
is also the year when relationships with particular people or groups is strongest.
The ten-year-old usually gets on well with parents but needs more time alone.
Personal talents begin to show.

In the United States twenty is the average age for the first marriage for women,
although probably only a third marry at this age because they want to; the others
marry because of social pressure. The human brain is at its finest at twenty. It is
the age when people can vote in Denmark, Japan, Norway and Switzerland. And
in Japan it is the minimum age for buying alcohol. **20**

30 For optimists thirty is one of the happiest ages, for pessimists it marks the end
of feeling young. At this age you need to take a little more care with your body
than when you were younger. Young people who enjoyed an all-night party at
twenty will feel much worse the next day at thirty. In Britain it is the youngest
you can become a bishop.

Forty is the year of the 'middle-aged', although nobody who is forty wants to
admit the fact. Bob Hope said that you are middle-aged when your age starts to
show around your middle. In fact, the body starts to get smaller at forty and
continues to do so until you die. **40**

50 Fifty is an age when old friendships get closer and relationships with colleagues
and relatives warmer. According to old proverbs, fifty is the age when you
should be rich. George Orwell said, 'At fifty everyone has the face he deserves.'
People need to wear glasses and some food loses its strong taste.

Adapted from The Book of Ages, by Desmond Morris

18

GENERAL COMMENTS

Average age

This lesson continues the theme of describing character begun in Lesson 17.

Making comparisons (1)

This is the first of three lessons about making comparisons. This lesson focuses on the comparative and superlative forms of adjectives. The rules for this are a little complicated and if this is the first time your students have done this, you may need to give further practice using material from *Reward* Practice Book and the Resource Pack. There is also a more detailed explanation in the Grammar review at the back of the Student's Book.

Photo: possible questions

Who are these people? What's their relation to each other?
Do you have any family photos like this?
Do they look like interesting or pleasant people?
What nationality are they? How can you tell?

VOCABULARY

1 Aim: to present the words in the vocabulary box.

● Ask the students to choose five words from the box to describe themselves. Then ask one or two of them to tell the class about their choice of words. It is best to choose the more extrovert students for this task and it is important for the atmosphere to remain lighthearted.

● Ask the students to think of other words to describe themselves. They can use their dictionaries for this activity.

● Ask the students to work in pairs and to discuss their choice of words.

● Ask one or two pairs about their choice of words and discuss these choices with the rest of the class. Once again, it is important to choose students who are not shy about discussing their personalities.

2 Aim: to practise using the new words.

● This activity is a game which is not intended to be taken too seriously. The communication activity will reveal that the words the students chose to describe the first person show the kind of person they'd like to be. The words they chose for the second person show how they think other people see them. The words they chose to describe the third person show their true character!

● Ask the students to follow the instructions, choosing three famous people and three sets of adjectives to describe these people.

● Before you tell the students to turn to the Communication activity, ask one or two people to tell you the names of the people and the adjectives they chose. Write them on the board.

● Now tell them to turn to Communication activity 13 on page 100.

READING

1 Aim: to read and react to a passage.

● The passage concerns a subject which everyone has some experience of. It is their own general knowledge and experience which provides the reading comprehension check. You may want to let the students use a dictionary.

2 Aim: to discuss reactions to the passage.

● When the students have read the passage, ask them to discuss their reactions and decide which is the most surprising piece of information. If, at any stage, you think the students may need a comparative or superlative adjective, work through the *Grammar and functions* section before doing this activity.

● Broaden the discussion by asking the students if they know of any exceptional old or young people.

GRAMMAR AND FUNCTIONS

1 Aim: to focus on the formation of comparative and superlative adjectives.

● Ask the students to read the information about making comparisons in the grammar box.

● Ask the students to do the activity.

> **Answers**
> kind kinder kindest
> nice nicer nicest
> lazy lazier laziest
> sad sadder saddest
> careful more careful most careful
>
> The rule for forming the comparative and superlative forms of short adjectives ending in:
> -e: add -r, -st
> -y: drop the -y and add -ier, -iest
> **a vowel and a consonant:** double the consonant and add -er, -est
>
> The rule for forming the comparative and superlative forms of longer adjectives:
> *more, most* + adjective

● The students may want to look at the explanation in the Grammar review at this stage.

2 Aim: to focus on the formation of comparative and superlative adjectives.

> **Answers**
> cold colder coldest
> imaginative more imaginative most imaginative
> intelligent more intelligent most intelligent
> fit fitter fittest
> tidy tidier tidiest
> beautiful more beautiful most beautiful
> polite more polite most polite
> patient more patient most patient
> young younger youngest
> funny funnier funniest
> nervous more nervous most nervous
> warm warmer warmest
> old older oldest

3 Aim: to practise using comparative and superlative adjectives.

● Ask the students to continue discussing their reactions to the passage using comparative and superlative adjectives. Return to any points made during the discussion in *Reading* activity 2 and express them differently using comparative and superlative adjectives.

4 Aim: to practise speaking and using comparative and superlative adjectives.

● Ask the students to continue their discussion about the best age to do things. If you like, you can extend this activity so that the students not only suggest the best age to do things, but also suggest the reasons why.

● Ask the students to write sentences comparing their opinions.

SPEAKING AND WRITING

1 Aim: to prepare to talk about exceptional people, places and things.

● The students can choose from the adjectives in *Grammar* activity 2, or from the adjectives in the vocabulary box to fill in the blanks in the questions.

2 Aim: to practise talking about exceptional people, places and things.

● Ask the students to work in groups, asking their questions and answering other people's questions. They should write down the answers if possible.

3 Aim: to practise writing about exceptional people, places and things.

● Ask the students to write a few sentences describing other people's answers to their questions.

● If time is short, the students can do this activity for homework.

2 Work in pairs. Which is the most surprising piece of information in the passage?

GRAMMAR AND FUNCTIONS

> **Making comparisons (1)**
> **Comparative adjectives**
> **You form the comparative of most adjectives by adding -er, -r, -ier or more + adjective.**
> kind**er** nic**er** laz**ier** sad**der** **more** careful
> *Fifty is an age when old friendships get **closer**.*
>
> **Superlative adjectives**
> **You form the superlative of adjectives with -est, -st, -iest or most + adjective.**
> kind**est** nic**est** laz**iest** sad**dest** **most** careful
> *For optimists thirty is one of **the happiest** ages.*
> **There are some irregular comparative and superlative forms.**
> good better best   bad worse worst

1 Look at the comparative and superlative adjectives in the box above. Write down the adjective they come from and the comparative and superlative forms.
kind   kinder   kindest

What's the rule for forming the comparative and superlative forms of short adjectives ending in -e, -y, a vowel + consonant?

What's the rule for forming the comparative and superlative forms of longer adjectives?

2 Make comparative and superlative adjectives from the following.

cold   imaginative   intelligent   fit   tidy   beautiful
polite   patient   young   funny   nervous   warm   old

3 Work in pairs and compare the information in *Average age* with your own experience.
In my country, I think that many women are older when they get married.

4 Work in groups of two or three. Talk about the best age for doing the following things.

– getting married   – buying a home
– having children   – leaving your parents' home
– going to university   – learning to drive
– leaving school   – learning a foreign language

I think it's best to get married at twenty-five.

Now write sentences comparing your opinions.
We all think it's best to leave school at sixteen.
Jerome thinks it's best to learn a foreign language at six, but we think it's best when you're older.

SPEAKING AND WRITING

1 Choose a superlative adjective you made in *Grammar* activity 2 to complete these questions about exceptional people, places and things.

– Who is the _____ person you know?
– What is the _____ thing you own?
– What is the _____ time of the year?
– Where is the _____ place you know?
– What is the _____ country for a holiday?
– Who is the _____ person to be with at a party?

Who is the funniest person you know?

2 Work in groups of three or four. Ask and answer the questions you wrote in 1 about exceptional people, places and things.
I think the funniest person I know is Dietrich.

3 Write sentences comparing your partners' answers to your questions.
Hans thinks the funniest person he knows is Dietrich.

19 | *Dressing up*

Making comparisons (2): *more than, less than, as...as*

VOCABULARY

1 Look at these words for clothes. Which do you wear? Put the words in four lists under these headings: *always, often, sometimes, never*.

> blouse coat dress hat jacket jeans shirt shoes
> skirt socks suit sweater swimsuit tie tights
> trainers trousers T-shirt underwear

always: trousers often: jeans sometimes...

Think of other clothes you wear in winter, in summer, for work and at home and add them to your lists.

2 Choose suitable adjectives from the list below to describe your own clothes. Add them to the four lists you made in 1.

> black blue brown casual formal green grey
> orange pink red smart white yellow

always: dark trousers often: smart jeans sometimes...

3 Work in pairs. Ask and say what you always, often, sometimes and never wear.
Do you often wear trousers, Erina?
Yes. I usually wear dark trousers.

READING

1 Read *Dressing up*, which is about clothing in Kuwait, Sweden and India, and find out if it says anything about:

– clothes for work
– traditional dress
– young people's fashions

Check your answers with another student.

In Kuwait, men and women wear their traditional dress most of the time. For men, it is a long robe and a cloth covering the head. For women it's similar and they wear a veil. Foreign male visitors usually wear lightweight cotton trousers and white shirts with short or long sleeves. Men often wear sandals during the day but never in the office. They wear a jacket and a tie for social occasions, but when it's really hot, it's usual to take off the jacket. Foreign women visitors usually wear long, loose clothes which cover their neck and arms.

The Swedish are very interested in clothes and are less formal than they were. People usually dress well in public and wear bright colours. In Sweden the winters are very cold, so overcoats and ski jackets are very common. Men wear business suits for work, with a shirt and a tie and women often wear trousers. People often carry a spare pair of shoes because you need boots outside. Children and teenagers are more casual than their parents. For school, they wear blue jeans and T-shirts.

Traditional dress in India for women is the *sari* and for men the *achkan* suit. The sari has its own distinctive style depending on which part of India it comes from – every region has its own special colours, decoration and style. The men wear their heavy and expensive *achkan* suits on formal occasions but for less formal occasions they wear the *kurtha* suit, a long shirt and loose trousers, which is not as heavy as the *achkan*. Indian people wear lighter colours as they grow older, and at funerals, white is the usual colour to wear.

Many people wear western-style clothes. For work they wear smart clothes, but not suits and ties. Women usually wear trousers and blouses but not dresses. Young people are as casual as young people all over the world with their jeans and T-shirts.

2 Work in pairs. Ask and say what clothes people in your country wear. Talk about clothes for work, clothes at home, traditional dress, young people's fashions.
Men wear suits for work. What do you wear?

3 Are the conventions for clothing in the passage different from conventions in your country?
In my country men don't usually wear sandals.

19

GENERAL COMMENTS

Making comparisons (2)

This lesson is the second of three lessons about making comparisons. The first was Lesson 18 and the third is Lesson 35. This lesson deals with the words which you use with adjectives to make comparisons, eg *than, as,* etc.

Clothes

The words for clothes and colours in the vocabulary box are likely to be the most useful ones for most students. However, the vocabulary field is a broad one and you may wish to extend the coverage beyond the items suggested.

Cross-cultural training

The choice of Kuwait, Sweden and India to illustrate styles of clothing in different countries is, to a certain extent, an arbitrary one, but they were felt to be different enough to create the opportunity for cross-cultural comparison. In this context, these are sample cultures with which students can compare their own. It is not the intention to give the students a large amount of culture-specific information.

Photos: possible questions

This lesson is about clothes. Can you say what nationality the people in the photos are from their clothes? What are typical clothes in your country? What do people from other countries typically wear? When do you wear national or regional dress in your country? How often do you wear it?

VOCABULARY

1 Aim: to focus on the words in the vocabulary box.

● This activity is designed to encourage the categorisation of the words according to personal criteria. This will help make the words more relevant to the student and promote the learning process. Write the four categories on the board and start categorising the words for yourself.

● Choose a student and guess how he or she would categorise the words. Find out if he/she agrees with you.

● Ask the students to put the words under the headings for themselves.

● You may like to ask people who know each other well to group the words for each other, and then find out if their partner agrees.

● Ask the students to add words to the list, and to group them under the new headings: *in winter, in summer, for work* and *at home.*

2 Aim: to present the words in the vocabulary box.

● Continue the process of grouping the words according to personal criteria.

● Find out how many people in the class like the different colours. Which is the most and least popular colour?

3 Aim: to practise using the new vocabulary.

● Ask the students to work in pairs and to talk about the clothes they and other people wear.

● Broaden the vocabulary work to a general discussion about clothes the students wear in different circumstances. Lead the discussion with the following questions: *Do all the students wear the same clothes in general, or is there anyone who is more or less formal? Do people like to blend in or stand out with the clothes they wear? Which is the most popular item of clothing?* You can do this activity in pairs and then finish it by finding out about the opinions of the class as a whole.

READING

1 Aim: to read for specific information.

● Explain that the reading passage is about clothing styles in Kuwait, Sweden and India. Ask the students to say which nationality they think is shown in the photo.

● Ask the students to read and find out about clothes for work, clothes for home, traditional dress and young people's fashions in the three countries.

> **Answers**
> **Kuwait**
> – clothes for work: men never wear sandals in the office
> – traditional dress: men: a long robe and a cloth covering the head; women: similar with a veil.
> **Sweden**
> – clothes for work: men: business suits, shirt and tie; women often wear trousers.
> – young people's fashions: jeans and T-shirts.
> **India**
> – clothes for work: smart clothes, but not suits and ties.
> – traditional dress: *sari* for women and the *achkan* suit for men on formal occasions, the *kurtha* suit for less formal occasions.
> – young people's fashions: jeans and T-shirts.

● Ask the students to compare their charts in pairs.

2 Aim: to discuss clothing conventions.

● Ask the class to talk about clothing for work, traditional dress and young people's fashions in their country. You may want to give them some extra vocabulary at this stage. If they have traditional dress, how often and on which occasions do they wear it?

3 Aim: to compare clothing conventions.

● Continue the discussion with the class as a whole by asking them to compare conventions of clothing with those in Kuwait, Sweden and India. Which of the three countries has similar conventions of clothing to the students' own countries?

● Ask the students to share any impressions of national styles of dress. Can you tell a person's nationality from the way they dress?

GRAMMAR AND FUNCTIONS

● Ask the students to read the explanation in the grammar and functions box and then do the exercises.

1 **Aim: to focus on the use of *as*, *than* and *from* in making comparisons.**

● Ask the students to read the information about making comparisons in the grammar and functions box.

● Ask them to do the activity.

> **Answers**
> 1 He's much smarter **than** I am.
> 2 She's **as** intelligent **as** he is.
> 3 Her clothes are different **from** mine.
> 4 She's got the same shoes **as** I have.
> 5 Sandals are less common here **than** in Kuwait.
> 6 Children are more casual **than** their parents.

2 **Aim: to focus on *more* and *less*.**

> **Answers**
> 1 Yes, he's more formal than she is.
> 2 Yes, it's less quiet now than it was.
> 3 Yes, they're less expensive here than at home.
> 4 Yes, he's more pessimistic than she is.
> 5 Yes, it's less difficult to get good clothes here.
> 6 Yes, he's more nervous than she is.

3 **Aim: to practise making comparisons.**

● Ask the students to write sentences comparing what they wear with other people.

SOUNDS

1 **Aim: to focus on /ə/ and /ɪ/.**

● This activity is designed to draw the students' attention to the weak syllable in comparative adjectives, which is usually /ə/ and superlative adjectives, where it is usually /ɪ/. Write /ə/ and /ɪ/ on the board. Ask the students to say the words aloud and to point to the phoneme they use.

● ▭ Play the tape.

> **Answers**
> /ə/: smaller bigger closer happier funnier
> /ɪ/: smallest biggest closest happiest funniest

2 **Aim: to focus on the unstressed pronunciation of *than, as* and *from*.**

● ▭ This activity is designed to draw the students' attention to the unstressed pronunciation of *than, as* and *from*. Play the tape and pause after each sentence. Ask them to repeat the sentences.

3 **Aim: to focus on contrastive stress when disagreeing.**

● ▭ There have already been some activities with contrastive stress in pre-intermediate *Reward*. This activity focuses on the stress and intonation pattern used when disagreeing over making comparisons. Play the tape.

● Ask the students to work in pairs and to disagree with the statements. Rewind and play the tape again, pausing each time so they can check their answers.

LISTENING AND SPEAKING

1 **Aim: to prepare for listening.**

● Ask the students to read the statements and decide if they are true or false for their country. Ask one or two students for their reactions.

2 **Aim: to listen for main ideas.**

● ▭ Explain to the students that they are going to listen to a British person talking about conventions of clothing in Britain, and that they should decide if the statements in activity 1 are true or false for Britain. Suggest that they tick the statements which are true and put a cross by the statements which are false.

> **Answers**
> 1 The weather is usually rather cold.   T
> 2 It's difficult to buy good clothes.   F
> 3 Good clothes are very expensive.   T
> 4 People are quite formal.   F
> 5 Many people are quite small.   F
> 6 The quality of clothes design is good.   T

● You may want to play the passage a second time.

3 **Aim: to practise making comparisons.**

● This activity is designed to give some final practice in using comparative and superlative adjectives and making comparisons. Ask the students to discuss their answers to the questions.

● If there is no time, you can ask the students to write their answers to the questions for homework.

GRAMMAR AND FUNCTIONS

> **Making comparisons (2): *more than, less than, as...as***
> *Children wear **more** casual clothes **than** their parents.*
> *They're **less** formal **than** they were.*
> *My father wears cheap**er** clothes **than** my mother.*
> *They're **as** casual **as** teenagers are all over the world.*
> *Dresses are **not as** popular **as** in Western countries.*
> ***It's the same as** my country.*
> ***It's different from** my country.*

1 Complete these sentences with *as, than* or *from*.

1 He's much smarter ___ I am.
2 She's ___ intelligent ___ he is.
3 Her clothes are different ___ mine.
4 She's got the same shoes ___ I have.
5 Sandals are less common here ___ in Kuwait.
6 Children are more casual ___ their parents.

2 Agree with these statements using *more* or *less* and the adjective in brackets.

1 He's less casual than she is. (formal)
2 It's more noisy now than it was. (quiet)
3 Clothes are cheaper here than at home. (expensive)
4 He's less optimistic than she is. (pessimistic)
5 It's easier to get good clothes here. (difficult)
6 He's less confident than she is. (nervous)

1 Yes, he's more formal than she is.

3 Write four sentences comparing what you wear and your appearance with other people. Use comparative adjectives.

I wear more colourful clothes than my father.

SOUNDS

1 Say these words aloud. Is the underlined sound /ə/ or /ɪ/?

small<u>e</u>r small<u>e</u>st bigg<u>e</u>r bigg<u>e</u>st clos<u>e</u>r clos<u>e</u>st happi<u>e</u>r happi<u>e</u>st funni<u>e</u>r funni<u>e</u>st

🔲 Listen and check.

2 🔲 Listen to the sentences in *Grammar and functions* activity 1. Notice that *than* is pronounced /ðən/, *as* is pronounced /əz/ and *from* is pronounced /frəm/.

Now say the sentences aloud.

3 🔲 Listen to someone disagreeing with the statements in *Grammar* activity 2.

Now work in pairs and disagree with the statements. Stress *more* or *less*.

LISTENING AND SPEAKING

1 Read these statements. Decide if they are true or false for your country.

	My Country	Britain
1 The weather is usually rather cold.		
2 It's difficult to buy good clothes.		
3 Good clothes are very expensive.		
4 People are quite formal.		
5 Many people are quite small.		
6 The quality of clothes design is good.		

2 🔲 Listen to Graham, from Britain, talking about clothing. Does he think the statements in 1 are true or false?

3 Work in pairs and answer the questions.

In your country...

– is the weather hotter or colder than in Britain?
– is it easier or more difficult to buy good clothes?
– are clothes cheaper or more expensive?
– are people more formal or more casual?
– are people smaller or larger than the British?
– is the quality of clothes design better or worse?

We have cold weather too. In my country it's more difficult to buy good clothes.

20 | *Memorable journeys*

Talking about journey time, distance, speed and prices

VOCABULARY AND LISTENING

1 How do you say these numbers?

505
  a five hundred five
  b five hundred and five

478
  a four hundred and seventy eight
  b four hundred seventy eight

3,563
  a three thousand and five hundred sixty three
  b three thousand, five hundred and sixty three

45,781
  a forty five thousand, seven hundred and eighty one
  b forty five thousand, seven hundred eighty and one

🔈 Now listen and check.

2 Say these numbers aloud.

346 678 2,345 18,664
24,589 123,456 202 54,566
3,481 407 10,020

🔈 Now listen and check.

3 Work in pairs and look at the photo. Use these words to describe what you can see.

> arrive border cost desert distance drive driver gallon • gas station get highway hill leave mile motel mountain move home passenger petrol police patrol reach set off speed limit take ticket truck turn off

Would you like to be one of the passengers?

4 🔈 Listen to Sarah, an English woman, talking about a memorable journey she made in the USA. As you listen, look at the vocabulary box again and tick (✓) the words you hear.

5 Tick (✓) the information you heard.

Journey time	9 days	19 days	90 days
Distance	250 miles	2,050 miles	2,500 miles
Price of petrol	25 cents a litre	50 cents a litre	55 cents a litre
Speed limit	50 miles an hour	55 miles an hour	70 miles an hour
Price of hotel rooms	$20 to $35 per person	$25 to $40 per person	$30 to $50 per person

🔈 Now listen again and check.

20

GENERAL COMMENTS

Numbers

Many speakers of a foreign language have difficulty in using numbers. Generally, it is more difficult to hear and write down numbers than to say them. With more extensive calculations, most people revert to their own language. So the aims of this lesson as far as numbers are concerned are modest. It is simply an opportunity to give some further practice in producing and receiving numbers. If your students need to be fluent in numbers, perhaps for professional reasons, you may have to supplement the work in *Reward* Student's Book with activities from the Practice Book and the Resource Packs. You may also want to create further practice by making your own exercises using those in the Student's Book as a model.

Accuracy and fluency

Some teachers feel it is important for their students to be as accurate as possible on every occasion, and will correct every error a student makes. While accuracy is important in many exercises in *Reward*, there are many discussion activities, especially in the *Speaking* sections, which are designed to promote fluency. On these occasions, it is better not to interrupt the communicative flow by correcting the students' mistakes. The criterion for correction on these occasions is intelligibility. If the student is intelligible, then it is better to make a note of the mistakes and draw the student's attention to them at the end of the activity.

Photo: possible questions

Where is this? How can you tell?
Is it like a street in your country? If so, where?
What can you see? Does it look like an interesting place?
What time of day is it? What's the weather like?
What do you think happened just before the photo was taken? What do you think happened just after it was taken?

VOCABULARY AND LISTENING

1 Aim: to practise saying longer numbers.
- 🔊 Ask the students to say the numbers.

- You may want to extend this activity by writing some more numbers on the board or asking students to work in pairs, writing numbers for each other and saying them aloud in turn.

> **Answers**
> 1 b 2 a 3 b 4 a
> You use *and* between hundreds and tens, or between hundreds and single figures.

2 Aim: to practise saying longer numbers.
- Ask the students to say the numbers aloud. As in activity 1 you may want to extend this activity.

- 🔊 Play the tape and ask the students to listen and check. You may need to pause the tape after each number for the students to repeat.

3 Aim: to present the words in the vocabulary box and to practise using them.
- If there is time, write the words on the board. Ask the students to look at the photo. Ask one or two students to use the words to describe what they can see. Tick the words as they use them.

- Ask the students if they would like to be one of the passengers in the journey across America.

4 Aim: to listen for main ideas.
- 🔊 Make sure everyone understands what the listening passage is going to be about. Then play the tape and ask the students to listen and tick the words in the vocabulary box as they hear them.

- Check the words they ticked.

- Ask a student to tell the class through which main towns the route passed.

5 Aim: to listen for specific information.
- There is a great deal of specific information concerning journey time, distance, speed and prices in the listening passage. Ask the students to try to remember which specific information they heard and underline it.

> **Answers**
> **Journey time:** 9 days
> **Distance:** 2,500 miles
> **Price of petrol:** 25 cents a litre
> **Speed limit:** 55 miles per hour
> **Price of hotel rooms:** $20 to $35 per person

- You can ask the students to try to remember other numbers which the speaker mentions.

- 🔊 Play the tape again and ask the students to check their answers.

SOUNDS

1 Aim: to focus on syllable stress in numbers.
- The stressed syllable in these numbers is very important, as it distinguishes between numbers in the teens and other numbers. Ask the students to underline the stressed syllable.

- 🔊 Play the tape and ask them to listen and check.

> **Answers**
> <u>thir</u>ty thir<u>teen</u> four<u>teen</u> <u>for</u>ty seven<u>teen</u> <u>seven</u>ty
> nine<u>teen</u> <u>nine</u>ty thirteen <u>dollars</u> fourteen
> <u>kilo</u>metres seventeen <u>hours</u> nineteen <u>miles</u>
> With 20, 30, 40, 50, 60 70, 80, 90 the first syllable is stressed. With numbers from thirteen to nineteen the second syllable is stressed.
> With numbers from thirteen to nineteen followed by a noun, the stress on the second syllable is less clear.

2 Aim: to practise listening to and writing down numbers.

● 📼 This is likely to be a difficult exercise, so give the students plenty of time. You may decide to supplement it with other material.

> **Answers**
> 21 33 421 645 4,591 3,542 77,889 12,523
> 101,456 987,241

● Ask the students to say the numbers aloud. Check they use a falling intonation on the last number.

FUNCTIONS

1 Aim: to practise talking about time, distance, speed and prices.

● Ask the students to read the information about talking about journey time, distance, speed and prices in the functions box.

● Ask the students to check their answers to *Vocabulary and listening* activity 5. Encourage them to use the structures in the functions box.

2 Aim: to focus on asking questions about time, distance, speed and prices.

● Ask the students to think about places in their country which are:
- an hour away from each other by train, place, car or boat.
- 500 kilometres away from each other.
- half an hour away from each other by train, place, car or boat.
- 20 kilometres away from each other.
Ask them to write the questions.

● The students can then ask each other their questions. See if the answers correspond to what the questioners had in mind.

3 Aim: to practise talking about time and distance.

● Ask the students to choose a town in their country and to write sentences saying how far away it is and how long it takes to get there.

4 Aim: to practise asking questions about speed and prices.

● Write the following on the board: *speed limit in town speed limit on motorways, price of a train ticket to the next big city, price of petrol, price of a bus ticket in your town.* Ask the students to write questions saying how much these things cost or how fast you can go.

● Ask the students to work in pairs and ask and answer the questions.

SPEAKING

1 Aim: to practise talking about time, distance, speed and prices; to make comparisons.

● This activity encourages students to make comparisons which they learnt in Lessons 18 and 19. Ask them to say if the information about America which they ticked in the chart in *Vocabulary and listening* activity 5 is the same as or different from their country. Do this activity orally with the whole class.

2 Aim: to practise talking about time, distance, speed and prices.

● Ask the students to work in pairs and to talk about a memorable journey they have made. It doesn't need to be as significant as Sarah's journey across the southern United States of America. It could be a holiday journey or a memorable journey to work or school. Don't worry about correcting mistakes during this activity; this is an opportunity for fluency practice.

● Ask one or two students to describe their memorable journeys to the rest of the class.

3 Aim: to practise using superlative adjectives; to ask questions about height, size, population and temperature.

● Ask the students to look at the information about America and to write two questions for each piece of information.

> **Answers**
> | | |
> |---|---|
> | What's the tallest building? | How tall is it? |
> | What's the highest mountain? | How high is it? |
> | What's the longest river? | How long is it? |
> | What's the largest lake? | How large is it? |
> | What's the biggest city? | How many people live there? *or* How big is it? |
> | What's the hottest place? | How hot is it in summer? |
> | What's the coldest region? | How cold is it in winter? |

4 Aim: to practise talking about height, size, population and temperature.

● Ask the students to answer the questions they prepared in activity 3 with reference to their own country.

● If time is short, the students can write answers to the questions for homework.

SOUNDS

1 Say these words aloud. Underline the stressed syllable.

thirty thirteen fourteen forty seventeen seventy
nineteen ninety thirteen dollars fourteen kilometres
seventeen hours nineteen miles

[cassette icon] Listen and check. Which words are stressed on the first syllable? Which words are stressed on the second syllable?

2 [cassette icon] Listen and write down the numbers you hear.

Now say the numbers aloud.

FUNCTIONS

Talking about journey time, distance, speed and prices
Journey time
How long *does it take **by** car?* (It takes) nine days.
How long *does it take **by** train?* Five hours.
How long *does it take **on** foot?* It's a five-minute walk/drive/flight/journey.
Distance
How far *is it?* It's 2,500 miles **away**.
How far *is your school from your home?* Ten kilometres.
Speed
How fast *can you drive?* Fifty-five miles **per** hour. (mph) Ninety kilometres **an** hour. (km/h)
Prices
How much *is petrol?* (It's) $2 **a** gallon.
How much *does petrol cost?* (It costs) 25 cents **a** litre.
How much *are hotel rooms?* $25 **per** person **per** night.
How much *do hotel rooms cost?*

1 Work in pairs and check your answers to *Vocabulary and listening* activity 5.

2 Imagine these are the correct answers to questions about your country. Write the questions, using *How long* or *How far.*

1 An hour.
2 500 kilometres.
3 Half an hour.
4 20 kilometres.
5 Two weeks.
6 It's three kilometres away.
7 Ten hours.
8 It's a ten-minute drive.

1 An hour.   How long does it take to get from Buenos Aires to Cordoba by plane?

2 500 kilometres. How far is it from Istanbul to Izmir?

3 Think of a town in your country. Write sentences saying how far it is from where you are now and how long it takes to get there by different means of transport or on foot.

4 Write six questions about speed and prices in your country.
How fast can you drive in town?
How much does it cost to fly from Buenos Aires to Cordoba?

SPEAKING

1 Compare the information about America in *Vocabulary and listening* activity 5 with your country.
It takes less time to cross my country.
Petrol is more expensive in my country.

2 Think about a memorable journey by car across your country. What is the best route to take?

Now work in pairs and talk about your journeys.

3 Look at some more information about America.

Tallest building	Sears Tower, Chicago, 443 m
Highest mountain	Mount McKinley, Alaska, 6 194 m
Longest river	Mississippi River, 6 020 km
Largest lake	Lake Superior, 82 260 sq km
Biggest city	New York City, population 7.5 million
Hottest place	Death Valley, average temperature 50°C in July
Coldest region	Alaska, average temperatures -12°C in January

Write two questions for each piece of information.
What's the tallest building? How tall is it?

4 Think about answers to the questions about your country. It doesn't matter if you don't know the exact figures.

Now work in pairs and ask and answer questions about your country.
Monique, what's the tallest building in France?
It's the Eiffel Tower.
How tall is it?
It's over three hundred metres high.

Progress check ⬤ 16-20

VOCABULARY

1 Look at these international words.

> pizza rioja sushi disco rock television football
> restaurant concert cinema tennis theatre film
> ballet opera jazz stadium museum video

Put them in these groups: *food, sport, places, types of entertainment, music.*

If you know any more international words, add them to your groups of words.

2 Look at the endings used for these adjectives.

friend**ly** gener**ous** act**ive** dynam**ic** confid**ent**
thought**ful** temperament**al** nois**y** cap**able**

Now look at the adjectives in Lesson 18 again. Are there any with similar endings? Words with these endings are often adjectives.

3 Some words are used either for men or for women, but not both. Put a cross (✗) by the sentences which sound odd.

1 He's got a really pretty face.
2 She bought some patterned tights yesterday.
3 He wears a white blouse in the office.
4 She's a very handsome little girl.
5 He's got a warm night-dress for winter nights.
6 She has two pairs of blue jeans.

4 Decide if these items of clothing are usually worn by men or women, or both.

skirt bikini bra knickers underpants tights
boots swimsuit shirt shorts jacket sandals
pyjamas

5 Look at the vocabulary boxes in lessons 16 – 20 again. Choose words which are useful to you and group them under headings of your choice in your *Wordbank*.

GRAMMAR

1 Complete these sentences with *in, at* or *on*.

1 The football season starts ___ August and finishes ___ May.
2 The ballet is ___ the Apollo Theatre.
3 It's ___ Monday 22 June ___ 7.30pm.
4 The match is ___ 7.15pm ___ Saturday.
5 The Olympic Games ___ 1996 are ___ Atlanta ___ the USA.
6 The film starts ___ 3pm ___ Saturday.

2 Complete these sentences with *to* or *at*.

1 I like going ___ the theatre.
2 I'm working ___ home tomorrow.
3 Shall we meet ___ the cinema?
4 The football match is ___ the main stadium.
5 Would you like to take us ___ the museum?
6 Let's walk ___ the swimming pool.

3 Write questions about Frank.

Ask about:

family likeness age height colour of hair
colour of eyes looks

4 Write sentences saying what Frank looks like.

Family likeness: father
Age: 24
Height: 1m 78

Progress check 16–20

GENERAL COMMENTS

You can work through this Progress check in the order shown, or concentrate on areas which may have caused difficulty in Lessons 16 to 20. You can also let the students choose the activities which they would like to or feel the need to do.

VOCABULARY

1 Aim: to focus on international words.

● Explain that many words in English are used in other languages as well. Sometimes it's hard to know which language a word originally came from. Ask the students to group the words under the headings.

> **Answers**
> **food:** pizza, rioja, sushi
> **sport:** football, tennis
> **places:** disco, restaurant, cinema, theatre, stadium, museum
> **types of entertainment:** disco, television, concert, cinema, theatre, film, ballet, video
> **music:** disco, rock, opera, jazz

● Some words can go under other headings. Ask them to think of other words to go under the headings.

● Other possible groups of international words are science and technology and politics.

2 Aim: to focus on adjective suffixes.

● Ask the students to look back at Lesson 18 and write down the adjectives in the vocabulary box which end in *-y, -ous, -ive, -ic, -ent, -ful, -al, -ly* and *-able*.

> **Answers**
> careful, confident, funny, imaginative, intelligent, lazy, nervous, optimistic, patient, pessimistic, sensitive, serious, tidy, thoughtful

3 Aim: to present or revise male and female words.

● In Lesson 17 a difference was made between adjectives you usually use to describe men or women but not both. Sentences 1 and 4 will be revision, the rest may be new to the students.

> **Answer**
> 2 and 6 do not sound odd.

4 Aim: to present male and female words for clothes.

> **Answers**
> **Men:** underpants, shirt
> **Women:** skirt, bikini, bra, knickers, tights, swimsuit
> **Both:** boots, shorts, jacket, sandals, pyjamas

5 Aim: to help the students organise their vocabulary learning.

● Remind the students that they need to go back over their *Wordbanks* every so often. It also may be helpful to organise the words in different categories.

GRAMMAR

1 Aim: to revise prepositions of time and place.

> **Answers**
> 1 The football season starts **in** August and finishes **in** May.
> 2 The ballet is **at** the Apollo Theatre.
> 3 It's **on** Monday 22 June **at** 7.30pm.
> 4 The match is **at** 7.15pm **on** Saturday.
> 5 The Olympics Games **in** 1996 are **in** Atlanta **in** the USA.
> 6 The film starts **at** 3pm **on** Saturday.

2 Aim: to revise *to* and *at*.

> **Answers**
> 1 I like going **to** the theatre.
> 2 I'm working **at** home tomorrow.
> 3 Shall we meet **at** the cinema?
> 4 The football match is **at** the main stadium.
> 5 Would you like to take us **to** the museum?
> 6 Let's walk **to** the swimming pool.

3 Aim: to revise questions about people's appearance.

> **Answers**
> Who does he look like?
> How old is he?
> How tall is he?
> What colour is his hair?
> What colour are his eyes?
> What does he look like?

4 Aim: to revise talking about personal appearance.

> **Answers**
> He looks like his father.
> He's 24 years old.
> He's one metre seventy-eight centimetres tall.
> He's got fair hair.
> He's got blue eyes.

5 Aim: to revise talking about personal appearance.
- Ask the students to write their own sentences.

6 Aim: to revise talking about personal appearance.

Answers
1 She's got a very **pleasant** face.
2 He's got no hair. He's quite **bald**.
3 He has a **long** grey beard.
4 She's over two metres. She's quite **tall**.
5 He works as a model. He's very **good-looking**.
6 It's cold today. I'll wear a **coat**.

7 Aim: to revise comparative/superlative adjectives.

Answers
big bigger biggest
calm calmer calmest
careful more careful most careful
clever more clever most clever
confident more confident most confident
friendly more friendly/friendlier most
friendly/friendliest
generous more generous most generous
imaginative more imaginative most imaginative
informal more informal most informal
lazy lazier laziest
nervous more nervous most nervous
quiet quieter quietest
small smaller smallest
smart smarter smartest
thoughtful more thoughtful most thoughtful
tidy tidier tidiest
warm warmer warmest

8 Aim: to revise comparative form of adjectives.

Answers
1 No, they aren't. They're more expensive.
2 No, he isn't. He's shorter than Philip.
3 No, it isn't. It's colder than Brazil.
4 No, it isn't. It's more difficult to buy nice clothes in the winter.
5 No, she isn't. She's older than Penny.
6 No they aren't. They're more casual than the Germans.
7 No, he isn't. He's ruder than Jack.
8 No, he isn't. He's more hard-working than Joe.

9 Aim: to revise superlative adjectives.

Answers
1 Yes, she's the kindest person I know.
2 Yes, it's the most beautiful town I know.
3 Yes, he's the politest person I know.
4 Yes, she's the shortest person I know.
5 Yes, it's the most expensive dress I know.
6 Yes, it's the most powerful car I know.
7 Yes, he's the most handsome man I know.
8 Yes, she's the most sensitive person I know.

10 Aim: to revise talking about time, speed, distances and prices.

Answers
A How **far** is it to the nearest station?
B It's two kilometres **away.**
A How **long** does it take **by** car?
B It's **a** five-minute drive.
A How long does it **take** to walk?
B It's thirty minutes **by/on** foot.
A How **much** is a ticket to London?
B It **costs** £5.50.

SOUNDS

1 Aim: to focus on /ʊ/ and /uː/.
- Ask the students to say the words.
- ▭ Play the tape and ask the students to write the words in two columns.

Answers
/ʊ/: good book pullover took cook
/uː/: blue shoe boot suit

- In some regional accents in Britain there is little difference between the two phonemes.

2 Aim: to focus on /dʒ/.
- ▭ Play each word and then pause the tape; ask the students to repeat each word.

Answers
geography journalist soldier engineer teenager
job manager

3 Aim: to present simple transactional language for buying things.
- ▭ Ask the students to put the sentences in order.

Answers
c Can I help you?
e Yes, I'm looking for a sweater.
f How about this one? It suits you.
d It's too small. Have you got it in a bigger size?
b I'm sorry. This is the largest size we've got.
a Well, I think I'll leave it. Thank you.

- Remind the students that people in Britain try to sound polite and friendly in most situations.

SPEAKING AND WRITING

Aim: to do a mutual dictation involving speaking and writing.
- The students are going to do a mutual dictation, in which they dictate to each other alternate lines of a short story. Ask them to turn to the Communication activities and to follow the instructions.

- Ask the students to check their versions by showing each other the text in their instructions.

5 Write sentences about what you look like. Use the questions you wrote in 3 to help you.

6 Choose the best words and complete the sentences.

1 She's got a very ___ face.
   a curly   b tall   c pleasant
2 He's got no hair. he's quite ___ .
   a bald   b fair   c grey
3 He has a ___ grey beard.
   a round   b long   c square
4 She's over two metres. She's quite ___ .
   a short   b honest   c tall
5 He works as a model. He's very ___ .
   a good-looking   b careless   c bossy
6 It's cold today. I'll wear a ___ .
   a swimsuit   b T-shirt   c coat

7 Write the comparative and superlative forms of these adjectives.

big calm careful clever confident friendly
generous imaginative informal lazy nervous
quiet small smart thoughtful tidy warm

8 Disagree with these statements using the adjective in brackets.

1 Clothes are cheaper than food. (expensive)
2 Mike is taller than Philip. (short)
3 Britain is hotter than Brazil. (cold)
4 It's easier to buy nice clothes in the winter. (difficult)
5 Kate is younger than Penny. (old)
6 The British are more formal than the Germans. (casual)
7 Peter is more polite than Jack. (rude)
8 Graham is lazier than Joe. (hard-working)

1 No they aren't. They're more expensive.

9 Agree with these statements using the superlative form of the adjective.

1 She's very kind.
2 It's a very beautiful town.
3 He's very polite.
4 She's very short.
5 This dress is very expensive.
6 It's a very powerful car.
7 He's extremely handsome.
8 She's very sensitive.

1 Yes, she's the kindest person I know.

10 Complete the dialogue.

A How ___ is it to the nearest station?
B It's two kilometres ___ .
A How ___ does it take ___ car?
B It's ___ five-minute drive.
A How long does it ___ to walk?
B It's thirty minutes ___ foot.
A How ___ is a ticket to London?
B It ___ £5.50.

SOUNDS

1 Say these words aloud.

blue good book shoe boot suit
pullover took cook

Is the underlined sound /ʊ/ or /uː/? Put the words in two columns.

🔊 Listen and check.

2 Say these words aloud. Underline the /dʒ/ sound.

geography journalist soldier engineer teenager
job manager

🔊 Now listen and check.

3 Put these sentences in the correct order and make a dialogue.

a Well, I think I'll leave it. Thank you.
b I'm sorry. This is the largest size we've got.
c Can I help you?
d It's too small. Have you got it in a bigger size?
e Yes, I'm looking for a sweater.
f How about this one? It suits you.

🔊 Listen and check. Do you think the speakers sound polite and friendly?

Now work in pairs and say the sentences aloud. Try to sound polite and friendly.

SPEAKING AND WRITING

Work in pairs. You're going to recreate a story called *The Phantom of the Opera and the empty seat.*

Student A: Turn to Communication activity 16 on page 100.
Student B: Turn to Communication activity 3 on page 98.

Present perfect simple (1) for experiences

How are you keeping?

When someone says 'How are you?' do you reply 'Fine thanks, how are you?' or do you say 'I'm not feeling very well. I've had a bad cold, I've been off work, and now I've got a dreadful cough.'? Some people never seem to be ill, others have always got something wrong with them... or think they have.

Try the questionnaire and find out how you're keeping.

			Yes	No
1	Have you ever broken an arm or a leg?		☐	☐
2	Have you ever stayed at home because of illness?		☐	☐
3	Have you ever taken vitamin pills?		☐	☐
4	Have you ever given up any of the following because of your health?	smoking	☐	☐
		drinking	☐	☐
		coffee	☐	☐
		meat	☐	☐
		sunbathing	☐	☐
5	Have you ever taken up any of the following because of your health?	running	☐	☐
		swimming	☐	☐
		regular exercise	☐	☐
6	Have you ever had an accident while watching a sport?		☐	☐
7	Have you ever had an accident while playing a sport?		☐	☐
8	Have you ever had...?	a heart attack	☐	☐
		high blood pressure	☐	☐
		malaria	☐	☐
9	Have you ever had...?	flu	☐	☐
		a headache	☐	☐
		food poisoning	☐	☐
10	Have you ever become ill on holiday?		☐	☐
11	Have you ever worried about getting ill?		☐	☐
12	Have you ever stayed in hospital?		☐	☐
13	Have you ever looked up an illness in a medical dictionary?		☐	☐
14	How are you keeping?	Not so good.	☐	☐
		I've never felt better!	☐	☐

Mostly **Yes**: Either you've been unlucky with your health or you've become a hypochondriac. Relax! Life's too short to worry so much about your health.

Mostly **No**: You're very lucky ... so far. You're healthy and you don't worry much. But maybe you need to take better care of yourself – just in case.

READING

1 Work in pairs. When was the last time you were ill? Do you worry about staying well? Do you think you're fairly healthy?

2 How are you keeping? Read the questionnaire and find out.

VOCABULARY

1 Complete the diagrams with words for parts of the body.

> tooth   mouth   eye   shoulder
> finger   waist   knee   ankle
> toe   foot   throat   neck
> wrist   thumb   elbow   back

2 Can you name the other parts of the body?

3 Look at this list of parts of the body. Which part doesn't belong?

finger   ankle   thumb
wrist   elbow

'Ankle' is a part of the leg, the others are parts of the arm.

Write some more lists of parts of the body with one part which doesn't belong.

4 Group all the words under four headings: *head*, *body*, *arm*, and *leg*.

21

GENERAL COMMENTS

Present perfect simple (1)

This is the first of three lessons on the present perfect simple. This lesson concentrates on how the tense can be used to describe experiences. The key concept with this use is that we are not interested in when the action in the past was performed. As soon as we need to use an expression of past time, the past simple is required instead of the present perfect. The tense is a difficult one for students to master. They have to learn to manipulate the form, which involves past participles, and the concept. Some languages may have a tense which looks like the present perfect but which operates like the past simple.

How are you keeping?

The theme of the lesson focuses on two important vocabulary fields, parts of the body and illness. While the students may need this vocabulary one day in serious circumstances, the context here is intended to be lighthearted.

Cross-cultural awareness

Most British people use the questions *How are you?* and *How are you keeping?* as a greeting or a way of starting a conversation. It is not intended as a serious enquiry about someone's health, and should not be perceived as being unacceptably curious or personal. Ask your students if their native language has an opening gambit used in the same way.

Left-hand illustration: possible questions

How do you think the man in the drawing feels?
What do you think is wrong with him?
Where is he?
What has he got on his arm? What has he got in his mouth?
What do you think the theme of the lesson is?

READING

1 Aim: to prepare for reading.
● Ask the students to think about their health, and about staying fit. Ask different people what they do to stay fit, if anything. You may like to ask them to suggest ways of keeping fit and make a list of these suggestions on the board.

2 Aim: to read and react to a questionnaire.
● Questionnaires establish an immediate relationship with the reader and their reactions to it constitute the reading comprehension check. Ask the students to read and think about their answers to the questions.

● The students may feel ready to share their answers to the questions, but it is better for them to do the next stage in this activity sequence first in order to be sure they have suitable language to continue.

VOCABULARY

1 Aim: to present the words in the vocabulary box.
● The students may already know many of these basic items of vocabulary. The items which are already in position on the chart are likely to be the most common. The items which are listed and to be added to the diagram are the words which the students need to acquire. Begin the activity by pointing to different parts of your body and asking what they are in English.

● Then ask the students to complete the chart with the words. If possible, it would be better to draw a stick person on the board, and ask students to come up and name each part of the body. Ask them to help each other with identifying the parts of the body.

> **Answers**
> 1 toe 2 knee 3 neck 4 back 5 elbow 6 shoulder
> 7 waist 8 foot 9 thumb 10 wrists 11 finger
> 12 ankle 13 tooth 14 mouth 15 eye 16 throat

2 Aim: to find out the names of other parts of the body.
● Ask the students to name the other parts of the body. You can help them by giving them the words on the board, or by suggesting they look the words up in bilingual dictionaries. But of course, it is much better if they can name the parts of the body without help.

3 Aim: to practise using the words for parts of the body.
● Write the words shown in the book on the board. Ask the students which word doesn't belong. Write some or all of the following word lists, and continue the activity.
  eyes, ankle, nose, ear
  arm, elbow, thumb, toe
  waist, back, shoulder

● Ask the students to prepare some word lists like this with one word which doesn't belong.

● When the students are ready, they should work in pairs and show their word lists to each other. The partner should decide which is the word which doesn't belong.

4 Aim: to consolidate the acquisition of words to describe parts for the body.
● As a final activity in this sequence, ask the students to write down the words under the four headings.

> **Answers**
> **head:** tooth, mouth, eyes, throat, neck, nose, face, ear
> **body:** shoulder, waist, back
> **arm:** finger, wrist, thumb, elbow
> **leg:** knee, ankle, toe, foot

GRAMMAR

● Ask the students to read the explanation in the grammar box and then do the exercises.

1 Aim: to focus on the difference between the present perfect and the past simple.

● Ask the students to read the information about the present perfect simple in the grammar box.

● If necessary, stress the fact that when we use the present perfect to describe experiences we are not interested in when the experience happened. Ask the students to do the activity.

> **Answers**
> 1 Have you ever been in an ambulance?
> 3 Have you ever eaten *sauerkraut?*
> 4 Have you ever met a famous person?
> 5 Have you ever played tennis?
> 6 What did you have for dinner last night?

● You may want to ask if anyone knows what *sauerkraut* is (a dish made with pickled cabbage and which comes from Alsace in France and the western part of Germany).

2 Aim: to focus on past participles.

● This activity is designed to draw the students' attention to the fact that past participles are used in the formation of the present perfect. This may be a new concept for them. You may like to explain that many past participles are like the past simple form of the verb, and the exceptions are with certain irregular verbs.

> **Answers**
> been, had, said, eaten, played, taught, visited, lived, seen, worked, loved, known, paid, broken, met, made, tried, won, worn, sold

● Write the answers on the board and circle the past participles. Ask the students to use the list at the back of their Student's Book to compare these past participles with their past simple forms.

3 Aim: to practise using the present perfect to talk about experiences.

● Ask the students to work in pairs and to talk about their answers to the questionnaire. You may like to check one or two students' answers first. Make sure everyone forms the present perfect with a suitable past participle.

SPEAKING

1 Aim: to practise using the present perfect to talk about experiences.

● The focus of this activity has moved away from parts of the body and experience, and towards more general experiences. Encourage the students to talk about experiences using not only the verbs in *Grammar* activity 2 but also any other suitable verbs. Give them plenty of time to write their questions.

2 Aim: to practise using the present perfect to talk about experiences.

● Ask the students to go round asking and answering their questions about experiences.

● You may like to ask the students to write about other people's experiences for homework.

● At this stage of the lesson, you may find that the students are beginning to manipulate the structure accurately, but without being confident that they will do so on every occasion in the future. You may need to supplement their work with some material from *Reward* Practice Book or Resource Packs.

GRAMMAR

> **Present perfect simple (1) for experiences**
> You use the present perfect simple to talk about an action which happened at an indefinite time in the past. You often use it to talk about experiences with *ever* and *never*.
>
> *Have you **ever stayed** in hospital?* (= Do you have any experience of staying in hospitals?)
>
> *Yes, I **have**.* (= Yes, at some time in my life, but it's not important when.)
>
> *No, I **haven't**. I've **never stayed** in hospital.*
>
> Remember that if you ask for and give more information about these experiences, such as *when, how, why* and *how long*, you use the past simple.
>
> *When **did** you **stay** in hospital?   In 1975.*
>
> You form the present perfect simple with *has/have* + past participle. You usually use the contracted form *'ve* or *'s*.
>
> *I've **been** ill.   He**'s broken** his leg.*
>
> **Negative**
> *I **haven't had** malaria.   I've **never had** malaria.*
>
> **Question**
> *Have you **ever broken** your leg?*

1 Rewrite these questions with the present perfect or the past simple of the verb in brackets.

1 (be) you ever in an ambulance?
2 When (be) the last time you (be) ill?
3 (eat) you ever *sauerkraut?*
4 (meet) you ever a famous person?
5 (play) you ever tennis?
6 What (have) you for dinner last night?

2 Write the past participle of these verbs.

be   have   say   eat   play   teach   visit   live   see   work   love   know   pay   break   meet   make   try   win   wear   sell

3 Work in pairs. Ask and answer the questions in *How are you keeping?* If you or your partner answer yes, ask for or give extra information.
Have you ever broken your arm or leg?   No.
Have you ever been ill on holiday?   Yes.
When was that?   When we went to India in 1987.

SPEAKING

1 Look at the verbs in *Grammar* activity 2 and think of questions to ask people about their experiences. Write down seven or eight questions.
Have you ever met anyone famous?

2 Find out about the experiences of other people in your class using the questions you wrote in 1. Ask for and give extra information.
Have you ever met anyone famous?
Yes, I have. I've met the President.
Really? When did you meet her?
In 1987.

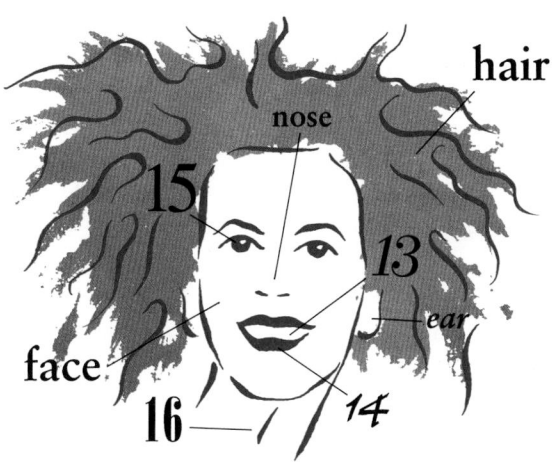

22 | *What's new with you?*

Present perfect simple (2) for past actions with present results

VOCABULARY AND LISTENING

1 You are going to hear Heide Meyer, a teacher in Berlin, talking about what's new with her. She's talking about changes in Germany and in her life since 1989. Which of these words do you expect to hear?

> election war government standard of living
> employment inflation finance police law and order
> freedom rights minister housing tourism education

2 Look at these questions about changes in Germany. Do you think Heide will answer *yes* or *no*?

1 Has there been a change of government?
2 Has the standard of living got better?
3 Have more people got jobs?
4 Has inflation gone down?
5 Have more tourists visited the country?

🔲 Now listen to Heide. Did you guess correctly?

3 🔲 Listen to Heide talking about changes in her own life. Tick (✓) the things that she mentions.

move to a new flat	☐	learn French	☐
finish her studies	☐	find a new job	☐
buy a new car	☐	lose weight	☐
have a baby	☐	stop smoking	☐
learn to cook	☐	get married	☐

4 Match the pairs of sentences about Heide.

1 She's had a baby.
2 She's got married.
3 They've moved to a new flat.
4 She's found a new job.
5 She's bought a new car.
6 She's given up smoking.

a She drives a Volkswagen now.
b It's in a very new building in Potsdam.
c She starts on Monday.
d She feels much healthier.
e She lives with her husband now.
f Their son is one month old.

🔲 Listen again and check.

SPEAKING

1 What's new with you? Think about your life a year or two ago and your life today. Can you think of anything that's different? Make notes.

last year	*today*
lived in London	*live in Bristol*
smoked twenty cigarettes a day	*don't smoke*

2 Work in pairs and talk about what's new with you.

22

GENERAL COMMENTS

Present perfect simple (2)

This is the second of three lessons on the present perfect simple tense. This lesson deals with the use of the present perfect to talk about past actions with present results. This is a very common use of the present perfect in British English although much less common in American English.

Vocabulary

The vocabulary area presents a few items to deal with more abstract concepts of politics and social affairs. You may decide that it would interest your students to expand this area further. As far as this lesson is concerned, the words in the box are simply the words required to give access to the listening material.

Accents

Heide speaks with a typical German accent. It is important to expose the students to as many different accents as possible, not just standard British and American ones. In general terms, the students are more likely to hear English spoken by a non-native speaker in real life than by a native speaker. Of course, you may decide to use a standard variety of English as a model for pronunciation, but you can still use non-native accents for receptive purposes. Listening to non-native accents is by no means a poor substitute for listening to native speakers. On the contrary, it is an important part of developing their overall competence in listening.

Left-hand photo

Soldiers helping a child over the Berlin Wall in November 1989.

Possible questions

Where are the people in the photo?
What are they doing? When did this happen?
Which way are we looking – east or west? (It's probable that the woman below is the child's mother, and she would have climbed over before she sent her child over, so we're probably looking eastwards, back where she came from.)
Can you remember what happened after this? (celebrations, the Wall came down, the unification of Germany)

Right-hand photo: possible questions

Who do you think this person is?
Where does she come from?
What do you think is new with her?

SPEAKING

1 **Aim: to talk about life today and life a year or two ago; to prepare for the distinction between the present perfect and the past simple.**
● This activity is important in establishing the conceptual framework for showing the difference between the present perfect and the past simple. Make a list on the board of things in your life which are different today. Think about your teaching circumstances and, if

appropriate, your personal life as well. There may have been some major changes in government, politics, etc. as well. Try to write at least four or five points on the board to show what you expect of the students.

● Ask the students to do the same with notes about their own lives.

2 **Aim: to practise talking about life today and life a year or two ago.**
● Ask the students to work in pairs and use the notes they made in activity 1 to talk about life today and life a year ago.

VOCABULARY AND LISTENING

1 **Aim: to present the words in the vocabulary box; to prepare for listening; to pre-teach some words from the listening passage.**
● This activity is an important stage in preparation for listening in that it encourages students to think about what they're going to listen to, and so reconstitutes a little of the context they would otherwise have if they were more actively participating in the conversation. Ask the students to think a little about the events in Germany in recent years and to predict which words from the box Heide is likely to use.

● Ask them to think also about other words they are likely to hear.

2 **Aim: to ask some concept questions; to listen for specific information.**
● To continue the process of preparing the students for listening, ask them to guess if Heide will say *yes* or *no* in response to the questions about the changes in Germany. Do this activity orally, and fairly quickly as the students should be keen to listen to the passage by now.

● ▭ Play the tape and ask the students to listen and check their answers.

3 **Aim: to practise listening for specific information.**
● ▭ Explain that the next part of the tape is concerned with changes to Heide's personal life. Play the tape and ask the students to tick the things she mentions.

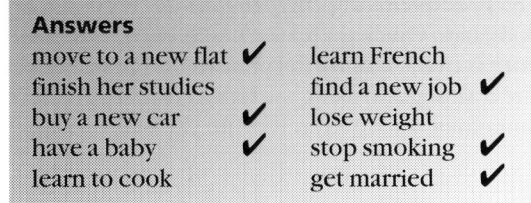

Answers

move to a new flat ✔	learn French
finish her studies	find a new job ✔
buy a new car ✔	lose weight
have a baby ✔	stop smoking ✔
learn to cook	get married ✔

4 **Aim: to focus on past actions and present results.**
● Ask the students to match the sentences.

Answers
1 f 2 e 3 b 4 c 5 a 6 d

● ▭ Play the tape again to check the answers.

GRAMMAR

1 Aim: to focus on the formation of past participles.
- Ask the students to read the information about the present perfect simple in the grammar box.
- Ask them to do the activity. They will have done some work in Lesson 21 on the formation of past participles.

> **Answers**
> | | | |
> |---|---|---|
> | change | changed | changed* |
> | move | moved | moved* |
> | leave | left | left* |
> | finish | finished | finished* |
> | find | found | found* |
> | get | got | got* (AmE: gotten) |
> | vote | voted | voted* |
> | buy | bought | bought* |
> | have | had | had* |
> | drink | drank | drunk |
> | read | read | read* |
> | ring | rang | rung |
> | write | wrote | written |
>
> The verbs marked * have the same form in the past simple and the past participle.

- You could draw the students' attention to the list of irregular verbs at the back of the book, where there is information about past participles which have the same form as their infinitive and/or their past simple forms.

2 Aim: to practise forming the present perfect.
- Ask the students to do the exercise. Explain that the word *now* indicates a change in the life of the person mentioned in each sentence, and suggests a present perfect in the preceding clause.

> **Answers**
> 1 I've bought a new car and now I drive to work.
> 2 He's moved house and now lives in London.
> 3 She's finished her book and is now watching TV.
> 4 We've drunk the whole bottle. It's empty.
> 5 I've found my bag. It's here.
> 6 She's written me a letter and I'm reading it now.

3 Aim: to practise using the present perfect.
- Ask them to check their answers to *Vocabulary and listening* activity 3, using the present perfect. Check they are forming and using the tense correctly. If there are difficulties, do the exercise with the whole class.

SOUNDS

1 Aim: to focus on the sound of the auxiliary 've and 's in the present perfect.
- Ask the students to tell you the past participles of the verbs in *Vocabulary and listening* activity 3.
- 🔲 Play the tape and pause after the first pair of sentences. Ask the students if they noticed how similar they sounded. Point out that you don't always hear *'ve* and *'s*. Repeat with the second pair of sentences, etc.

- 🔲 Remind the students, however, that the context will help them decide if the verb is present perfect or past simple. Play the tape and ask them to repeat the sentences.

2 Aim: to focus on the sound of the auxiliary in the present perfect.
- This activity is designed to show how the context helps to indicate if the verb is in the present perfect or the past simple. Ask the students to predict which verb form they're likely to hear by choosing the correct verb form in each sentence. Remind them that when there is a reference to a specific time in the past, the verb is likely to be in the past simple.
- 🔲 Play the tape and ask the students to underline the verb form they hear. Check that the verbs they underlined are correct.

> **Answers**
> She stayed late last Friday.
> They've finished their work and now they're going home.
> We visited America in 1983.
> He studied French five years ago.

WRITING

1 Aim: to read a letter to be used as a model for guided writing practice.
- This reading activity is designed to make the students read with some purpose the model they will later use to write about changes to their lives. Ask them to read and underline any verbs Heide uses which they can also use to describe changes in their lives during the last year.

2 Aim: to write sentences using the present perfect.
- Refer the students back to the notes they made in *Speaking* activity 1. Ask them to write full sentences.

3 Aim: to expand the sentences and add extra information.
- Ask the students to expand their sentences and to give more information about how things are now.

4 Aim: to join sentences using *and*.
- The intention is that the students write a series of connected sentences. Go round and help them decide if any of their sentences can be joined using *and* or other conjunctions, such as *but, because* and *so*.

5 Aim: to write an informal letter.
- Ask the students to think of someone they would like to write to and to use all the sentences they have written in a letter to this person. They should use Heide's letter as a model. Make sure they add their address and the date at the top of the letter, begin it *Dear* (name) and finish *Best wishes*.

- If time is short, you can ask the students to do this activity for homework.

GRAMMAR

> **Present perfect simple (2) for past actions with present results**
> **You use the present perfect simple to talk about a past action which has a result in the present. It is not important when the action happened. You often use it to describe changes.**
> *She's got married.*
> **(She wasn't married before. We don't know when she got married. She's married now.)**
> **You often use** *just* **to emphasise that something has happened very recently.**
> *She's just had a baby.*
> *They've just moved to a new flat.*

1 Look at these past participles.

moved   changed   left   finished
found   got   voted   bought   had
drunk   read   rung   written

Write down their infinitive and past simple forms.
move   moved   moved

Which past participle forms are the same as their past simple forms?

2 Write the verb in brackets in the present perfect form.

1 I (buy) a new car and now I drive to work.
2 He (move) house and now lives in London.
3 She (finish) her book and is now watching TV.
4 We (drink) the whole bottle. It's empty.
5 I (found) my bag. It's here.
6 She (write) me a letter and I'm reading it now.

3 Work in pairs. Check your answers to *Vocabulary and listening* activity 3. Ask and answer questions about Heide.
Has she moved to a new flat?
Yes, she has.

SOUNDS

1 🔲 Listen to these sentences. Notice how you don't always hear *'ve* or *'s*.

1 She's stayed late.   She stayed late.
2 They've finished their work.   They finished their work.
3 We've visited America.   We visited America.
4 He's studied French.   He studied French.

2 Now look at the sentences in context and underline the correct verb form.

1 *She's stayed/She stayed* late last Friday.
2 *They've finished/They finished* their work and now they're going home.
3 *We've visited/We visited* America in 1983.
4 *He's studied/He studied* French five years ago.

🔲 Listen and check. Say the sentences aloud.

WRITING

1 Read this letter from Heide to her friend in America and underline any verbs which you can use to describe changes in your life during the last year.

> 1 December
>
> Dear Stacey,
>
> How are you? I've got a lot of news for you. Things have changed quite a lot since I last wrote to you. I've got married, and I now live with my husband. We've moved to a different part of Berlin and we now live in Potsdam. I've stopped smoking and I feel much healthier now. I've found a new job and I can work from home – which is good because the most important news is I've just had a baby. He's one month old and very beautiful.
>
> Here are some recent photos. It would be nice to hear from you.
>
> Best wishes,
>
> *Heide*

2 Look at the notes you made in *Speaking* activity 1. Write full sentences which describe the changes in your life. Use the present perfect.
I've moved to Bristol. I've stopped smoking.

3 Write down some extra information about how your life is now.
I've moved to Bristol. I like it here.

4 Join the two sentences with *and*.
I've moved to Bristol and I like it here very much.

5 Now write a letter to a friend describing how things have changed for you in the last year. Use Heide's letter to help you.

23 | *It's a holiday*

Present perfect simple (3): *for* and *since*

VOCABULARY AND LISTENING

1 Work in pairs. Make a list of important events or festivals in your town or country. (Try not to choose religious occasions.)
Carnival, the Palio, April Fool's Day

2 Look at the words in the box. Which of these words can you use to describe what happens?

barbecue   celebrate   dancing fair   fireworks   flag   holiday king   meal   parade   party picnic   president   prime minister queen   race   soldier   speech

Are there any other words you can use?
music, bands...

3 Work in two groups. You are going to hear Barry, who is Australian, talking about an important event in Australia.

Group A: Turn to Communication activity 5 on page 98.
Group B: Turn to Communication activity 18 on page 101.

4 Now work with someone from another group and answer the questions in the *Australia* column.

	Australia	Your country
What's it called?		
When does it take place?		
Where does it take place?		
When did it first take place?		
What happens?		
Is it a public holiday?		
Are there any other interesting features?		

5 🔲 Listen again and check your answers to 4. Can you add any extra information?

6 Choose an important day in your country and complete the column *Your country*.

23

GENERAL COMMENTS

Present perfect simple (3)

This is the third lesson in a series of three which focus on the use of the present perfect simple. By now, the students may be forming the tense accurately, and using it with some degree of fluency. This lesson deals with the common use of the present perfect with *for* and *since* and focuses on verbs that are not usually used in the continuous form. Other uses of the present perfect, notably with *yet* and *already*, and the present perfect continuous will be covered in intermediate *Reward*.

Top right-hand photo: possible questions

What can you see in the photo?
Do you have fireworks in your country?
What do they celebrate?

Bottom left-hand photo: possible questions

What can you see in the photo?
Where do you think the people are?
Do you think it might be a scene from your country?

VOCABULARY AND LISTENING

1 Aim: to introduce the theme of the lesson.
● Ask the students to work in pairs and make a list of important events or festivals in their country. Suggest they choose national rather than religious occasions, as the latter will be covered in Lesson 34. You may like to do this activity orally with the whole class; if so, write the names of the important occasions on the board. Try to leave these names on the board as they will be useful for *Grammar* activity 4.

2 Aim: to present the words in the vocabulary box; to pre-teach some important vocabulary in the listening passage.
● Ask the students to choose the words they can use to describe what happens during the important occasions you have listed in activity 1.

● In the *Writing* section, they will be asked to write about an important event or festival, so use this opportunity to supply them with the necessary vocabulary. You can group the new items under the names of the occasions on the board.

3 Aim: to listen for specific information.
● This is a jigsaw listening activity. Divide the class into two groups, A and B, and ask each group to follow its instructions in the relevant Communication activity.

● 🔲 Play the tape.

● Allow enough time for every member of each group to find out the same information, which they should write down.

4 Aim: to exchange information; to complete the chart.

● Ask the students to work with someone from another group and to exchange information to complete the chart. They are only allowed to reveal the information they have written down in activity 3.

Answers	
What's it called?	The Melbourne Cup.
When does it take place?	The first Tuesday in November at 2.40pm.
Where does it take place?	Flemington race course, Melbourne, Australia.
When did it first take place?	1874.
What happens?	People from all over Australia come to Melbourne for the day with picnics, and they bet on the horse they think will win.
Is it a public holiday?	Only in the city of Melbourne and the state of Victoria.
Are there any other interesting features?	Everyone in Australia stops work to listen.

5 Aim: to listen for specific information.
● 🔲 Play the tape again, so the students can check their answers.

● When they're ready, they can work in pairs and see if they can add any information. Check this stage with the whole class.

6 Aim: to practise talking about important events and festivals.
● The chart, and possibly the answers, will give the students the necessary framework and language to talk about an important day in their country. This is an important stage in the preparation for writing later in the lesson.

GRAMMAR

**1 Aim: to focus on the difference between *for* and
since.**

● Ask the students to read the information about the
present perfect simple in the grammar box.

● Ask the students to complete the sentences.

> **Answers**
> 1 They've had a parade **since** 1988.
> 2 There have been fireworks **for** ten years.
> 3 It's been a national holiday **since** 1993.
> 4 The president has given a speech every year **for**
>   five years.
> 5 She's been Queen **since** March.
> 6 We've celebrated Independence Day **for** ten
>   years.

**2 Aim: to practise forming the present perfect with
verbs that you do not usually use in the
continuous form.**

● These verbs belong to a category known as *stative
verbs*. To focus on these verbs at this stage will avoid
complications at a later stage when the present perfect
continuous is presented as a tense which can be used
with *for* and *since*. You may like to do this activity on
the board, writing down something you know, like,
want, etc.

**3 Aim: to practise asking and answering questions
using the present perfect with *for* and *since*.**

● Ask the students to work in pairs. They should use the
notes they wrote in activity 2 as prompts for questions
and answers using the target language.

**4 Aim: to practise using the present perfect with *for*
and *since*.**

● Ask students to say how long their countries have
celebrated the important occasions they noted down in
Vocabulary and listening activity 1. These occasions
may still be on the board for you to point to. Ask them to
make statements using *for* and *since*.

SOUNDS

Aim: to focus on the weak form pronunciation of *for*.

● 📼 Draw the students' attention to the weak form
pronunciation of *for*. Play the tape and ask them to
repeat the sentences. Pause after each sentence.

WRITING

**1 Aim: to provide a purpose for reading the model
writing passage.**

● Ask the students to read about Guy Fawkes Night and
to answer the questions. The passage is going to be used
as a model for guided writing practice in activity 2.

> **Answers**
> a For nearly 400 years.
> b The attempt to kill King James I.
> c There are fireworks and fires in gardens and parks.
> d Sausages, baked potatoes.

**2 Aim: to use the model passage to write a passage
about another important national occasion.**

● Ask the students to write about their own important
event or festival.

● If time is short, they may like to do this activity for
homework.

GRAMMAR

> **Present perfect simple (3): *for* and *since***
> You use the present perfect to talk about an action or state which began in the past and continues to the present.
> You use *for* to talk about the length of time.
> *The 1st October **has been** our national day **for a hundred years**.*
> You use *since* to say when the action or state began.
> *We've had a President **since 1888**.*

1 Complete these sentences with *for* or *since*.

1 They've had a parade ___ 1988.
2 There have been fireworks ___ ten years.
3 It's been a national holiday ___ 1993.
4 The president has given a speech every year ___ five years.
5 She's been Queen ___ March.
6 We've celebrated Independence Day ___ ten years.

2 On a piece of paper, write down something or someone you:

– know   – like   – want   – need
– hate   – own   – believe

know – Karen,   like – sport...

3 Work in pairs and show the list of people and things you wrote in 2. Ask and say how long.
How long have you known Karen?
Since January.
How long have you liked sport?
For ten years.

4 Look at the list of important days or occasions you made in *Vocabulary and Listening* activity 1. Say how long they have been important days or occasions. Use *for*.
Independence Day has been an important occasion for 20 years.

Now say since when your country has celebrated them. Use *since*.
We've celebrated Independence Day since 1864.

SOUNDS

[cassette icon] Listen and repeat these sentences.
Pronounce *for* /fə/.

1 They've celebrated it for seventy years.
2 She's lived here for six months.
3 They've had fireworks on Bastille Day for many years.
4 He's worked at the bank for ten years.

WRITING

1 Read this description of an important occasion in Britain, Guy Fawkes Night. Find the answers to these questions.

a How long has it been important?
b What event does it celebrate?
c What happens during the celebration these days?
d What do people have to eat or drink?

> We've celebrated Guy Fawkes Night on November 5th every year since 1605. Guy Fawkes was an English Catholic who wanted to kill the Protestant King James I. He tried to destroy the Houses of Parliament in London with a bomb. The plan failed when one of the Catholics warned a relative not to go to Parliament that day. Soldiers arrested Guy Fawkes and he was executed. These days we celebrate Guy Fawkes night with fireworks and a large fire in gardens or in parks. The celebration starts when it's dark. On the fire we put a 'guy', which is made with old clothes and straw and looks like Guy Fawkes. Then the fireworks begin and we burn the guy on the fire. There's usually warm food like sausages and potatoes baked in the fire. I don't know if we celebrate Guy Fawkes because he failed or because he very nearly succeeded.

2 Write sentences describing an important national occasion in your country. Use the questions above and your notes in the chart in *Vocabulary and listening* activity 6 to help you.

We've celebrated Bastille Day since 1789...

24 | *Divided by a common language?*

Defining relative clauses: *who, which/that* and *where*

Divided by a common language?

George Bernard Shaw said that America and Britain were two nations divided by a common language. But how different is British English from American English? Some British and American people gave their definitions for some common words.

British		American
'Something that you burn for heating and cooking.'	*gas*	'Something you put in your car to make it go.'
'A school which is private.'	*public school*	'A school that is open to everyone.'
'A path which passes under a road.'	*subway*	'A railway which runs under the ground.'
'Something that you wear under your trousers.'	*pants*	'Something which you wear to cover your legs, over your underpants.'
'Clothing which you wear under your shirt.'	*vest*	'Clothing that you wear over your shirt and under your jacket.'
'A list of things that you have bought or eaten in a restaurant and which tells you how much to pay.'	*bill*	'Money which is made of paper.'
'Long sticks of potato which you cook in deep oil and eat hot with a meal.'	*chips*	'Very thin slices of fried potato which you eat cold before a meal or as a snack.'

Confused? British and American English have lots of words which look the same but have different meanings. Nobody ever gets into serious trouble if they make a mistake, although you may get a strange look if you ask for the wrong clothes. But things get even more complicated! Here are some American English words which the British don't use at all.

druggist	someone who sells medicine in a shop.
parking lot	a place where you park the car.
drugstore	a shop where you can buy medicine, beauty products, school supplies, small things to eat.
main street	the street in a town where all the shops are.
stop lights	lights which control the traffic.
faucet	something you turn on and off to control water in a bath or a basin.
elevator	a device which carries people from one floor to another in a building.

But most of the differences between British English and American English are minor and are only concerned with vocabulary, spelling and pronunciation. You can usually understand what words mean from the context. Good luck (British English) or break a leg (American English)!

24

GENERAL COMMENTS

British and American English

This lesson covers some of the main differences between British English and American English. Inevitably, it is a fairly brief overview and will leave out more than it can include. However, it is important to stress that the main differences are to do with vocabulary. American English and British English accents are mutually intelligible, and the differences in spelling and grammar usually present few problems of comprehension. The lesson focuses on some of these vocabulary differences, and makes the student aware that other differences may exist. It should be added that the context can usually resolve any confusion which may arise.

Illustrations: possible questions

What do the pictures show?
Do you think they illustrate something from Great Britain or from the USA?
(Clockwise from the top-left hand corner: a truck (US), Spiderman (US), a Formula 1 racing car (GB), (just underneath it, above the grammar box) a post box (GB), a cricket player (GB), a football player (US), a football player (GB), a mail box (US), Big Ben (GB), (just above it) an American car (US), the space shuttle (US), the Statue of Liberty (US), an Aston Martin car (GB))
Can you think of any other typically British or American people or things?

SPEAKING AND READING

1 Aim: to practise talking about attitudes towards British and American English; to prepare for reading.

● Many students have a natural preference towards British or American English. This activity encourage the students to confront their preference towards and their prejudice against British or American English. There is, of course, no right or wrong answer. Encourage your students to explain why they prefer one or the other. It may be better to do this activity orally with the whole class.

2 Aim: to read and infer.

● Ask the students to read the statements and then read the passage to find out if the statements are true or false. The answer is not always obvious; students may have to infer what the writer means.

> **Answers**
> 1 False 2 True 3 True 4 True

3 Aim: to check comprehension.

● Ask the students to discuss how many words they recognised or knew already. You may like to ask them to write down some of the most useful words in their *Wordbank*.

GRAMMAR

1 Aim: to focus on defining relative clauses.

● Ask the students to read the information about defining relative clauses in the grammar box.

● Ask the students to do the exercise.

> **Answers**
> 1 A hat is something **which/that** you wear on your head.
> 2 A post office is a place **where** you can buy stamps and post letters.
> 3 A journalist is someone **who** writes for a newspaper.
> 4 A swimsuit is something **which/that** you wear when you go swimming.
> 5 A disco is a place **where** you go to dance.
> 6 A hairbrush is something **which/that** you use to brush your hair.

2 Aim: to practise writing defining relative clauses.

● Do the activity orally, and help the students with any necessary vocabulary.

> **Possible answers**
> A kitchen is a place where you cook food.
> A supermarket is a place where you can buy shopping.
> A language teacher is someone who teaches language.
> A fridge is a place where you keep food and drink cold.
> A pilot is someone who flies a plane.
> Shoes are something you wear on your feet.
> A restaurant is a place where you can have lunch or dinner.
> A pub is a place where you can have a drink and meet friends.

● Ask the students to write the definitions.

● There is more work on defining relative clauses with objects in Lesson 25.

SOUNDS

1 Aim: to compare British and American pronunciation.

● 🔲 Many students can recognise an American or a British accent, but are not aware why they sound different. This activity gives students an opportunity to compare the two with each other. Play the tape and ask the students to listen to an American speaker and then to a British speaker.

2 Aim: to distinguish between an American and a British accent.

● 🔲 Ask the students to listen and guess if the speaker is British or American. Play the tape.

Answer
The speaker is an American.

3 Aim: to present the basic differences of British and American pronunciation.

● 🔲 This activity can only cover some of the main differences between standard British and American pronunciation. Regional accents may vary even more. Simply play the tape and ask students to listen. If they want to repeat the words in a British then an American way, they can, but you need to decide or agree with them which accent you wish them to model their own accent on.

VOCABULARY AND LISTENING

1 Aim: to present the British and American words in the vocabulary boxes.

● Whether the students choose to acquire British or American English, it is still useful at least for receptive purposes to be aware of some words which may cause confusion. The words in the boxes are the words featured in the reading passage.

Answers	
US	**GB**
bill	bank note
chips	crisps
druggist	chemist
faucet	tap
french fries	chips
gas	petrol
main street	high street
vest	waistcoat
pants	trousers
stop lights	traffic lights
parking lot	car park
subway	underground
public school	state school

The following American English words have different meanings in British English: *gas, pants, subway, vest.*

2 Aim: to check the answers to activity 1, and to listen for specific information.

● 🔲 Ask the students to listen and check their answers to activity 1.

3 Aim: to help the students acquire new vocabulary by categorising according to personal choice.

● Ask the students to put the words in the box into their own categories. They may already have suitable categories in the *Wordbanks*, or they may need to create them specially for this activity.

● This activity can be done for homework.

SPEAKING AND READING

1 Which is more useful to you – British English, American English or another type of English? Work in pairs and say why.

2 Read *Divided by a common language* and decide if these statements are true or false according to the passage.

1 British English and American English are two very different languages.
2 Some words have different meanings in British and American English.
3 Some words are used in one type of English but not in the other.
4 If you don't understand a word, you can usually guess the meaning.

3 Which of the words in the passage did you recognise or know already?

GRAMMAR

> Defining relative clauses: *who, which/that* and *where*
> **You use a defining relative clause to define people, things and places.**
>
> **You use *who* for people.**
> *A druggist is someone **who** sells medicine in a shop.*
>
> **You use *which* for things.**
> *A subway is a railway **which** runs under the ground.*
>
> **You often use *that* instead of *who* or *which*.**
> *A druggist is someone **that** sells medicine in a shop.*
> *A subway is a railway **that** runs under the ground.*
>
> **You use *where* for places.**
> *A parking lot is a place **where** you park your car.*

1 Complete the sentences with *who, which/that* or *where*.

1 A hat is something ___ you wear on your head.
2 A post office is a place ___ you can buy stamps and post letters.
3 A journalist is someone ___ writes for a newspaper.
4 A swimsuit is something ___ you wear when you go swimming.
5 A disco is a place ___ you go to dance.
6 A hairbrush is something ___ you use to brush your hair.

2 Write definitions for these words.

kitchen supermarket language teacher fridge pilot shoes restaurant pub

SOUNDS

1 🔲 Listen to the first paragraph from the passage, spoken first by an American and then by a British person.

Now listen to the second paragraph. Is the speaker American or British?

2 The sound of some words in American English is different from British English.

🔲 Look at the underlined sound and listen to some of the main differences.

fa<u>th</u>er mo<u>th</u>er grandfa<u>th</u>er sist<u>er</u>

<u>o</u>pera c<u>o</u>st c<u>o</u>ncert f<u>o</u>g

bo<u>tt</u>le thir<u>t</u>y dir<u>t</u>y bu<u>tt</u>er

d<u>a</u>nce b<u>a</u>th c<u>a</u>n't pl<u>a</u>nt

n<u>ew</u> T<u>ue</u>sday n<u>u</u>clear t<u>u</u>ne

VOCABULARY AND LISTENING

1 Match the American English words with the British English words in the boxes below.

> bill chips druggist faucet french fries gas
> main street vest pants stop lights parking lot
> subway public school

> trousers car park state school traffic lights
> underground waistcoat chemist chips crisps
> high street bank note petrol tap

Check the passage to remind yourself which American English words in the box have different meanings in British English.

2 🔲 Listen to a British person and an American talking about the words and check your answers to 1.

3 Put the words in 1 into categories of your choice. Use a dictionary to find out if there are any other British and American words which can go in each category.

What's it called in English?

Describing things when you don't know the word

VOCABULARY AND SPEAKING

1 Choose five or six of the words in the box and think of something you can describe with each one.

> cloth cotton curved glass hard heavy high
> leather light long low metal narrow nylon
> oblong oval paper plastic round rubber short
> soft square stone wide wood wool

cotton: a shirt   oval: egg

2 Work in pairs. Tell your partner one of the things you chose. He/she must guess the word to describe it.

A: Shirt.   B: Cotton?
A: That's right.

3 Work in pairs. Decide which questions the words in the vocabulary box go with.

– What does it feel like?   – What's it made of?
– What shape is it?   – What size is it?

4 Look at some words you can use to describe something if you don't know the English word.

> liquid machine powder stuff thing tool

Think of things you can describe with these words.
toothpaste: stuff

5 Match the objects in the pictures with their names in the box below.

> wallet vacuum cleaner glue towel matches string
> sun-tan lotion soap tin opener sponge bag sweater
> hairdryer camera window cleaner shoe polish
> wine glass   pad of paper

LISTENING

1 Look at the objects in the pictures. Think of words to describe what they look like, what they're made of, what shape they are etc.

2 📼 Listen to people describing five of the objects in the pictures, which they don't know the English word for. Write the name of the object each speaker describes.

Speaker 1 _____ Speaker 2 _____ Speaker 3 _____
Speaker 4 _____ Speaker 5 _____

25

GENERAL COMMENTS

What's it called in English?

The theme of this lesson is how to make yourself understood if you don't know the word for something, a survival skill for circumstances with which your students may be very familiar. An ability to describe something's purpose or appearance in the absence of the word for the thing itself will greatly contribute to the student's fluency. Students may have difficulty with prepositions coming at the end of a sentence, eg *It's something to write on.* *It's for drinking wine out of* because it is a construction which doesn't often occur in other languages.

VOCABULARY AND SPEAKING

1 Aim: to present the words in the vocabulary box.

● Write the words for four or five very simple objects on the board, eg *table, book, pen, tyre, curtains.* Ask the students to look at the words in the box and to suggest words they could use to describe the objects.

● Ask the students to choose five or six words in the box and think of things they can describe with each one.

2 Aim: to practise using the words in the vocabulary box.

● Ask the students to work in pairs and to say one of the things they chose in activity 1. The other student must guess the word or words used to describe it.

3 Aim: to categorise the words in the vocabulary box.

● The process of categorising the words will reinforce the learning process.

> **Answers**
> **What does it feel like?** hard, heavy, light, soft
> **What's it made of?** cloth, cotton, glass, leather, metal, nylon, paper, plastic, rubber, stone, wood, wool
> **What shape is it?** curved, oblong, oval, round, square
> **What size is it?** high, long, low, narrow, short, wide

4 Aim: to present the words in the vocabulary box.

● The words are particularly useful when you're describing the function of something.

> **Possible answers**
> **liquid:** water, oil
> **machine:** dishwasher, lawnmower
> **powder:** washing up powder, soap powder
> **stuff:** toothpaste, lipstick, shampoo
> **thing:** cooker, television
> **tool:** hammer, spoon

● Explain that a machine is something that works on its own, and does something a human can do. For example, you wouldn't call a *television* a machine.

5 Aim: to present all the words in the vocabulary box.

● Ask the students to match the words with the items in the pictures.

LISTENING

1 Aim: to prepare for listening.

● Ask the students to look at the objects in the pictures and to find words in the vocabulary boxes to describe them. You may want to write the words for the objects on the board and ask the students to suggest suitable words.

2 Aim: to listen for main ideas.

● 🔲 The speakers in the listening passages are all non-native speaker who don't know the word for one of the objects in the pictures. This can happen even to advanced speakers of English. This is a productive comprehension check in that the students have to produce language to demonstrate comprehension, but they only have to write a single word each time; alternatively, you can do the task orally. You can also stop the tape after each description to allow the students plenty of time.

> **Answers**
> 1 glue
> 2 towel
> 3 matches
> 4 sun-tan lotion
> 5 vacuum cleaner

3 Aim: to present the target language.

● Each description shows an example of the different constructions used for describing things when you don't know the word.

> **Answers**
> a sweater b shoe polish c hairdryer d camera
> e window cleaner f bag g wine glass h pad of paper

FUNCTIONS

1 Aim: to focus on the prepositions at the end of the sentence, and on defining relative clauses.

● Remind the students if necessary that they learnt about defining relative clauses in Lesson 24. Ask them to read the information about describing something when you don't know the word in the functions box.

● Ask the students to do the exercise.

> **Answers**
> 1 A chair is a thing to sit **on**.
> 2 A mug is for drinking **out of**.
> 3 A frying pan is for cooking things **in**.
> 4 A towel is for drying yourself **with**.
> 5 An oven is for cooking things **in**.
> 6 A plant pot is a thing for putting plants **in**.

2 Aim: to practise describing things.

● You may like to do this orally, as the students may need help with suitable vocabulary.

> **Answers**
> A wallet is oblong, made of leather and is for keeping money in.
> A vacuum cleaner is a machine for cleaning carpets.
> Glue is stuff for sticking things together.
> A towel is oblong, made of a kind of cloth and you use it to dry yourself.
> Matches are small pieces of wood which you use to set light to things.

● When the students have done the exercise orally, ask them to write down the definitions.

3 Aim: to practise describing things.

● If the preceding exercises have been done successfully, ask the students to describe the purpose of the things mentioned on their own.

> **Answers**
> 1 A sleeping bag is for sleeping in.
> 2 An envelope is a thing for putting letters in.
> 3 Washing-up powder is stuff for washing the dishes in a dishwasher.
> 4 A dishwasher is a machine for washing the dishes.
> 5 A fridge is a thing for keeping food and drink cold in.
> 6 A pen is a tool for writing with.
> 7 A knife is a tool you use to cut things with.
> 8 A tea pot is a thing for making tea in.
> 9 A telephone is a thing you use to talk to people at a distance.
> 10 A typewriter is a machine you use to type things with.

SOUNDS

1 Aim: to focus on consonant clusters.

● Some students find it difficult to pronounce two or more consonants in a cluster. Ask them to say the words aloud and to merge the consonant clusters together. Be careful that they do not separate them or put in extra vowels to make them easier to say.

● ▭ Play the tape and pause after each word. Ask the students to listen and check.

2 Aim: to focus on links between words.

● ▭ In connected speech many words run into each other. If you have time, write these sentences on the board. Play the tape and pause after each sentence. Ask the students to listen and notice which words run into each other. Mark the links on the board.

> **Answers**
> Is it heavy?
> An egg is oval.
> A fridge is cold inside.
> It's for opening tins.
> It's something to write on.
> To open, pull it out.

SPEAKING

1 Aim: to focus on communication strategies.

● This activity sequence is designed to help students explore communication strategies and to be aware that a successful communicator does not always have to be a competent linguist. Ask the students to think about the objects in the pictures and to decide which is the best way to refer to them if they don't know the word for them.

2 Aim: to practise communication strategies.

● The game is designed to help students practise different ways of making themselves understood. You will need between five and ten minutes to do it.

● First, ask one student from each group to come up to you. Whisper the first item in the list of words below to these students. Make sure the others in the group don't hear.

> *1 a sleeping bag 2 an umbrella 3 a suitcase*
> *4 a shower 5 a traveller's cheque 6 scissors*
> *7 a passport 8 a lettuce 9 a watch 10 an aspirin*

● The players return to their groups and describe, draw, mime or point to the object. They mustn't say what it is. When someone guesses correctly, this student comes to you and whispers the word. If it is correct, you tell him/her the next word. You continue until a group has guessed all the words. If time is short, you can leave out some of the words.

3 Read these descriptions of some more objects in the pictures. Decide which object each one describes.

a It's a thing you wear to keep warm.
b It's for polishing shoes.
c It's a machine to dry your hair with.
d You use it to take pictures with.
e It's stuff to clean windows with.
f It's for carrying shopping in.
g It's for drinking wine out of.
h It's something to write on.

FUNCTIONS

> **Describing things when you don't know the word**
> *It's a thing you wear **to keep** warm.*
> *It's **for** polish**ing** shoes.*
> *It's **for** carry**ing** shopping **in**.*
> *It's a machine **to** dry your hair **with**.*
> ***You use** it **to** take pictures **with**.*
> *It's **stuff to** clean windows **with**.*
> *It's **for** drink**ing** wine **out of**.*
> *It's **something** to write **on**.*
>
> **You can use *to* + infinitive to describe the purpose of something.**
> *You use a cassette player **to play** cassettes.*
>
> **You can also use *(to be) for* + -*ing* to describe the purpose of something.**
> *A cassette player **is for play*ing** cassettes.*

1 Complete these sentences with *in, out of, with* and *on*.

1 A chair is a thing to sit ___ .
2 A mug is for drinking ___ .
3 A frying pan is for cooking things ___ .
4 A towel is for drying yourself ___ .
5 An oven is for cooking things ___ .
6 A plant pot is a thing for putting plants ___ .

2 Look at the five pictures of objects which have not yet been described. Write sentences describing them using as many expressions from this lesson as possible.

3 Describe the purpose of these things.

1 a sleeping bag
2 an envelope
3 washing-up powder
4 a dishwasher
5 a fridge
6 a pen
7 a knife
8 a tea pot
9 a telephone
10 a typewriter

SOUNDS

1 Say these words aloud. Make sure you pronounce the underlined consonant groups carefully.

pra<u>ct</u>ical <u>tr</u>ansport news<u>p</u>aper car<u>db</u>oard
too<u>thp</u>aste di<u>shw</u>asher cal<u>c</u>ulator

🔊 Now listen and check.

2 🔊 Listen and notice the words the speaker links.

1 Is it heavy?
2 An egg is oval.
3 A fridge is cold inside.
4 It's for opening tins.
5 It's something to write on.
6 To open, pull it out.

Now say the sentences aloud.

SPEAKING

1 Work in pairs. There are several ways to refer to something if you don't know the English word. You can describe it, draw it or mime it. You can also point if you can see one.

Look at the objects in the pictures. Which is the best way to refer to them?
I think I would draw a wine glass.

2 Work in groups of three or four. You're going to play *What's it called in English?* Here are the rules.

> # What's it called in English?
>
> AIM: for one person in each group to refer to a word by describing it, drawing it, miming it or pointing to it. The first group to guess all the words is the winner.
>
> *1* One student from each group comes to the teacher who gives them a word. They must describe it, draw it, mime it or point to it. They mustn't say the word!
>
> *2* The student who guesses the word correctly goes to the teacher and tells him/her what the word is. If it is correct, the teacher gives him/her another word.
>
> *3* The student acts out the word, and the game continues until a group has guessed all the words. The first group to finish is the winner.

Progress check 21-25

VOCABULARY

1 Write the verbs which come from these nouns. Underline the suffixes for each noun.

government employment election education information reading writing

govern govern<u>ment</u>

Now write nouns which come from these verbs, using the suffixes you underlined.

entertain refresh teach greet exhibit connect manage

When you write new nouns in your *Wordbank* you may like to group them according to their suffix.

2 *Get* is a common verb in spoken English. Here are some of its meanings.

1 *get* + past participle = *become*   *They've got married.*
2 *get* + adjective = *become*   *I'm getting angry.*
3 *get* + object = *fetch, obtain*   *Could I get you a drink?*
4 *get* + object = *catch, take*   *I get a train at 7.15.*
5 *get* + object = *receive*   *We got the news yesterday.*
6 *get* + to + object = *arrive in/at*   *I got to work at 8am.*
7 *get* + preposition has the idea   *We get up at 7.30.*
   of movement, sometimes   *He got in through the window.*
   with difficulty.

Look at these sentences and decide which meaning of *get* they show. Rewrite them replacing *get* with another verb.

a It's getting dark.
b We got a call from her.
c The train gets to London at 6 o'clock.
d He got the tickets at the box office.
e The child got down from the table.
f She gets worried very easily.
g The bus to get is the number 39.

It's a good idea to note the different meanings of *get* as you come across them.

3 Look at the vocabulary boxes in lessons 21 – 25 again. Choose words which are useful to you and group them under headings of your choice in your *Wordbank*.

GRAMMAR

1 Write the past participles of these verbs.

be become break bring come do find get go have learn lose put read say sit sleep take think understand write

2 Write short answers to these questions.

Have you ever...

1 ...visited the USA?
2 ...learnt a musical instrument?
3 ...been on TV?
4 ...met a rock star?
5 ...tasted English beer?
6 ...eaten English cheese?
7 ...heard Beethoven's Fifth Symphony?
8 ...written a letter to a newspaper?

3 You haven't seen your friend Jenny for five years. Ask her questions using the words below.

1 find a new job
2 get married
3 stop smoking
4 buy a motorcycle
5 visit USA
6 write a book
7 travel around Europe
8 learn to cook

1 Have you found a new job?

Progress check 21–25

GENERAL COMMENTS

You can work through this Progress check in the order shown, or concentrate on areas which may have caused difficulty in Lessons 21 to 25. You can also let the students choose the activities which they would like to or feel the need to do.

VOCABULARY

1 Aim: to focus on noun suffixes.
● Your students may like to use a dictionary for this activity. Mention that words with the suffixes below are likely to be nouns.

> **Answers**
> govern employ elect educate inform read write
>
> govern*ment* employ*ment* elect*ion* educat*ion* informat*ion* read*ing* writ*ing*
>
> entertainment refreshment teaching greeting exhibition connection management

2 Aim: to focus on a few uses of *get*.
● As the introduction says, *get* is a very common word in English, and it is not possible to cover all of its meanings. Here are some of the meanings which are most suitable at this level.

> **Answers**
> a2 b5 c6 d3 e7 f1 g4

3 Aim: to help organise the students' vocabulary learning.
● It may be a suitable moment to take a look at your students' *Wordbanks* in their Practice Books to see if they are completing them successfully.

GRAMMAR

1 Aim: to revise past participles.

> **Answers**
> been, become, broken, brought, come, done, found, got, gone, had, learnt, lost, put, read, said, sat, slept, taken, thought, understood, written

2 Aim: to revise short answers in the present perfect.
● Do this activity orally first of all and check the students answer using *Yes, I have* or *No, I haven't*.

● Ask the students to write their answers to this activity.

3 Aim: to revise asking questions using the present perfect.

> **Answers**
> 1 Have you found a new job?
> 2 Have you got married?
> 3 Have you stopped smoking?
> 4 Have you bought a motorcycle?
> 5 Have you visited the USA?
> 6 Have you written a book?
> 7 Have you travelled around Europe?
> 8 Have you learnt to cook?

4 Aim: to revise writing sentences using the present perfect.

Answers
1 She's found a new job.
2 She hasn't got married.
3 She's stopped smoking.
4 She hasn't bought a motorcycle.
5 She's visited the USA.
6 She hasn't written a book.
7 She's travelled round Europe.
8 She hasn't learnt to cook.

5 Aim: to revise the difference between the present perfect and the simple past.

Answers
1 I've never driven a car in my life.
2 I left my secondary school in 1989.
3 We haven't met before. My name's John.
4 Where did you buy your coat?
5 What did you pay for it?
6 How long have you been here now?
7 Last Sunday I got up at eleven.
8 Have you seen the latest Harrison Ford film?

6 Aim: to revise the use of *for* and *since*.
● The students should answer these questions for themselves.

7 Aim: to revise saying what you use things to do.

Answers
1 You use an umbrella to shelter from the rain.
2 You use sunglasses to protect your eyes from the sun.
3 You use sticky tape to stick paper with.
4 You use an overcoat to keep warm in.
5 You use a saw for cutting wood and metal with.
6 You use water for washing, cooking and drinking.

8 Aim: to revise saying what things are for.

Answers
1 A garage is for keeping your car in.
2 A bath is for washing yourself in.
3 A wardrobe is for keeping your clothes in.
4 A toothbrush is for cleaning your teeth with.
5 A door handle is for opening the door with.
6 A kettle is for boiling water in.

SOUNDS

1 Aim: to focus on /ɜː/ and /ə/.
● 🔲 This activity focuses on stressed and unstressed sounds. Ask the students to put the words in two columns, and then play the tape to listen and check.

Answers
/ɜː/: surfing fireworks learnt heard university word
/ə/: primary economics lecture chosen

2 Aim: to focus on /ɔː/ and /ʌ/.
● 🔲 /ɔː/ is a difficult sound for non-native speakers to acquire. This activity is to encourage a very simple distinction between /ɔː/ and /ʌ/. Ask the students to listen to the tape and to repeat each word. Play the tape and pause after each word.

● Ask them to put the words in two columns.

Answers
/ɔː/: sport awful walking performance four
/ʌ/: up summer bus pub sunny

SPEAKING AND WRITING

1 Aim: to practise speaking about experiences.
● Ask the students to look at the experiences and to think of other experiences. Do this activity orally with the whole class, and write the activities the students suggest on the board.

2 Aim: to practise speaking about experiences.
● Ask the students to go round the class finding people who have done the things in activity 1. Ask the students to ask for extra information, using the past simple.

3 Aim: to practise writing about experiences.
● The students may like to do this activity for homework.

4 Here are Jenny's answers to the questions in 3. Write full answers.

1 Yes.	2 No.	3 Yes.	4 No.
5 Yes.	6 No.	7 Yes.	8 No.

1 Yes. She's found a new job.

5 Complete these sentences with the present perfect or the past simple form of the verb in brackets.

1 I never (drive) a car in my life.
2 I (leave) my secondary school in 1989.
3 We not (meet) before. My name's John.
4 Where you (buy) your coat?
5 What you (pay) for it?
6 How long you (be) here now?
7 Last Sunday I (get up) at eleven.
8 You (see) the latest Harrison Ford film?

6 Answer the questions. Use *for* and *since* in turn.

How long have you...

1 ...been a student of English?
2 ...lived in your home?
3 ...known your best friend?
4 ...liked your favourite food?
5 ...been in class?
6 ...had the clothes you're wearing?

7 Say what you use these things to do.

1 an umbrella	4 an overcoat
2 sunglasses	5 a saw
3 sticky tape	6 water

8 Say what these things are for.

1 a garage	4 a toothbrush
2 a bath	5 a door handle
3 a wardrobe	6 a kettle

SOUNDS

1 Say these words aloud.

surfing fireworks primary learnt economics
lecture heard university chosen word

Is the underlined sound /ɜ:/ or /ə/? Put the words in two columns.

 Listen and check.

2 Say these words aloud.

sport up awful walking summer bus pub
sunny performance four

Is the underlined sound /ɔ:/ or /ʌ/? Put the words in two columns.

 Now listen and check.

SPEAKING AND WRITING

1 Look at these experiences.

– visit the Great Wall of China
– play tennis
– drive a Porsche
– cook dinner for six friends
– swim in a river
– see a play by Shakespeare
– write a diary
– read *Time* magazine

Think of two more experiences and add them to the list.

2 Now find people in the class who have done the things in 1. Ask two extra questions about these experiences.
Have you ever visited the Great Wall of China?
Yes, I have.
Really! When was that? In 1989, when I was twelve.
What was it like? Fantastic!

3 Write a paragraph describing what you have learnt about the other students.
Two people have visited the Great Wall of China. Paco went in 1989, when he was twelve...

26 Safety first

Modal verbs; *must* for obligation; *mustn't* for prohibition

SPEAKING

Work in pairs and look at these safety instructions. Who do you think is speaking and where?

1 'You must fasten your seat belt.'
2 'You mustn't lean out of the window.'
3 'You must put out your cigarette.'
4 'You mustn't leave them on their own.'
5 'You must wear strong shoes.'
6 'You mustn't play with matches.'
7 'You must stay in your seat until we stop.'
8 'You must wear a helmet.'

GRAMMAR

> **Modal verbs**
>
> **Modal verbs:**
> – have the same form for all persons
> – don't take the auxiliary *do*
> – take an infinitive without *to*.
>
> *Must* for obligation, *mustn't* for prohibition
> ***Must*** is a modal verb. You use ***must*** to talk about something you're obliged or strongly advised to do. You often use it when you talk about safety instructions.
> *You **must** fasten your seat belt.*
>
> You use ***mustn't*** to talk about something you aren't allowed to do or you're strongly advised not to do.
> *You **mustn't** lean out of the window.*
>
> For strong prohibition you use ***must never***.
> *You **must never** walk on the railway line.*
>
> ***Must*** and ***have to*** have almost the same meaning. You usually use ***must*** when the obligation comes from one of the speakers.
> *I usually forget her birthday. I **must** remember this year.*
> *The baby's asleep. You **must** be quiet.*
>
> You usually use ***have to*** when the obligation comes from a third person. You often use it when you talk about rules.
> *The government says you **have to** do military service.*
> *You **have to** show a cheque card when you pay by cheque.*

1 Think of five or six things you must do to make your life safer.
I must stop smoking.
I must mend the brakes on my car.

2 Complete these sentences with *must* or *mustn't*.

1 You ___ eat less to lose weight.
2 You ___ wear a seat belt while the plane is taking off.
3 You ___ travel on a train without a ticket.
4 You ___ ride a bicycle on a motorway.
5 You ___ always lock your car.
6 You ___ drive at more than 55 km/h in town.

3 Think of other rules or safety instructions for these situations.

– in a plane – in a car – in the street – on a boat
– in the mountains – in the desert

You mustn't smoke during take-off and landing.

26

GENERAL COMMENTS

Modal verbs

The focus of this lesson is modal auxiliary verbs. Modal verbs have a number of points in common, which are mentioned in the grammar box. Most of them have a number of different meanings; for example, in this lesson, the use of *must* for obligation is presented, but it can also be used to talk about deductions. Lessons 27 to 30 continue the work on modal verbs.

Must and *have*

In the grammar box, it is suggested that *must* and *have to* have almost the same meaning. Students may have difficulty in appreciating the difference between the two forms. *Have to* is often used when you talk about a regulation, because this suggests that the obligation comes from a third person. You can use *must* to talk about a regulation, but it suggests a strong suggestion, as if the obligation comes from the speaker and not from some outside source. *Mustn't* can be used to talk about a regulation and a strong suggestion.

SPEAKING

1 Aim: to present *must* and *mustn't*.

● In these sentences you can replace *must* with *have to* if you want to stress that the source of the obligation is a third person, such as a rule. Stress the idea of safety instructions rather than rules for this activity. Write the sentences on the board and ask the students to say who is speaking and to whom. There are many possibilities. You may like to discuss these possibilities after the students have read the distinction between *have to* and *must* in the grammar box.

> **Possible answers**
> 1 A parent to a child in a car.
> 2 A parent to a child in a car or a train.
> 3 A flight attendant to a passenger on a plane.
> 4 A parent to a babysitter.
> 5 A guide to walkers.
> 6 A parent to a child.
> 7 A flight attendant to passengers on a plane.
> 8 A builder to a visitor on a building site.

GRAMMAR

1 Aim: to focus on the use of *must* to talk about safety instructions.

● Ask the students to read the information about modal verbs and *must, mustn't* and *have to* in the grammar box.

● Ask the students to think of five or six things which will make their lives safer. You may like to suggest these areas to think about: *at home, in the street, money and valuables, the environment, transport.* Then ask them to write sentences using *must.*

● Ask the students to work in pairs and to show each other the sentences they have written. Ask them if they written similar sentences.

2 Aim: to focus on *must* and *mustn't* to talk about regulations.

● Stress that this activity focuses on obligation from a third person, such as a regulation.

> **Answers**
> 1 You **must** eat less to lose weight.
> 2 You **must** wear a seat belt while the plane is taking off.
> 3 You **mustn't** travel on a train without a ticket.
> 4 You **mustn't** ride a bicycle on a motorway.
> 5 You **must** always lock your car.
> 6 You **mustn't** drive at more than 55 km/h in town.

3 Aim: to practise using *must* and *mustn't*.

● Ask the students to write some more sentences using *must, mustn't* and *have to.*

● When they are ready, ask the students to suggest a few sentences and draw attention to the distinction between *must* and *have to.*

READING

1 Aim: to read for main ideas and to practise talking about prohibition.

● The style of the illustrations and the text is humorous but the serious point is clearly made. This is a very short reading passage, but it illustrates the language point quite well. Ask the students to read the rules and match them with the pictures.

> **Answers**
> 1 F 2 A 3 B 4 E 5 D 6 C

● You may want to explain a few items of vocabulary such as *platform, skateboard, stick out.*

2 Aim: to react to a text.

● Ask the students to say which drawings they like best.

3 Aim: to practice using *must* and *mustn't*.

● All of the sentences can be rewritten with *must never* and *mustn't*. Ask the students to extend the activity by writing sentences saying what you must do, eg 1 *You must get off your bicycle and walk.*

> **Answers**
> 1 You mustn't ride a bicycle on a station platform. You mustn't use a skateboard there either.
> 2 You must never stick your head out of a window of a moving train.
> 3 You mustn't throw anything out of the windows.
> 4 You must never open a door before the train has stopped.
> 5 You must never go onto a railway line or walk along it.
> 6 You must never put anything on the railway line.

SOUNDS

1 Aim: to focus on strong and weak pronunciation of *must* and *mustn't*.

● When *must* and *mustn't* are followed by other words, the /t/ is weakened and may disappear completely. Ask the students to say the words in the list.

● 🔲 Ask the students to listen and decide which they hear, /mʌs/, /mʌst/, /mʌsn/ or /mʌsnt/.

> **Answers**
> 1 a 2 c 3 b 4 d 5 a 6 a 7 c 8 d

2 Aim: to focus on strong stress and intonation for insisting.

● 🔲 Explain to the students that obligation and prohibition can be stressed by strong intonation. Ask the students to listen to speaker B insisting. Play the tape.

● Ask the students to work in pairs and repeat the sentences. Make sure speaker B uses strong stress and intonation to insist.

● 🔲 Play the tape and pause after every sentence. Ask the students to repeat.

VOCABULARY AND LISTENING

1 Aim: to present the words in the vocabulary box.

● Ask the students to match the words in the box and the situations. Some words can go with more than one situation.

> **Answers**
> **on a motorway:** drive, fast, lane, speed, traffic
> **in a train:** carriage, first class, second class, ticket inspector
> **at the border:** customs, duty-free, guard
> **in the street:** bicycle, pavement, policeman, ride, traffic

2 Aim: to practise talking about obligation and prohibition.

● The rest of this activity sequence focuses on rules, even though the rules may be imposed for reasons of safety. Ask the students to suggest some rules for the situations in activity 1. You may like to do this orally.

> **Suggested answers**
> **on a motorway:** You mustn't drive too fast.
> **in a train:** You mustn't lean out of the window.
> **at the border:** You have to show your passport.
> **in the street:** You mustn't walk in the road.

3 Aim: to listen for main ideas.

● 🔲 Ask the students to listen and decide what the situations are.

> **Answers**
> **Dialogue 1:** on a motorway
> **Dialogue 2:** in the street
> **Dialogue 3:** in a train.

● Some students may be surprised that you mustn't ride a bicycle on the pavement in Britain.

4 Aim: to practise using *have to* and *mustn't*, and to listen for specific information.

● Ask the students to think about what the people have to or mustn't do.

> **Answers**
> **Dialogue 1:** He mustn't drive so fast. He mustn't use his car phone when he is driving.
> **Dialogue 2:** He mustn't ride through a red light. He has to have lights at night. He mustn't ride on the pavement.
> **Dialogue 3:** He mustn't sit in a first class carriage with a second class ticket.

● 🔲 Play the tape again. Pause after each dialogue and check the answers.

5 Aim: to practise using *must, mustn't* and *have to*.

● Ask the students to think about other rules and safety instructions. They may like to write sentences about this for homework.

READING

1 Look at the drawings for a guide to railway safety. Match them with the rules below.

1 Do not ride a bicycle on a station platform. Do not use a skateboard there either.
2 Never, never, never stick your head out of a window of a moving train.
3 Do not throw anything out of the windows.
4 Never open a door before the train has stopped.
5 Never go onto a railway line or walk along it.
6 Never put anything on the railway line.

2 Work in pairs and say which drawing you like best.

3 Rewrite the sentences in 1 saying what you *mustn't* do.
1 You mustn't ride a bicycle on a station platform.

SOUNDS

1 Say these words.

a /mʌs/ b /mʌst/ c /mʌsn/ d /mʌsnt/

🔲 Listen to these sentences and decide if you hear *a, b, c* or *d*.

1 You must go now.
2 You mustn't be late.
3 Yes, you must.
4 No, you mustn't.
5 You must visit us.
6 We must have lunch.
7 He mustn't say that.
8 You mustn't eat much.

Now say the sentences aloud.

2 🔲 Listen to the way speaker *B* uses a strong stress and intonation to sound insistent.

A I don't want to go to school. **B** You must go.
A Must I really go? **B** Yes, you must.
A I'll be late. **B** You mustn't be late.
A Well, I'm going to leave early. **B** No, you mustn't.

Now work in pairs and say the sentences aloud. Try to sound insistent.

VOCABULARY AND LISTENING

1 Match the words in the vocabulary box with the situations below.

– on a motorway – in a train
– at the border – in the street

bicycle carriage customs drive
duty-free fast first class guard
lane pavement police officer
ride second class speed
ticket inspector traffic

2 Work in pairs and say what you have to or mustn't do in the situations in 1.

3 🔲 Listen to three conversations and decide what the situation is. Choose from the situations in 1.

4 Work in pairs and say what the people in the conversations have to or mustn't do.
He mustn't drive so fast.

🔲 Listen again and check.

5 Talk about rules or safety instructions you have at your language class or school. How many can you think of?
You mustn't drop litter.
You have to do your homework.

27 The Skylight

Can, could (1) for ability

SPEAKING

1 Find someone in the class who can:

- swim a hundred metres
- speak a foreign language (not English)
- drive a car
- stay awake all day
- stay up all night
- write with their left hand
- ride a bicycle
- use a computer
- cook
- play a musical instrument

Can you swim a hundred metres? Yes, I can. No, I can't.

2 Choose the last person you spoke to and find two things you can both do, and two things neither of you can do.

GRAMMAR

Can, could (1) for ability

Can is a modal verb. It has the same form for all persons and you don't use the auxiliary *do* in questions and negatives. You use *can* to express general ability, something you are able to do on most occasions.
I **can** *swim a hundred metres.*

Negative	**Questions**	**Short answers**
I can't ride a bicycle.	*Can you swim?*	*Yes, I can.*
		No, I can't.

You can also use *be able to* to express general ability. It is more formal than *can* and is mostly used in the future or the past tense.

You use *could* and *couldn't* or *wasn't/weren't able to* in the past to express general ability.
When I was five, I **could** *swim, but I* **couldn't** *write my own name.*

1 Think about when you were five years old. Which things in *Speaking* activity 1 could you do? Write sentences.
I could swim, but I couldn't cook.

2 Find out what other people could or couldn't do when they were five years old.

LISTENING AND SPEAKING

1 Work in pairs. You are going to hear a story in four parts, adapted from *The Skylight* by Penelope Mortimer. Here are some key words from the first part. What do you think the story is about? (The words are in the right order.)

taxi woman five-year-old boy
sleep house summer
mountains hot afraid suitcases
shutters doors closed locked

2 📼 Listen to part 1 of the story and check your answers to 1.

3 Here are some questions about part 2 of the story. Try and guess the answers.

1 Why didn't they go in the house?
2 What did she see on the roof?
3 How did she get up there?
4 Did she get through the skylight?
5 Who climbed through the skylight?
6 What did she ask him to do when he was in the house?
7 Did he do it?

4 📼 Listen to part 2 of the story and check your answers to the questions in 3.

27

GENERAL COMMENTS

Can and *could* (1) for ability

This use of *can* is likely to have been covered at an earlier stage in the students' English lessons. In this context, *can* means *know how to* and refers to a general ability. You can also use *can* to express possibility, permission and prohibition. Some languages do not use the same verb to express these different functions. The use of *could* is complex. You can use it in affirmative sentences to talk about general ability in the past, eg *I could read when I was six,* but not usually for a particular ability in the past. In this case, you use *was/were able to,* eg *There was no traffic on the way to the station so I was able to catch my train.* The exception is when *could* is followed by *see* and *hear,* eg *It was cloudy and I could see it was raining in the distance.* You can use *couldn't* in negative sentences to talk about a general or a particular ability in the past, eg *I couldn't swim when I was five. I'm sorry, I couldn't get here earlier.*

Pedagogic and real-life motivation

Choosing material which you think the students will find interesting is one way of creating motivation. *The Skylight* was chosen for its interest and dramatic tension, but in the classroom even strong material needs the pedagogic motivation of tasks to captivate the students' interest until real-life motivation takes over. The story is divided into four parts, because the length of a listening passage can compromise motivation, and each part has a different task to be done while listening, except for the fourth part, when it is to be hoped that students will be listening because they are genuinely interested and real-life motivation has taken over.

Illustration: possible questions

What can you see? Where do you think it is? Does it look like a scene from your country? What time of year do you think it is? What time of day is it? Would you like to go there on holiday? Would you like to go there to live? What kind of house can you see? How many rooms does it have? What furniture and fittings does it have?

SPEAKING

1 Aim: to practise using *can* for general ability.

● Ask one or two students which things they can do. Some of them may appear strange, but will be useful when the students use *could* in *Grammar* activity 1.

● Ask the students to go round and find people who can or can't do the things mentioned in the list.

● Ask the students to find other things they can or can't do. You may want to write the following prompts on the board: *read Japanese, speak German, tell lies, repair your car, sleep ten hours, run a kilometre, stay up all night.* Explain that you can use *So can I / Nor can I* if you can or can't do the same things and *Can you? I can't* if you can't do the same things.

GRAMMAR

1 Aim: to focus on *could* for general ability in the past.

● Ask the students to read the information about *can* in the grammar box.

● Ask the students to think about what they could do in the past when they were five years old, and to write some sentences.

2 Aim: to practise using *could* for general ability in the past.

● Ask the students to go round asking and saying what they could do in the past.

● Find out what things people could do in the past but can't do in the present.

LISTENING AND SPEAKING

1 Aim: to prepare for listening and to pre-teach some important words from the story.

● Explain that the story is about a mother and her five-year-old son, and they are going on holiday. A *skylight* is a small window in a roof. You can see one in the illustration.

● Ask the students to work in pairs and to use the words to predict what is going to happen in the story. Because this activity is designed to pre-teach some important vocabulary, they may need to use their dictionaries or ask you to explain any unfamiliar words. Try to limit the explanations to five or six words.

2 Aim: to practise listening for main ideas.

● 🔲 This activity is designed to help the students check their predictions. They are not asked to understand every detail. Make quite sure that everyone knows the meaning of the word *skylight.* Play the tape.

● Ask one or two students to re-tell the story. Check that everyone has understood the main points. If necessary, play the tape a second time.

3 Aim: to check comprehension and to prepare for listening.

● Give the students time to read the questions.

● Ask the students to predict what is going to happen in the second part of the story. Do this activity orally with the class as a whole.

4 Aim: to listen for main ideas.

● 🔲 Ask the students to listen to the second part of the story and to check their answers to the questions in activity 3. Play the tape.

5 Aim: to prepare for listening.
- Ask the students to complete the missing words and phrases from the transcript of part three of the story. They can do this individually or in pairs.

> **Answers**
> name, help, car, stop, cry, farm, round, road, door, happened, car, house, lane

6 Aim: to listen for specific information.
- 🔲 Ask the students to listen and check their answers to activity 5. Play the tape.

7 Aim: to prepare for listening; to predict the end of the story.
- Ask the students to work in pairs and to predict the end of the story.

- Ask the students to tell the rest of the class what they think is going to happen.

8 Aim: to pre-teach some important vocabulary; to listen for main ideas.
- Check that everyone understands the key items of vocabulary.

- 🔲 Play the tape.

- Ask the students how they would feel in the woman's position at the end of the story.

VOCABULARY

1 Aim: to focus on the words in the vocabulary box.
- Most of the listening activities accompanying the story were concerned with the main ideas, and no special attention was given to the vocabulary. The words in the box represent the items which are most useful at this level but you may want to exploit more vocabulary from the story. Ask the students to reconstitute the story using the words in the box. Ask one student to start telling the story. If anyone thinks he/she has missed something out, they can ask a question. If the first person can answer the question, he/she continues telling the story. If he/she can't answer the question, the person who asked the question takes over.

2 Aim: to encourage vocabulary acquisition according to personal categories.
- Ask the students to think of categories of their own choice which they can use to organise the new words. They may like to put the words into different categories in their *Wordbanks*.

- If time is short, the students may like to do this for homework.

5 Here's part 3 of the story with some words and phrases missing. Can you guess what they are?

It was now dark. She went down again and ran round the house, shouting his _____ . Something has happened to him, I must go for _____ . She ran to the road and when she saw the lights of a _____ , she waved her arms to _____ it. She started to _____ . It was a long time before the three men understood.

'But how can we get in? We have no tools,' they said.

'There's a _____ back there. Will you take me?' They let her into the car.

'Turn _____ . It's back there on the left. There it is!' They turned off the _____ She got out of the car and ran to the front _____ .

A small woman in trousers opened the door. 'My dear, what's _____ ?'

'You're English?'

She told her the story. Another woman appeared. 'Yvonne,' Miss Jardine said, 'Get some tools, a hammer and an axe.'

They all got into the _____ and went back to the ____ . They drove up the _____ and stopped. She ran to the house, calling 'Johnny? Johnny?'

6 🔊 Listen to part 3 of the story and check your answers to 5.

7 Work in pairs. What do you think happens at the end of the story?

8 Before you listen to part 4 of the story, check you understand these words.

axe   hammer   toys   smash   thumb

🔊 Now listen to part 4 of the story and find out what happens.

VOCABULARY

1 Here are some words from the story. Check you know what they mean. (You can use a dictionary.) Try to remember where they came in the story.

> hill sleep asleep narrow stones grass suitcase
> square skylight roof shutter door lock knock
> ladder lower climb shout hurry wave farm tools
> axe hammer toys smash thumb whisper

2 Can you think of categories to group the words? Think of other words which go with them.

28 Breaking the rules?

Can, can't (2) for permission and prohibition

READING

1 Work in pairs. Look at the signs in the photos. Do you understand what they mean? Where do you think the signs are? Choose from these situations.

- in the street
- in a bar
- in a park
- in a shop
- in a church
- in a cinema
- in a restaurant

2 Read *Breaking the rules?* and answer the questions with *yes* or *no* for your country.

Breaking the rules?

These days there are rules everywhere we go and it's hard to obey them all. But what about the rules and customs we don't know about when we're visitors to a town or a country? Are you breaking the rules?

· No Bicycles Please ·

1 You're waiting to cross the street at a pedestrian crossing. The lights are red for pedestrians, but there are no cars. Can you cross?

2 You're in a large city park which has lots of trees and grass. It's a beautiful summer's day and the sun is shining. Can you walk on the grass?

3 It's half past three on a Sunday afternoon, and you're thirsty. Can you buy a nice, cold glass of beer?

4 You're with some very young children and you'd like to have a drink in a bar. Can they go in with you?

5 You're in a restaurant but you don't know what the food is like. Can you walk into the kitchen and have a look?

6 You want to buy a postcard in the newsagent's or tobacconist's shop. Can you also get stamps there?

7 You're in the cinema and you'd like a cigarette. There are no signs which say 'no smoking'. So can you smoke?

8 You're in a cafe and you'd like a drink. Can you get it at the bar and pay for it there?

9 You feel quite ill, but you don't have any medical insurance or much money. Can you make an appointment to see the doctor?

10 You want to take the bus. Can you buy a single ticket before you get on?

28
GENERAL COMMENTS

Can and *can't* (2) for permission and prohibition

This is the second lesson on *can*. Some students may be confused that you can use the same verb for talking about permission and prohibition and for describing general ability. If so, write *can* on the board and underneath write *know how to* and *be allowed to*. Some students may have used *can't* in Lesson 26 for prohibition. It has the same meaning as *mustn't* in this context, and the only reason it was not introduced in Lesson 26 was in order not to overload the students with too many structures.

Cross-cultural awareness

If you are teaching a mono-cultural group, the answers to the questions in *Breaking the rules?* are all likely to be very similar. In the listening passage, a British person talks about the rules and customs in Britain in order to create the opportunity of cross-cultural comparison. It is not assumed that the students need to acquire this information about Britain because they are going to visit the country. This is simply an opportunity to draw attention to the fact that people do things differently in different countries, and that it should not be assumed that the rules and customs in one country can be transferred to another cultural context.

READING

1 Aim: to prepare for reading.

● The photos show signs of rules and customs in different countries. Ask the students where they might see the signs.

> **Answers**
> No dogs: in a park
> No bicycles please: in a park, or on a building in the street
> Please keep off the grass: in a park
> Please do not obstruct: in the street
> No parking in front of these gates: in a park, or in the street
> Don't walk: in the street

2 Aim: to read and answer the questions.

● The passage is a kind of questionnaire, in which direct interaction is established with the reader. Ask the students to read and think about the answers for their country.

● With a multi-cultural group, you may want to discuss the answers to the questions immediately. With a mono-cultural group, in which all the answers are likely to be similar, explain that there will be an opportunity in the *Listening* section to compare rules and customs in their country with those in Britain.

GRAMMAR

1 Aim: to practise using *can* and *can't*.

● Ask the students to read the information about *can* and *can't* in the grammar box.

● Ask the students to use *can* or *can't* to make true statements about their countries.

2 Aim: to practise using *can* and *can't*.

● Ask the students to discuss their answers to activity 1 in pairs.

● You may like to check the answers with the whole class.

SOUNDS

1 Aim: to focus on strong and weak pronunciation of *can* and *can't*.

● When *can* is at the end of the sentence, it is pronounced /kæn/. When it is in the middle of a sentence, it is unstressed and pronounced /kən/. *Can't* is always pronounced /kɑːnt/. Ask the students to say the sentences aloud and decide if the sound is /æ/, /ə/ or /ɑː/.

● 🔲 Ask the students to listen and check their answers.

> **Answers**
> Can I come in.
> The underlined sound is /ə/.
> Yes, you can. You can sit down there.
> The first underlined sound is /æ/, the second is /ə/.
> Can I smoke?
> The underlined sound is /ə/.
> You can't smoke in here, but you can smoke outside.
> The first underlined sound is /æ/, the second is /ə/.

2 Aim: to focus on the American pronunciation of *can* and *can't*.

● 🔲 In spoken American English, it is very often difficult to distinguish between *can* and *can't* as they are both pronounced with the letter *a* in the strong form /æ/ and, as in English, the /t/ sound may be assimilated with the next sound. If the word is at the end of the sentence, the /t/ ending may be clearer. Ask the students to listen and tick the sentence they hear.

> **Answers**
> You can't smoke.
> You can walk on the grass.
> You can't cross now.
> You can go.

LISTENING

1 Aim: to listen for main ideas.

● 🔲 Explain to the students that they are going to listen to a British person answering the questions in *Breaking the rules?* Remind the students that they may not understand every word, but they should understand enough to do the task successfully.

> **Answers**
> 1 yes   2 yes   3 no   4 no   5 no   6 yes   7 yes   8 yes
> 9 yes   10 no

2 Aim: to practise talking about rules and customs.

● Make sure you check the answers to the questions with the students.

● Ask the students to work in pairs and say if the rules and customs are the same in Britain as they are in their countries.

● Ask the whole class for their reactions. Is there anything they find surprising? Go through each point and ask them to remember as much detail as possible.

● 🔲 Play the tape a second time.

VOCABULARY AND SPEAKING

1 Aim: to practise using *can* and *can't* for permission and prohibition; to present the words in the vocabulary box.

● Explain that the words in the vocabulary box are all situations in which there may be some special rules and customs. Write sentences such as *Men can't wear hats in church. You can walk across the road at a pedestrian crossing* on the board Ask the students to make sentences with *can* and *can't* using the words.

2 Aim: to practise using *can* and *can't* for permission and prohibition.

● Ask the class what rules and customs there are in their countries. You may like to ask them to write sentences about this for homework.

GRAMMAR

> **Can, can't (2) for permission and prohibition**
> **Can** is a modal verb. It has the same form for all persons
> and you don't use the auxiliary **do** in questions and
> negatives. You use **can** + infinitive to talk about what
> you're allowed to do or what it is possible to do.
> *You* **can** *cross when the light is green.*
>
> You use **can't** to talk about what you're not allowed to do
> or what it is not possible to do.
> *You* **can't** *cross when the light is red.*
>
> **Questions**   **Short answers**
> **Can** *you cross when the light is green?*   *Yes, you* **can**.
> **Can** *you cross when the light is red?*   *No, you* **can't**.
>
> **Can't** and **mustn't** mean the same.
> *You* **mustn't** *drive a car if you're only sixteen.*
> *You* **can't** *drive a car if you're only sixteen.*

1 Complete these sentences with *can* or *can't* and make
true statements about your country.

1 You ___ drive a car when you're sixteen.
2 You ___ go into a bar when you're fourteen.
3 You ___ get married when you're sixteen.
4 You ___ wash your car on Sundays.
5 You ___ visit the USA without a visa.
6 You ___ get cheap housing.

2 Work in pairs and check your answers to the questions
in *Breaking the rules?*.
In my country you can't walk on the grass in parks.

SOUNDS

1 Work in pairs. Say these sentences aloud.

'Can I come in?'
'Yes, you c<u>a</u>n. You c<u>a</u>n sit down there.'
'Can I smoke?'
'You c<u>a</u>n't smoke in here, but you c<u>a</u>n smoke outside.'

Is the underlined sound /æ/, /ə/ or /ɑː/?

[cassette] Listen and check.

2 In American English it is sometimes difficult to hear the
difference between *can* and *can't*.

[cassette] Listen and tick (✓) the word you hear.

1 You *can/can't* smoke.
2 You *can/can't* walk on the grass.
3 You *can/can't* cross now.
4 You *can/can't* go.

LISTENING

1 [cassette] Listen to Jane, a British woman, answering the
questions in *Breaking the rules?*. Put a tick (✓) if the
answer is *yes* and a cross (✗) if the answer is *no*.

2 Work in pairs and check your answers. Are the rules
and customs the same as in your country?

[cassette] Listen again and check.

VOCABULARY AND SPEAKING

1 Look at the words in the box. Which of the following
things can you and can't you do in the situations?

– read a newspaper – wear shoes – wear a hat
– go swimming – sunbathe – watch TV
– go for a walk – cross the road
– ride a motorbike – wear jeans – listen to music
– smoke – eat – drink – go to sleep – talk
– sing – take your dog – feed the animals

> escalator lift hotel bar petrol station church mosque
> hospital level crossing office shop beach mountain
> canal library stadium traffic lights pedestrian crossing
> park museum prison aeroplane bus train zoo

2 Talk about other rules or customs in your country?
*In my country, children can go into a bar but they
can't drink alcohol.*

29 | *Warning: flying is bad for your health*

Should **and** *shouldn't for advice*

READING

1 Look at the title of the passage. Why do you think flying may be bad for your health?

2 Read *Warning: flying is bad for your health* and find out if it is safe for you to fly.

3 Write down the reasons why flying is bad for people's health.
Less oxygen, changes of pressure…

Flying is the safest way to travel … or is it? Some doctors think the aeroplane is a dangerous place, especially for the old or the unhealthy.

Although the aeroplane is pressurised, there is less oxygen than on the ground. Anyone with heart disease or a lung problem notices the difference much sooner. Even healthy people find it difficult to concentrate after hours of breathing less oxygen than usual. So anyone who has had a heart attack should not fly for at least two weeks after the attack. After an operation, you should stay on the ground for at least ten days.

Because of changes of pressure, pregnant women shouldn't take a flight lasting more than four hours after their thirty-fifth week or a shorter flight after 36 weeks. People with bad colds will probably get earache during take-off and landing.

Even if you feel well when you get on the plane, you will possibly feel ill when you get off. Sitting on a plane for many hours – especially in economy class where there isn't very much leg room – gives everyone aches and pains, so you should take some exercise, especially on long flights.

Most of the air you breathe is recycled so you will possibly catch a cold or flu from one of the other passengers.

Flying also causes dehydration. If you drink or eat too much, you'll wake up feeling ill. Everyone needs to drink more in the air, but you shouldn't drink alcohol because it makes you even more thirsty.

The most common problem is jet lag. But there isn't much you can do to prevent it. You should change to your new time zone as soon as possible and you shouldn't sleep if it's still daylight.

Crowded airports, long queues and delays cause stress and high blood pressure. So, be careful! Flying is the safest way to travel, but is it the healthiest?

Adapted from *The Independent on Sunday*

29

GENERAL COMMENTS

Should

Like *must, should* can be used to talk about obligation, particularly a moral obligation, eg *I must ring my mother. I should tell her my news.* But *should* is slightly less forceful. You use *should* to make suggestions, give advice and talk about one's duty, perhaps in a subjective way, eg *You really shouldn't work so hard. You should spend more time with your family.*

Vocabulary load per lesson

It is important for the students to realise that in real-life language situations, they will have to confront many items of vocabulary which they don't understand. In the classroom, a certain amount of vocabulary needs to be explained in order to gain access to the main ideas of a passage or to perform a given task. But it is vital not to use a reading passage as a resource for vocabulary building. Students must be encouraged to accept that some of the gaps in their understanding will not be filled by the dictionary or by your own explanations. Try to limit the number of words you explain or the students write down in a lesson. Most students will have difficulty in retaining more than ten or twelve new words per lesson. In pre-intermediate *Reward* the vocabulary boxes focus on about twenty words per lesson, but it would be difficult to retain all of them in one go.

Photo: possible questions

What can you see in the aeroplane?
What kind of aeroplane is it?
Who do you think the people are?
Where do you think they're going?
Do you like flying?
What do you like most/least about flying?
What's your favourite form of transport?

READING

1 Aim: to prepare for reading.

● Ask the students to think about the title of the passage and to predict why flying may be bad for your health.

● Ask the students to predict any words that are likely to be in the passage.

● You may decide to pre-teach some difficult items of vocabulary before they read. Limit them to no more than six or seven.

2 Aim: to read for specific information.

● The task is for the students to read the passage and find out how the information concerns them personally.

3 Aim: to read for specific information.

● Ask the students to re-read the passage and to focus on all the reasons why flying is bad for people's health. You may like to do this activity orally with the whole class and write a list of the reasons on the board.

Answers
less oxygen, changes of pressure, sitting still, recycled air, dehydration, jet lag, stress and high blood pressure

GRAMMAR

1 Aim: to practise giving advice.
- Ask the students to read the information about *should* and *shouldn't* in the grammar box.

- Ask them to do the exercise.

> **Answers**
> 1 You should take some aspirin.
> 2 You should see a doctor.
> 3 You should go to the dentist.
> 4 You should take some medicine.
> 5 You should take some exercise.
> 6 You should lie down.

2 Aim: to practise giving advice.
- The students will have to re-read the passage again in order to get the exact answers for this task.

> **Answers**
> People who have recently had an operation shouldn't fly for ten days.
> Pregnant women shouldn't take long flights after their thirty-fifth week.
> Pregnant women shouldn't fly at all after their thirty-sixth week.
> You should take some exercise during the flight.
> You shouldn't eat or drink too much.
> You shouldn't drink alcohol.
> You should change to your new time zone as soon as possible.
> You shouldn't sleep if it's still daylight.

VOCABULARY

1 Aim: to focus on the words in the vocabulary box; to practise giving advice.
- Make sure the students understand that *to be sick* in British English means *to vomit* and in American English *to be ill.*

> **Answers**
> If you are ill, you should go to the doctor.
> If you are sick, you should lie down and you shouldn't eat anything.
> If you are sore, you should rest or get some medicine.
> If you are thirsty, you should have something to drink.
> If you are tired, you should go to bed.

2 Aim: to focus on the words in the vocabulary box.
- Ask the students to group the words under the headings.

> **Answers**
> **parts of the body:** arm, ear, head, heart, leg, lung, nose, stomach, throat
> **complaints:** ache, blood pressure, cold, disease, flu, hangover, insect bite, jet lag, pain, sunstroke, temperature
> **Words which go together:** earache, headache, stomach ache, heart disease, lung disease

- You may like to revise some of the vocabulary presented in Lesson 21 on parts of the body.

LISTENING

1 Aim: to prepare for listening.
- The listening passage is quite complex and dense, so it is a good idea to help the students predict as much of the passage as possible, and then listen in order to check their predictions.

2 Aim: to listen for specific information.
- ▭ Play the tape and ask students to check their answers to activity 1.

> **Answers**
>
> | complaint | advice |
> | --- | --- |
> | jet lag | Drink lots of water. |
> | | Try to stay awake. |
> | stomach upsets | Don't eat uncooked food. |
> | | Drink only bottled or boiled water. |
> | insect bites | Take an insect repellent. |
> | | Cover your arms and legs. |
> | sunstroke | Wear a hat. |
> | | Don't spend all day in the sun. |

3 Aim: to practise speaking and to listen for specific information.
- Ask the students to try to remember as much detail as possible about the doctor's advice. Ask the following questions and elicit suitable answers:
> *Why should you drink lots of water?*
> *Why should you try to stay awake?*
> *Why shouldn't you eat uncooked food?*
> *Why should you drink only bottled or boiled water?*
> *Why should you take an insect repellent?*
> *Why should you cover your arms and legs?*
> *Why should you wear a hat?*
> *Why shouldn't you spend all day in the sun?*

- ▭ Play the tape again and ask them to listen and check.

SPEAKING

Aim: to practise speaking.
- Ask the students to think about dangers to health in their country and to make a list of things you should or shouldn't do. They may like to discuss their advice, or write it down for homework.

GRAMMAR

> *Should* and *shouldn't* for advice
>
> *Should* **is a modal verb. It has the same form for all
> persons and you don't use the auxiliary** *do* **in
> questions and negatives. You use** *should* **+ infinitive
> to give advice.**
>
> You **should** *take some exercise on long flights.*
> You **shouldn't** *drink alcohol.*
>
> | **Questions** | **Short answers** |
> |---|---|
> | *Should* I take some exercise? | Yes, you *should*. |
> | *Should* I drink alcohol? | No, you *shouldn't*. |
>
> *Should(n't)* **and** *ought(n't) to* **mean the same.**
> You **ought to** *take some exercise on long flights.*
> You **oughtn't to** *drink alcohol.*

1 Give advice to these people. Use these words and
phrases.

see a doctor   go to the dentist   take some aspirin
lie down   take some exercise   take some medicine

1 I've got a headache.
2 My arm hurts.
3 I've got toothache.
4 I've got a sore throat.
5 My legs are stiff.
6 I've got a temperature.

2 Look at the article again and give advice to people
who travel by air.

People with heart disease should be careful.

VOCABULARY

1 Work in pairs. Say what you should do if you are:

– ill   – sick   – sore   – thirsty   – tired

2 Look at the words and expressions in the box and write
them under two headings: *parts of the body* and
complaints.

ache   arm   high blood pressure   cold   disease   ear   flu
hangover   head   heart   insect bite   jet lag   leg   lung
nose   pain   stomach   sunstroke   temperature   throat

Which words go together to make a complaint?
headache...

LISTENING

1 You are going to hear a doctor talk about what you
should and shouldn't do to stay healthy when you
travel. Match the complaints with the advice.

complaint	advice
jet lag	Drink lots of water.
	Don't spend all day in the sun.
stomach upsets	Take an insect repellent.
	Cover your arms and legs.
insect bites	Wear a hat.
	Drink only bottled or boiled water.
sunstroke	Don't eat uncooked food.
	Try to stay awake.

2 🔊 Listen to the doctor and check if you were right.

3 Try and remember any other advice the doctor gives.

🔊 Listen again and check.

SPEAKING

Work in pairs. Talk about possible dangers to your
health in your country. Think about the following and
say what you should or shouldn't do to stay healthy.

– the environment   – the climate   – animals
– the livestyle   – dangerous sports

*You shouldn't go skiing on your own when it's snowing
heavily.*

Asking for permission; asking people to do things; offering

1 **You're a guest in someone's home. You'd like a cigarette. What do you say?**
a 'Is it all right if I smoke?'
b 'Would you like a cigarette?'
c Nothing and light up.

2 **A friend suggests you have dinner together in a certain restaurant. At the end of the meal, the waiter brings the check. What do you say?**
a 'I'll pay this.'
b 'Shall we share this?'
c Nothing. Your friend suggested dinner, and you expect him to pay.

3 **You're visiting a friend when the phone rings. What do you expect her to say to the caller?**
a 'It's great to hear from you. Hold on while I get a chair.'
b 'Would you mind if I called you back? I've got a visitor here at the moment.'
c Nothing. It's rude to pick up the phone when you've got guests.

4 **It's late and your neighbours are playing very loud music. What do you say to them?**
a 'Turn down the music!'
b 'Could you turn the music down, please?'
c Nothing. You call the police.

5 **You meet a Ms Esther Craig for the first time. You don't know how to address her. What do you say?**
a 'What do I call you?'
b 'Can I call you Esther?'
c 'Would you mind telling me what to call you?'

6 **You meet someone at a party and get on very well. As she leaves, she says, 'Nice meeting you. We must do lunch sometime.' What do you say to her?**
a 'Great! Shall we make a date?'
b 'Would you mind giving me your phone number?'
c 'That's a great idea. Bye!'

7 **You'd like your friend to lend you a book. What do you say?**
a 'Lend it to me, will you?'
b 'It's much too expensive for me to buy.'
c 'Would you mind lending it to me?'

8 **You're at the information office at the railway station and you want to know some train times. What do you say?**
a 'When's the next train to New York?'
b 'Tell me when the next train to New York is!'
c 'I wonder if you could tell me when the next train to New York is.'

9 **Your host serves you food you don't like. You eat it, but then the host offers you more. What do you say?**
a 'It was very nice, but no thank you. I've had enough.'
b 'Yes, would you mind if I only had a little?'
c 'No way. What was that, anyway?'

10 **As you're leaving a shop the assistant says, 'Have a nice day!' What do you say?**
a 'Thank you. Same to you. Bye.'
b 'Have a nice day yourself!'
c 'No, thanks. I've made other arrangements.'

70

30

GENERAL COMMENTS

Asking for permission; asking people to do things; offering

The language functions in this lesson use many of the modal verbs and other structures which have already been presented in pre-intermediate *Reward*. The grammar will, therefore, be largely revision but the language functions may be new. For some students, it is a common mistake to use an infinitive after a modal. Remind them, if necessary, that modal verbs do not take an infinitive.

Cross-cultural training

The aim of the questionnaire is not to encourage students to adopt British or American customs, but to focus on the students' own way of doing things, to compare this with customs and behaviour in other countries, and to draw attention to any differences, particularly anything which may cause offence or be perceived as offensive. Even in a mono-cultural group, there are likely to be some different answers to these questions, so take the opportunity of reflecting on their own culture, customs and behaviour. The subject is likely to generate a great deal of discussion.

Illustration: possible questions

What can you see in the drawing?
Where are the people? In a restaurant or at someone's house?
Who do you think the two people are? Which is the host and which is the guest?
What are the people doing?

READING AND LISTENING

1 Aim: to read and answer a questionnaire.

● Ask the students to read the questionnaire and answer the questions. Tell the students that the questionnaire is in American English, and includes words like *check* (Question 2) meaning *bill* in British English.

● There are a number of aspects of behaviour and customs which need commenting on:
1 In some countries nowadays, people have very strong feelings against smoking. In Britain and America, smokers are much less common than they used to be, and the unwritten rules about smoking are particularly strong when you're a guest in someone's home. It is a situation in which the guest is expected to conform to the rules of the house, not the other way round.
2 It is sometimes difficult to decide who pays for a restaurant bill. The students should simply be aware that the convention over paying may be different from what they expect.
3 In certain countries, a telephone caller can take precedence over a visitor, who may be offended by having to wait to regain the host's attention.
4 Relations between neighbours can vary from neighbourhood to neighbourhood as well as from country to country.

5 In certain countries, such as Britain and America, it's very common to use first names even with people you don't know very well. In other countries, first names are only used between members of the same family.
6 Invitations may not always be meant as a definite arrangement to get together again, but more as a vague desire to meet again in the future. This is especially the case with the invitation in this question.
7 In Britain and America, the politeness formulae are very important, so the more elaborate request is more suitable.
8 In Britain it is important to be polite even in transactional situations. Other countries may consider it unnecessary or even insincere to use politeness formulae in such circumstances.
9 Once again, it is important not to be rude in this situation in many countries.
10 The parting salutation *Have a nice day* is primarily American usage, and British English has no real equivalent.

2 Aim: to listen for main ideas.

● 🔊 Play the tape and ask the students to listen and find out how the two speakers answer the questionnaire.

> **Answers**
> 1a 2b 3b 4b 5b 6b 7c 8a 9a 10a

3 Aim: to practise speaking and to listen for specific information.

● Ask the students to think about what the speakers said for each answer. Do this activity orally with the whole of the class.

● 🔊 Play the tape again for the students to check if they remembered correctly.

4 Aim: to practise speaking and to compare different customs.

● Ask the students to work together and to compare their answers with what they heard on the tape.

FUNCTIONS

1 Aim: to focus on polite questions.

● Ask the students to read the information in the functions box about asking for permission and asking people to do things.

● Ask the students to do the exercise.

> **Answers**
> 1 Would you mind if I left early?
> 2 Would you mind helping me?
> 3 I wonder if you can tell me where the station is.
> 4 Could you tell me what your phone number is?
> 5 Is it all right if I open the window?

2 Aim: to focus on polite replies.

> **Possible answers**
> 1 Yes, of course.
> 2 No, not at all.
> 3 I'm sorry, I'm afraid I can't.
> 4 Yes, of course.
> 5 I'm sorry, I'm afraid not.

SOUNDS

Aim: to focus on polite intonation.

● 📼 Explain that intonation can also be used to make you sound polite in English. Ask the students to listen and decide if the speaker sounds polite or rude.

> **Answers**
> 1 Is it all right if I smoke? [polite]
> 2 Could you turn the music down, please? [rude]
> 3 Would you mind if I called you back? [polite]
> 4 Would you mind telling me what to call you? [rude]
> 5 Would you mind giving me your phone number? [polite]

● Ask the students to say the sentences politely.

VOCABULARY AND SPEAKING

1 Aim: to present the words in the vocabulary box.

● Explain that the words come from this lesson and ask the students to remember which words they went with.

2 Aim: to use the new words in this lesson.

● Ask the students to go round the class asking for permission to do things or asking people to do things. Make sure they use expressions from the functions box.

T 71

READING AND LISTENING

1 Read the questionnaire *Doing things the right way*. What do you do or say in your country?

2 🔲 Listen to two Americans talking about the questionnaire and tick (✓) their answers.

3 Work in pairs. Try to remember what the speakers said about each possible answer.

🔲 Listen again and check.

4 Compare your answers to the questionnaire with the speakers' answers. Do you do things the same way?

FUNCTIONS

Asking for permission	Asking people to do things
Can *I smoke?*	**Can** *you speak louder, please?*
Could *I leave now?*	**Could** *you help me?*
May *I call you Esther?*	**Could** *you tell me* **where** *the station is?*
Is it all right if *I smoke?*	**I wonder if** *you could help me.*
Would you mind if *I borrowed it?*	**Would you mind** *lending it to me.*

Could is a little more formal than **can**. **May** is very formal.

Agreeing	Refusing
Yes, of course. Yes, go ahead.	*(I'm sorry,) I'm afraid I can't/ you can't.*
By all means.	*(I'm sorry,) I'm afraid not.*

In response to *Would you mind...?*

No, of course not. No, go ahead.	*(I'm sorry,) I'm afraid I do (mind).*

Offering

	Accepting	Refusing
Shall *I do that?*	*That's very kind of you.*	*No, it's all right, thank you.*
I'll do that, shall I?	*Thank you.*	*No, I'll do it.*

1 Rewrite these questions using the phrase in brackets.

1 Can I leave early. (Would you mind...?)
2 Can you help me? (Would you mind...?)
3 Where's the station? (I wonder if...?)
4 What's your phone number? (Could you tell me...?)
5 Can I open the window? (Is it all right if...?)

1 Would you mind if I left early?

2 Write suitable responses to the questions in 2.

1 Agree.   2 Agree.   3 Refuse.   4 Agree.   5 Refuse.

1 Yes, of course.

SOUNDS

🔲 Listen to these sentences. Does the speaker sound rude or polite?

1 Is it all right if I smoke?
2 Could you turn the music down, please?
3 Would you mind if I called you back?
4 Would you mind telling me what to call you?
5 Would you mind giving me your phone number?

Now say the sentences aloud. Try to sound polite.

VOCABULARY AND SPEAKING

1 Here are some new words from this lesson. Check you know what they mean. (You can use a dictionary.) Can you remember the sentence where you saw them?

share   hold on   neighbour turn down   get on with someone date   phone number   lend borrow   rude   polite   formal informal   manners

2 Go round the class asking for permission to do things or asking people to do things. Be polite.
Is it all right if I open the window?
Yes, of course.

Progress check 26-30

VOCABULARY

1 Some adjectives and nouns often go together. For example:

high temperature   red light   sore throat

Match the adjectives in box A with the nouns in box B. Some adjectives can go with more than one noun.

A	alcoholic bad duty-free first-class heavy high-speed narrow next-door steep

B	cigarettes cold drink hill manners neighbour path suitcase ticket train

Think of other nouns which the adjectives often go with, and other adjectives which the nouns often go with.

2 Look at these words and find out:

– how many meanings they have
– how many parts of speech they can be.

head   complaint   cold   back   neck   ache   sore   arm

3 Complete these sentences with the words in 2. There are some extra words.

1 He held the bottle by the ___ .
2 I'm very angry. I'd like to make a ___ .
3 I'm tired. Let's go ___ home.
4 I feel ___ . Put the heating on.
5 She was the ___ of the department.
6 My legs ___ .

4 Here are some strategies you can use when you are reading and you come across a word you don't understand.

– Decide what part of speech the word is.
– Decide if the word is important to the general sense of the passage.
– Try and guess the general sense of the word.
– Read on and confirm or revise your guess.

5 It is useful to learn new words by associating them with other words. For example:

sunstroke – temperature, headache, lie down
thirsty – drink, water, hungry

Look at the vocabulary boxes in lessons 26 – 30 again. Choose words which are useful to you and group them under headings of your choice in your *Wordbank*. Then think of two or three words that you associate with each one.

GRAMMAR

1 Complete these sentences with *must* or *mustn't*.

1 You ___ wear a warm coat, because it's cold.
2 You ___ be rude to your teacher.
3 You ___ stand too close to the edge of the platform.
4 You ___ be late for an important meeting.
5 You ___ drive slowly, as I'm very nervous.
6 You ___ make too much noise in a library.
7 You ___ do your homework every day.
8 You ___ feed the animals in the zoo.

2 Write sentences describing three things you must do in your English classes and three things you mustn't do.

You must speak English.

3 Write five things you couldn't do when you were ten years old but you can do now.

I couldn't speak English, but I can now.

4 Say if you can or can't do these things when you fly.

Can you...

1 ...arrive at the airport just before the plane leaves?
2 ...stand up during take-off?
3 ...smoke during landing?
4 ...have a drink during the flight?
5 ...visit the pilot?
6 ...listen to your personal stereo?
7 ...use a mobile phone?
8 ...buy a ticket at the airport?

1 No, you can't.

Progress check 26–30

GENERAL COMMENTS

You can work through this Progress check in the order shown, or concentrate on areas which may have caused difficulty in Lessons 26 to 30. You can also let the students choose the activities which they would like to or feel the need to do.

Photo

You may like to explain that the photo is of a sign on an old bridge in Dorset, in the south of England, which threatens to punish anyone damaging the bridge by sending them to Australia, where the British government used to send its prisoners.

VOCABULARY

1 Aim: to focus on collocations.

● The collocations in this activity have all occurred in Lessons 26 to 30. Ask the students to match the adjectives and the nouns.

> **Answers**
> alcoholic drink, bad cold, bad manners, duty-free drink, first-class ticket, heavy cold, heavy suitcase, high-speed train, narrow path, next-door neighbour, steep path, steep hill

● Ask the students to use a dictionary and to find other collocations with these adjectives and nouns.

2 Aim: to focus on headwords and meanings.

● Explain that a word may have several different meanings as well as being different parts of speech. There are likely to be different numbers of meanings depending on the dictionary you use. Ask the students if they know what meanings the words have and if they can be different parts of speech.

● Ask the students to look the words up in their dictionaries and count the number of meanings and check the parts of speech.

3 Aim: to use some of the meanings of the words in activity 2.

> **Answers**
> 1 He held the bottle by the **neck.**
> 2 I'm very angry. I'd like to make a **complaint.**
> 3 I'm tired. Let's go **back** home.
> 4 I feel **cold.** Put the heating on.
> 5 She was the **head** of the department.
> 6 My legs **ache.**

4 Aim: to focus on strategies for dealing with unfamiliar words.

● Students need to develop the strategies for dealing with unfamiliar vocabulary in the relatively reassuring context of the classroom. Ask them to read through these strategies and to use them when they are reading a passage with some unfamiliar vocabulary.

5 Aim: to organise vocabulary records in the Wordbank.

● It may be necessary to remind your students that they need to record new vocabulary not just once but in a number of different categories in order to help the process of acquisition.

GRAMMAR

1 Aim: to revise *must* and *mustn't*.

> **Answers**
> 1 You **must** wear a warm coat, because it's cold.
> 2 You **mustn't** be rude to your teacher.
> 3 You **mustn't** stand too close to the edge of the platform.
> 4 You **mustn't** be late for an important meeting.
> 5 You **must** drive slowly, as I'm very nervous.
> 6 You **mustn't** make too much noise in a library.
> 7 You **must** do your homework every day.
> 8 You **mustn't** feed the animals in the zoo.

2 Aim: to revise *must* and *mustn't*.

● You may like to remind your students that *must* in this context is a strong suggestion rather than a rule.

3 Aim: to revise *couldn't* and *can*.

● Remind the students that the use of *can* and *could* here is to talk about a general ability.

4 Aim: to revise *can* and *can't*.

● This use of *can* and *can't* is to talk about permission and prohibition.

> **Answers**
> 1 No, you can't.
> 2 No, you can't.
> 3 No, you can't.
> 4 Yes, you can.
> 5 Yes, you can.
> 6 Yes, you can.
> 7 No, you can't.
> 8 Yes, you can.

5 Aim: to revise *should(n't)* and *ought(n't)*.

Possible answers
1 You should go to bed early.
2 You ought to play tennis or go swimming.
3 You shouldn't stay in bed all day.
4 You oughtn't to play football at your age.
5 You should learn to speak her language.
6 You ought to go to Britain or America.
7 You should join a club.
8 You oughtn't to work so much.

6 Aim: to revise *can, should, have to* and *can't*.

Answers
1 You **can't** park a car on the pavement.
2 You **can** cross the road at traffic lights.
3 He **should** wear a tie.
4 You **have to** pass your driving test.
5 You **can't** drink and drive.
6 She **should** see a doctor.
7 Men **have to** do military service in many countries.
8 You **can** go to night-clubs when you're sixteen.

7 Aim: to revise making polite requests.
● There are many other possible answers which you may like to discuss with the students.

Possible answers
1 Could you turn it off, please?
2 Would you mind if I smoked?
3 I wonder if you could help me, please.
4 Could you tell me where the police station is?
5 Is it all right if I borrow your book?
6 Would you mind lending it to me?
7 I wonder if you could tell me what the date is, please.
8 Can you speak up, please?

SOUNDS

1 Aim: to focus on /əʊ/ and /ɔɪ/.
● /əʊ/ is a difficult sound for many language students, and at this stage it isn't necessary to get it absolutely right.

Answers
/əʊ/: wrote vote know only telephone home photo
/ɔɪ/: boy noise royal unemployment

● 📼 Play the tape and pause after each word. Ask the students to say the words aloud.

2 Aim: to focus on /ʃ/, /tʃ/, and /dʒ/.
● Ask the students to put the words in three columns. You can do this activity orally with the whole class.

Answers
/ʃ/: shoe station pressure situation
/tʃ/: teacher temperature
/dʒ/: oxygen passengers stranger

● Draw their attention to the different spellings for the same sound.

● 📼 Play the tape again and pause after every word. Ask the students to repeat each word.

3 Aim: to focus on polite and friendly intonation.
● 📼 Ask the students to say these sentences in a polite and friendly way. Then ask them to say them in a rude and unfriendly way.

● Ask one or two students to say one of the sentences either in a polite and friendly way or in a rude and unfriendly way. The rest of class should decide which he/she intended.

SPEAKING AND WRITING

1 Aim: to revise the language presented in lessons 26 to 30 and to practise speaking.
● Write the prompts on the board and ask the students to suggest some advice for foreign visitors to their country.

● Ask the students to work in small groups and to prepare some more advice. They should write this down.

2 Aim: to revise the language presented in lessons 26 to 30 and to practise writing.
● Ask the students in their groups to write some rules that foreign visitors should be aware of.

● They can pass their rules to other groups and receive a new set of rules if they all come from the same country.

3 Aim: to practise speaking.
● Ask the groups in turn to talk about their advice and rules for visitors to their country. If they all come from the same country, choose the five most useful pieces of advice and the five most important rules. If they all come from different countries, choose the most surprising rules and advice for each country.

5 Reply to these people and give advice. Use *should(n't)* and *ought(n't) to* in turn.

1 I feel so tired.
2 I don't get much exercise.
3 I stay in bed all day on Sundays.
4 I've hurt my leg playing football.
5 My wife doesn't understand me.
6 I need to learn English quickly.
7 I don't have many friends.
8 I spend twelve hours a day at work.

1 You should go to bed early tonight.

6 Rewrite these sentences with *can, should, have to* or *can't*.

1 You aren't allowed to park a car on the pavement.
2 You are allowed to cross the road at traffic lights.
3 He ought to wear a tie.
4 You are obliged to pass your driving test.
5 You aren't allowed to drink and drive.
6 She ought to see a doctor.
7 Men are obliged to do military service in many countries.
8 You are allowed to go to night-clubs when you're sixteen.

1 You mustn't park a car on the pavement.

7 Rewrite these requests more politely.

1 Turn it off!
2 I want to smoke.
3 Help!
4 Where's the police station?
5 I want to borrow your book.
6 Lend it to me!
7 What's the date?
8 Speak up!

SOUNDS

1 Say these words aloud.

wr<u>o</u>te   v<u>o</u>te   kn<u>o</u>w   b<u>oy</u>   <u>o</u>nly   teleph<u>o</u>ne
h<u>o</u>me   n<u>oi</u>se   r<u>oy</u>al   ph<u>o</u>to   unempl<u>oy</u>ment

Is the underlined sound /əʊ/ or /ɔɪ/ ? Put the words in two columns.

[cassette] Listen and check.

2 Say these words aloud.

<u>sh</u>oe   sta<u>ti</u>on   tea<u>ch</u>er   pre<u>ss</u>ure   o<u>x</u>ygen   pa<u>ss</u>engers   tempera<u>t</u>ure   situa<u>ti</u>on   stran<u>g</u>er

Is the underlined sound /ʃ/, /tʃ/ or /dʒ/? Put the words in three columns.

[cassette] Listen and check.

3 Say these sentences aloud. Try to sound polite and friendly.

1 Could you turn down the music, please?
2 Could you be quiet please?
3 You must fasten your seat belt.
4 I'm sorry, but you can't smoke in here.
5 Can I come in?
6 Where's the nearest bank?

[cassette] Listen and check.

SPEAKING AND WRITING

1 Work in groups of two or three. Write some advice for foreign visitors to your country. Think about the following:

– where to go   – where not to go   – what to see   – what to buy
– what to do   – what to wear   – what to drink   – what to eat

2 Write some rules for foreign visitors to your country. Think about the following.

– what you must be careful of
– what you mustn't bring into the country
– what you must wear in certain places
– what you mustn't photograph
– what you must do when you enter someone's home
– what you mustn't do in the streets

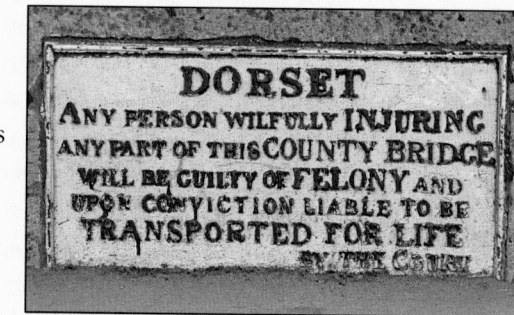

3 Compare your advice and rules with those of other groups. Choose the five most useful pieces of advice and the five most important rules.

31 *My strangest dream*

Past continuous (1) for interrupted actions: *when*

LISTENING AND SPEAKING

1 Work in pairs. You're going to hear an English woman talking about her strangest dream, which she called *The day the Queen came to tea*. First of all, think about these questions.

– How does the Queen of England usually travel?
– Do you think she does her own shopping?
– Do you think she ever visits people in their own homes?
– Does she ever do anything which is unexpected?

2 Look at the words below which come from the first part of the dream. What do you think it's about? (The words are in the right order).

sit fire knock front door surprised Queen crown
bag shops come in front room shopping tired
cup tea chocolate biscuits curtain material castle
redecorate burn down

3 🔈 Listen to the first part of the dream and tick (✓) the words as you hear them. Did you guess correctly in 2?

4 Here are some extracts from the second part of the dream. (They are in the wrong order.) What do you think happens next in the dream?

a '... and now she's going home.' ☐
b ...waved to me like she does on television. ☐
c It was Prince Philip. ☐
d Suddenly the phone rang. ☐
e 'Would you like a lift home?' ☐
f ... my husband arrived home from work. ☐
g '...I'll get the bus.' ☐
h and we carried on talking just as if we were old friends... ☐
i 'I'll be back at about five o'clock...' ☐
j 'Hello, Queen. Pleased to meet you.' ☐

5 🔈 Listen to the second part and number the extracts as you hear them. Did you guess correctly in 4?

GRAMMAR

> **Past continuous (1) for interrupted actions:** *when*
> **You form the past continuous with** *was/were* **+ present participle (verb + *-ing*).**
>
> *I **was** watch**ing** television.*
> *We **were** talk**ing** about the weather.*
>
> **You use the past continuous to talk about something that was in progress at a specific time in the past.**
>
> *What were you doing at nine o'clock yesterday morning?*
> *I was going to work.*
>
> **You also use the past continuous tense to talk about something that was in progress at the time something else happened or interrupted it. You join the two parts of the sentence with** *when*. **The verb in the** *when* **clause is usually in the past simple.**
>
> *I **was watching** television **when** there **was** a knock at the door.*
> *We **were talking** about the weather **when** the phone **rang**.*
>
> **Remember that you don't usually use these verbs in the continuous tenses:**
> *believe feel hear know like see smell sound taste think understand want*

74

31

GENERAL COMMENTS

Past continuous

The students' native languages may have a past continuous tense, but it may used in a different way to the English past continuous. A typical mistake would be to use the past continuous for more permanent actions, eg *She was spending her childhood in London,* or to use the past simple for background events instead of the past continuous, eg *I watched television when there was a knock at the door.* To explain the 'interrupted action' use of the past continuous, mentioned in the lower part of the grammar box, a time line might be useful.

I was watching television

↑

when there was a knock at the door

Illustration: possible questions

Who can you see in the picture?
Who is the person on the left?
Where are they?
What's happening?
Do you think this sort of event happens very often?

LISTENING AND SPEAKING

1 Aim: to prepare for listening.

● Explain that the passage the students are about to listen to is a lighthearted account of a dream. You may like to tell them it has been estimated that 80 per cent of English people admit to having had a dream about the Queen. Ask them to discuss their answers to the questions with you, and to think about her lifestyle.

2 Aim: to prepare for listening to the first part of the passage, and to pre-teach some important items of vocabulary from the listening passage.

● Explain that the passage is in two parts, and you are going to play the first part. The words in the list come from the first part. Ask the students to discuss the words in groups of three or four and to guess what the dream is about.

● Ask each group to tell the rest of the class their version of the dream. This activity can take about ten minutes to complete. The process of predicting the dream is important preparation for listening, so allow plenty of time.

3 Aim: to listen for main ideas.

● You may want to give the students some background information. Write on the board *Windsor Castle, Harrods, Philip* and elicit as much information about the words as you can.
For your information:
- The woman is talking about a dream she had in November 1992, shortly after Windsor Castle was badly damaged in a fire.
- Harrods is a very expensive shop in London.
- Prince Philip is the Queen's husband.

● 📼 Ask the students to listen to the dream and check if their predictions in activity 2 were correct. Play the tape.

● Ask the students to re-tell the story to check that everyone has understood what happened.

4 Aim: to prepare for listening to the last part of the passage.

● Ask the students to continue working in their groups and to try to predict what happens from the extracts. Make it quite clear that the extracts are in the wrong order.

5 Aim: to listen for specific information.

● 📼 Ask the students to listen to the last part and number the extracts as the woman says them. Play the tape.

Answers	
a '...and now she's going home.'	[6]
b '...waved to me like she does on television.'	[10]
c 'It was Prince Philip.'	[3]
d 'Suddenly the phone rang.'	[2]
e 'Would you like a lift home?'	[8]
f '...my husband arrived home from work.'	[5]
g '...I'll get the bus.'	[9]
h 'and we carried on talking just as if we were old friends...'	[1]
i 'I'll be back at about five o'clock...'	[4]
j 'Hello, Queen. Pleased to meet you.'	[7]

GRAMMAR

1 Aim: to join two sentences with *when*.
● Ask the students to read the information about the past continuous in the grammar box.

Answers
1 I was sitting in front of the fire...
   d ...when there was a knock at the door.
2 I was getting the tea ready...
   c ...when she called out to me.
3 She was looking for some curtains...
   e ...when she saw this lovely material.
4 She was finishing her tea...
   a ...when my husband arrived home from work.
5 She was going down the garden path...
   b ...when she stopped and waved.

2 Aim: to practise using the past continuous.
● Ask the students to work in pairs and ask and say what they were doing at the six times yesterday. Go round and check they are using the past continuous.

3 Aim: to focus on the difference between the past continuous and the past simple.

Answers
1 I understood what she said.
2 She was having dinner when the phone rang.
3 It was raining when he got into his car.
4 He made me a sandwich because I was hungry.

VOCABULARY AND WRITING

1 Aim: to focus on verbs and prepositions.
● This activity focuses for the first time on multi-part verbs. However, a more complete explanation of how they are used is to be found in Progress check lessons 31–35.

Answers
call out, come in, get out, look for, pick up, put up, show into, stand up, turn to, wave to

● There are a large number of verb + preposition combinations. You may like to write these common combinations on the board:
be up, call in, get into, go into, go out, look at, pick out, put into, show up, stand on, turn into, wave at

● Ask the students to write sentences using verb + preposition combinations. Make sure they put the noun/pronoun in the right position. If they don't, explain that there will be more work on verb + preposition combinations in the Progress check. It is not the intention to cover multi-part verbs fully in this lesson.

2 Aim: to focus on compound nouns and adjective + noun combinations.
● Ask the students to try to remember the combinations they heard in the passage.

Answers
chocolate biscuit, tea cup, front door, garden path, front room
Other possible combinations: front garden, tea room, front path, garden room

3 Aim: to present the words in the vocabulary box.
● These words will not present too many problems of comprehension, but the use of the comma after them is important.

Possible answers
1 We had a wonderful holiday. **Unfortunately,** we had to go back to work.
2 It was late at night and I was asleep. **Suddenly,** I heard a strange noise.
3 The guests stayed very late. **Finally,** they said goodbye and left.
4 There was a car accident. **Fortunately,** no one was hurt.
5 I lost my wallet. **To my surprise,** someone found it and gave it back.

4 Aim: to practise writing, using the past continuous and the words in activity 3.
● Ask the students to rewrite the dream, using the past continuous, the words and extracts earlier in the lesson, and the words in activity 3.

● If time is short, this would be a good activity for homework.

5 Aim: to practise speaking or writing.
● This activity is designed to give students the opportunity to write or talk about their dreams.

1 Join the two parts of the sentence from the dream with *when*.

1 I was sitting in front of the fire...
2 I was getting the tea ready...
3 She was looking for some curtains...
4 She was finishing her tea...
5 She was going down the garden path...

a my husband arrived home from work.
b she stopped and waved.
c she called out to me.
d there was a knock at the door.
e she saw this lovely material.

2 Work in pairs. Ask and say what you were doing at these times yesterday.

1 7.15am   2 8.35am   3 9.30am   4 1pm   5 5.15am   6 7.45pm

What were you doing at seven fifteen yesterday morning?
I was getting up. What were you doing?
I was having breakfast.

3 Choose the best verb.

1 I *understood/was understanding* what she said.
2 She *had/was having* dinner when the phone *rang/was ringing*.
3 It *rained/was raining* when he *got/was getting* into his car.
4 He *made/was making* me a sandwich because I was hungry.

VOCABULARY AND WRITING

1 In the dream you heard there were a number of verbs followed by a preposition. Match the verbs and the prepositions you heard. Try to remember the sentence you heard them in.

Verbs	be call come get go look pick put show stand
	turn wave
Prepositions	to on in in front of at for into out up

Are there other verb + preposition combinations above? Choose four or five and use them in sentences.

2 The dream also included some two-word nouns (compound nouns) or an adjective and a noun combination.
tea cup, front door

How many combinations did you hear with these words?

biscuit chocolate cup door front garden path room tea

Are there any other combinations you can make with the words in the box?

3 You often use these words and expressions in stories.

| suddenly   fortunately |
| unfortunately   finally |
| to my surprise |

Complete the sentences with the words or expressions.

1 We had a wonderful holiday. ___ we had to go back to work.
2 It was late at night and I was asleep. ___ I heard a strange noise.
3 The guests stayed very late. ___ they said goodbye and left.
4 There was a car accident. ___ no one was hurt.
5 I lost my wallet. ___ someone found it and gave it back.

4 Rewrite the dream of *The day the Queen came to tea*. Use the words and extracts in *Listening* activities 2 and 4, and some of the words and expressions in 3 above.
I was at home, watching television. Suddenly there was a knock at the front door.

5 Many dreams belong to one of the following common types.

– flying   – taking a test
– meeting someone important
– missing a train or plane
– giving a speech in public

Have you ever had one of these types of dream? Can you remember your strangest dream? Talk or write about it, using the past simple and past continuous tenses and some of the words or expressions in 3.

32 | *Time travellers*

Past continuous (2) : *while* and *when*

VOCABULARY AND READING

1 You're going to read a story called *Time travellers*. Here are some of the words from the story. Check you understand what they mean. What do you expect the story to be about?

> sightseeing grounds palace crowd favourite atmosphere strange shiver
> discover eighteenth century lose one's way cottage sculpture uneasy path
> evil angry disappear bridge lawn grass anniversary prison invade
> messenger warn guard

2 Read the story and find out why it's called *Time travellers*.

3 Work in pairs. What do you think happened to the two women?

4 Read the story again and decide where these sentences go. The sentences are in the order in which they appear in the story.

  a While they were visiting the Palace, one of the women had an idea.
  b While they were walking through the Palace grounds, the atmosphere suddenly changed.
  c While they were wondering which way to go, they saw a wooden hut under some trees.
  d While they walking past, she stared at them in a very royal way.
  e While she was relaxing at the Petit Trianon, an angry crowd of people from Paris invaded the Palace.

GRAMMAR

> **Past continuous (2):** *while* and *when*
> **You can use** *while* **+ past continuous to talk about something that was in progress at the time something else happened or interrupted it. You need a comma at the end of the** *while* **clause.**
> *While they **were visiting** the Palace, one of the women had an idea.*
>
> **You can also put the** *while* **clause at the end of the sentence. You don't need a comma.**
> *One of the women had an idea **while** they **were visiting** the Palace.*
>
> **You can also use** *when* **in the past simple clause.**
> *They were visiting the Palace **when** one of the women **had** an idea.*
>
> **You use** *when* **+ past simple to describe two things which happened one after the other. The second verb is often in the past simple.**
> *When the women **got** closer, they **saw** some people in eighteenth-century clothes.*
>
> **You can also say:**
> *The women **saw** some people in eighteenth-century clothes **when** they **got** closer.*

1 Write full answers to these questions.

  1 Who did they notice while they were walking to the Petit Trianon?
  2 What happened while the evil-looking man was staring at the two women?
  3 Who did they see while they were crossing the lawn in front of the Petit Trianon.

2 Rewrite your answers in 1 using *when* + past simple.

3 Write full answers to these questions.

  1 What did the two men do when they asked them the way?
  2 How did Miss Jourdain feel when she saw the woman and the young girl?
  3 What did they see when they crossed the bridge?
  4 What did they do when the tour was over?

SPEAKING

1 Work in pairs. Would you like to travel in time? What time would you like to travel to? Talk about the people you'd like to meet and things you'd like to do.

2 Talk about your time travelling with the rest of your class. Find someone who'd like to travel with you. Decide who has got the strangest reasons for travelling.

32
GENERAL COMMENTS

Past continuous (2): *while* and *when*

This is the second lesson on the past continuous and focuses on its use with *while* and *when*. A time line may be helpful again:

> **While they were visiting the Palace,**
>
> ↑
>
> **one of the women had an idea.**

> **They were visiting the Palace,**
>
> ↑
>
> **when one of the women had an idea.**

Time travellers

The passage describes a well-documented story of how two English women claimed to have been transported over 200 years back in time during a visit to Versailles in 1901. Elicit from the students information about Versailles, the Petit Trianon and Marie-Antoinette. For your information:
– Versailles is a palace outside Paris where King Louis XVI of France lived until he was executed in 1791 during the French Revolution. It has been a popular tourist attraction for many years.
– The Petit Trianon is a smaller building within the grounds of Versailles.
– Marie-Antoinette was Louis' wife and Queen.

Illustration: possible questions

Who can you see in the picture?
Where are they? What are they doing?
What period do you think they come from?
Could they be people from your country?

VOCABULARY AND READING

1 Aim: to pre-teach the words in the vocabulary box; to prepare for reading.
● The words in the box are of limited use for the student at pre-intermediate level. They are nevertheless important as items to be pre-taught, which will allow easier access to the passage. Ask the students to explain any items they know, and then explain the meaning of any they are unfamiliar with.

● Ask the students to guess what the passage might be about.

2 Aim: to read for main ideas.
● It is important for the students to understand the storyline, so this activity is a simple task to encourage reading for main ideas.

3 Aim: to check comprehension.
● When the students have finished reading, ask someone to re-tell the story to check everyone has understood.

● Ask the students to discuss what happened to the women.

4 Aim: to read and focus on text organisation; to present the target structure.
● Give the students plenty of time to do this activity, and then check their answers orally.

> **Answers**
> a ...and there were crowds of people. **While they were visiting the Palace, one of the women had an idea.** 'Why don't we walk to the Petit Trianon?'
> b ...Queen Marie-Antoinette's favourite places. **While they were walking through the Palace grounds, the atmosphere suddenly changed.** Suddenly everything felt strange.
> c While they were wondering which way to go, they saw a wooden hut under some trees. **...the path branched left and right. / An evil-looking man was sitting on the grass...**
> d ...sitting alone on the grass. **While they were walking past, she stared at them in a very royal way.** They went up the steps onto the terrace...
> e ...when Marie Antoinette went to prison. **While she was relaxing at the Petit Trianon, an angry crowd of people from Paris invaded the Palace.** Perhaps the man running to tell them...

GRAMMAR

1 Aim: to focus on interrupted actions with *while* + past continuous.

● Ask the students to read the information about the past continuous in the grammar box.

● Ask the students to do the exercise. For the most useful practice in manipulating this structure, make sure the students write full sentences.

Answers

1 While they were walking to the Petit Trianon, they noticed people wearing clothes from the eighteenth century.
2 While the evil-looking man was staring at the two women, a second man ran up to them and called out.
3 While they were crossing the lawn in front of the Petit Trianon, they saw a beautiful woman wearing a summer dress.

2 Aim: to focus on interrupted actions with *when* + past simple.

● Encourage the students to write full answers.

Answers

1 They were walking to the Petit Trianon when they noticed people wearing clothes from the eighteenth century.
2 The evil-looking man was staring at the two women when a second man ran up to them and called out.
3 They were crossing the lawn in front of the Petit Trianon when they saw a beautiful woman wearing a summer dress.

3 Aim: to focus on two consecutive actions using *when* + past simple.

● Ask the students to do the exercise.

Answers

1 When the women asked them the way, the men looked at the women in surprise.
2 When she saw the woman and the young girl, Miss Jourdain felt uneasy.
3 When they crossed the bridge they saw the Petit Trianon.
4 When the tour was over, they returned to Versailles for tea.

SPEAKING

1 Aim: to practise speaking.

● Ask the students if they believe the two Englishwomen's story. Do they think it is possible to travel in time?

● Ask one or two students where they would like to travel to in time.

● Ask them to work in pairs and to discuss their time travelling. They should talk about the time and the people they would like to meet.

2 Aim: to practise speaking.

● Ask the students to tell the rest of the class about their time travelling. Find out if anyone else wants to travel with them. As a class, decide who has the strangest reasons for travelling.

Time travellers

One hot August day in 1901, two young English women were sightseeing at the Palace of Versailles near Paris. It was extremely hot, and there were crowds of people. 'Why don't we walk to the Petit Trianon?' suggested Miss Anne Moberly. Her friend, Miss Eleanor Jourdain agreed. The Petit Trianon was one of Queen Marie-Antoinette's favourite places. Suddenly everything felt strange. The sun was still shining but Miss Moberly shivered. In the distance they noticed some people. But when the women got closer they saw the people were wearing clothes from the eighteenth century.

They went up to two men wearing long green coats and red three-cornered hats, who were talking to each other. 'Excuse me, gentlemen, but we have lost our way. Could you tell us how to get to the Petit Trianon?' The men turned and looked at the two women in surprise. One of them said, 'Go straight down this path, madam,' and carried on talking.

Down the path was a small cottage. While they were passing it, they looked in through an open door. Inside there was a woman and a young girl. Both were wearing eighteenth-century clothes and were standing very still, like sculptures. Miss Jourdain felt uneasy because she knew she wasn't dreaming.

They came to a point where the path branched left and right. An evil-looking man was sitting on the grass in front, staring angrily at the two women. Suddenly a second man ran up to them and called out to them, 'Don't go that way. Return to the house.' He looked over his shoulder and disappeared again.

The two women took the path the man suggested and crossed a little bridge. Beyond the trees, they could see the Petit Trianon. While they were crossing the lawn in front of the house, they saw a beautiful woman wearing a summer dress, sitting alone on the grass. They went up the steps onto the terrace and joined a group of tourists who were visiting the building. When the tour was over, they returned to Versailles for tea.

While they were walking round the palace, neither lady talked about the strange event. Then a week later, Miss Jourdain suddenly said, 'Do you think the lady in the summer dress was Marie-Antoinette?' The day they were there was 10 August, the anniversary of the day in 1789 when Marie-Antoinette went to prison. Perhaps the man running to tell them to return to the house was a messenger who was coming to warn the Queen about the danger. The two men in green coats and red hats were probably members of the Swiss guard who looked after the Queen while she was staying at the Petit Trianon, and the woman and young girl were possibly the wife and daughter of one of the royal gardeners. And was the evil-looking man the Count de Vaudreuil, who told the crowd where Marie-Antoinette was?

On a second visit to Versailles, Miss Jourdain discovered that the wooden hut was no longer there, nor was the bridge, although old maps showed they were once there.

So what really happened on that hot August day in 1901? Both Miss Jourdain and Miss Moberly believed that, somehow, they travelled back to the Court of Versailles in August 1789. Or was it just a dream caused by the heat of the day?

33 | *Is there a future for us?*

Expressions of quantity (2): *too much/many, not enough, fewer, less* and *more*

READING

1 Read *Is there a future for us?* and underline anything you think is a good suggestion and correct anything you think is wrong.

2 Match the words from the passage in list A with their meanings in list B.

A fumes
muck up
squashed
tandem
greedy

B wanting more than it needs
damage
flat
unpleasant smoke or gas
bike with two seats

3 Which of these statements about the environment do you think Henrietta and Eryn agree with?

1 There's too much noise.
2 The air in the cities isn't clean enough.
3 There isn't enough farmland.
4 There aren't enough clean rivers.
5 We should use less fuel.
6 There are too many people.
7 The sea is too polluted.
8 We should give more money to poor people.
9 We should cut down fewer trees.
10 We should build more houses.

4 Put a tick (✓) by the statements you agree with.

Is there a future for us?

Henrietta (aged 8): The biggest problem with the environment is the ozone layer; there's a hole, and it's getting bigger. It's made by cars and aeroplanes – things which give off fumes. There should be a law that you aren't allowed to go to school by car because the fumes go up into the ozone layer and muck it up.

Eryn (also aged 8): The ozone layer's like a piece of paper covering a rock. It's meant to protect us. I'm scared the hole will get bigger and move around the world and people will get cancer. I think about it sometimes in bed. That could happen in our time. We should stop using aerosols, because they get squashed and all the poisons come out. And we could use horses instead of cars.

Henrietta: We could get tandems, and longer bikes, so children could ride on the back. Cars should be very, very expensive. I heard a woman on the radio saying she didn't care about the ozone layer because she wouldn't be alive, but it's our family who will die.

Eryn: You also get bad pollution from burning down the rainforest. People should leave things alone if they don't own them. There should be fines for destroying the rainforest. And we should give money to poor people in Africa and places.

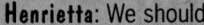

Henrietta: We should spread out the people evenly. We could say, 'Put your hands up all those who want to live in Africa.' And then we could spread out the food. There's enough to go round. We could easily grow it. England's a very rich and greedy country.

Eryn: We use up far more of the earth than people in Africa, so it's a good idea for the whole world to discuss the environment, but not prime ministers, because they always vote for their own side. The Queen should go instead.

Adapted from *The Independent on Sunday*

33
GENERAL COMMENTS

Expressions of quantity (2)
This is the second of two lessons in pre-intermediate *Reward* which focus on expressions of quantity, the first one being countables and uncountables in Lesson 15. You may want to ask the students to re-read the explanation in the grammar box of that lesson, as much of the information there will be relevant to this lesson as well. The structures should be simpler than those in Lessons 31 and 32, and should provide a change of tempo in the course.

Illustrations
The illustrations are of stamps designed by children for the Royal Mail in Britain.

Possible questions
What do the drawings show?
Who do you think did them?
What forms of pollution do they show?
Can you think of any other forms of pollution? (noise, oil spills in the sea, traffic)

READING

1 Aim: to read for specific information.
- Explain that the passage is a transcript of two eight-year-old children talking about the environment.

- Ask the students to read the passage and to note down or underline anything they think is a good suggestion and write a correction for anything which is wrong.

- When the students have finished, check their answers orally.

2 Aim: to focus on some difficult vocabulary and so to gain easier access to the text.
- There may have been some vocabulary difficulties, so ask the students to match the words and their meanings.

> **Answers**
> fumes = unpleasant smoke or gas
> muck up = damage (a very informal word)
> squashed = flat
> tandem = bike with two seats
> greedy = wanting more than it needs

- Ask the students to write down this new vocabulary only if they think the words will be useful.

3 Aim: to present the target structures and to practise reading for specific information.
- The structures to be presented and practised in the *Grammar* section can all be found in these statements. Ask the students to decide which Henrietta and Eryn would agree with. For statements which they think the children would not have an opinion about, suggest that they write a question mark.

> **Possible answers**
> Henrietta and Eryn would probably agree with 6, 7, 8 and 9. In general, they are concerned about pollution and overcrowding.

4 Aim: to focus on the target structures.
- Ask the students to tick the statements they agree with. In case the students tick them all, ask them to choose the three most important statements.

- Ask the whole class about their choice of statements.

GRAMMAR

1 Aim: to focus on *too*, *much*, *many* and *enough*.
- Ask the students to read the information about expressions of quantity in the grammar box.

- Ask the students to do the exercise.

> **Answers**
> 1 It's too quiet for me in the countryside.
> 2 There aren't enough forests in the world.
> 3 The centre of my country is too flat. I prefer the mountains.
> 4 There are too many factories.
> 5 It isn't peaceful enough in the city.
> 6 There's too much pollution in the sea.

2 Aim: to focus on expressions of quantity.
- Ask the students to write sentences giving their opinions about the issues mentioned and using the expressions of quantity in the grammar box.

3 Aim: to practise using expressions of quantity.
- Ask the students to work in pairs and to talk about their opinions about the statements in *Reading* activity 3.

4 Aim: to practise using expressions of quantity.
- Ask the students to write sentences about the environment near their homes. You may need to give individual students some relevant vocabulary at this stage.

- At the end of the activity, write on the board the extra vocabulary you supplied.

VOCABULARY AND SPEAKING

1 Aim: to present the words in the vocabulary box.
- The words in the vocabulary box are all useful to talk about geographical locations, the environment and the countryside. Ask the students to write the words which can go together.

> **Answers**
> **beach:** peaceful, noisy, quiet
> **city:** noisy, hilly, industrial, poor, quiet, rich
> **coast:** flat, hilly, mountainous, industrial, noisy, peaceful, poor, rich
> **countryside:** flat, hilly, mountainous, peaceful, quiet
> **desert:** flat
> **factory:** noisy, quiet
> **farmland:** poor, hilly, quiet, peaceful
> **field:** flat, hilly, peaceful
> **forest:** peaceful, quiet
> **island:** flat, hilly, mountainous, peaceful, rich, poor
> **jungle:** noisy, peaceful, quiet
> **lake:** flat, peaceful, quiet
> **ocean:** noisy, peaceful
> **region:** mountainous, peaceful, quiet, rich, poor, rural, industrial
> **sea:** noisy, peaceful
> **town:** noisy, hilly, flat, industrial, poor, rich, rural,
> **village:** noisy, hilly, flat, industrial, poor, rich, rural

- If the students ask, you may have to explain that some words go together and others don't. For example, *countryside* suggests the physical aspect of a region, so *rich* and *poor* don't really go with it, and nor does *industrial* because if the countryside is industrial, it's no longer called countryside.

2 Aim: to present the words in the vocabulary box.
- Ask the students to talk about their countries using the words in the box. Do this as a class activity.

3 Aim: to practise speaking.
- Ask the students to talk about the best place to live in their countries. This can also be done as a class activity. If you do it in pairs, leave time for some class feedback.

4 Aim: to practise speaking.
- Ask the students to work in groups and make predictions about the environment. Remind them, if necessary, to use *will*. Encourage them to choose from the phrases provided.

5 Aim: to practise speaking.
- Ask the class for their predictions and write them on the board. Then ask the class to vote for the predictions which most people agree with.

GRAMMAR

> ### Expressions of quantity (2)
>
> **Too much/many** + noun
> There's **too much** noise.
> There are **too many** people.
>
> **Too** + adjective
> The sea is **too polluted**.
>
> **Not enough** + noun
> There isn't **enough** farmland.
> There aren't **enough** clean rivers.
>
> **Not** + adjective + **enough**
> The air isn't **clean enough**.
>
> **Fewer, less** and **more**
> You use *fewer* and *more* with countable nouns.
> In Britain there are **fewer** men than women. There are **more** women than men.
>
> You use *less* and *more* with uncountable nouns.
> There's **more** pollution these days. There's **less** clean air.

1 Complete these sentences with *too, much, many* and *enough*.

1 It's ___ quiet for me in the countryside.
2 There aren't ___ forests in the world.
3 The centre of my country is ___ flat. I prefer the mountains.
4 There are too ___ factories.
5 It isn't peaceful ___ in the city.
6 There's ___ much pollution in the sea.

2 Give your opinion about these issues. Write sentences with *fewer, less and more* beginning *We need...*

cars industry clean water pollution factories farmland fields forests
We need fewer cars.

3 Work in pairs. Find out if your partner agrees with the statements in *Reading* activity 3.

4 Write sentences saying what you don't like about the environment near your home.
There's too much noise.

VOCABULARY AND SPEAKING

1 Work in pairs and look at the words in the box below. Which adjectives and nouns go together?

> beach city cliff coast countryside desert factory farmland field flat forest
> hill hilly industrial industry island jungle lake mountain mountainous
> noisy ocean peaceful poor quiet region rich river rural sea town village

a quiet beach, a flat field

Which words in the box can you use to talk about your country? Can you think of other words to describe your country?

2 Describe different regions of your country using the words in 1 and the words below.

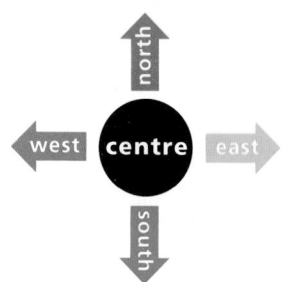

There's a desert in the centre of the country. It's quite rural in the north.

3 Work in pairs. In your opinion, where is the ideal place to live in your country?
I think it's probably somewhere on the West coast, not far from the mountains.

4 Work in groups of three or four and talk about what will happen to our environment in the future. Think about the following:

– the population
– the sea
– the sea level
– the countryside
– towns
– the climate
– the air

Choose from these phrases:

grow bigger grow smaller
become dirtier become cleaner
warm up cool down
improve deteriorate rise fall
get more crowded
get more deserted

The population will grow bigger.

5 Find five predictions about the environment which most people in the class agree with.

34 | *The Day of the Dead*

Present simple passive

The Day of the Dead

It's the end of October, and Mexico is preparing to celebrate the Day of the Dead on 1 November. In cities throughout the country for several weeks before the festival begins, street markets and shops are filled with symbols of death.

It's the highlight of the year of the year for all Mexicans: the day when dead spirits return to the land of the living. But there's nothing sad about this festival. It simply reflects the Indian belief that death is a natural part of life.

November 1 is known as the feast of All Saints and All Souls. The Day of the Dead is called *todos santos* or *dia de muertos* by the Mexicans and the festival usually involves two days of celebration on 31 October and 1 November. Mexicans believe that on the Day of the Dead the souls of dead relatives will return. The Indian festival became a Catholic one when the Spanish brought their religion to Mexico.

A feast is prepared for the dead with their favourite food and drink, cigarettes, sweets and fruit. A special kind of bread, known as *pan de muertos* ('bread of the dead') is baked, traditionally by the men – either the head of the family or the closest relative of the dead person. Today, however, the *pan de muertos* is often bought in markets. A bowl of water and a cloth is put on the table so that the spirits can wash their hands, and sometimes a favourite possession of the dead person is left.

The dead person is not usually seen when they return, but their spirit is felt by the family. After the festival, the food is given to the community, and the gifts are arranged around a wooden frame which is decorated with coloured papers, flowers and fruit.

The island of Janitzio is famous for its Day of the Dead celebrations, and has become a major tourist attraction. Just before midnight on 1 November, the lake which surrounds Janitzio is lit up by hundreds of torches. These show the route of the *lancha*s (small boats) which carry the families to the island. They go with their gifts to the cemetery where they will spend the night.

The cemetery is crowded not only with family and friends, but with tourists, photographers and even film crews. But later in the night, the tourists leave, and the families remain until morning. Through the night, attracted by the light of the candles and perfume of incense and flowers, the souls of the dead return once more to their families.

Adapted from *BBC World magazine*

34
GENERAL COMMENTS

The passive

Traditionally, the passive has been presented in the context of manufacturing processes. In fact, the structure occurs in many other contexts in every day language use. In this lesson, it can be found in a passage about the Day of the Dead in Mexico, in which we are interested in the events and the processes of the religious festival. Some languages do not have a passive form, and you may need to supplement the presentation in this lesson with material from the Practice Book and the Resource Pack. In the grammar box, reference is made to *by* as the agent, and you may like to add an explanation about *with* to be used for instruments which are used by the agent to perform the actions, eg *The bowl was filled with flowers.*

Reading

It is very common for a passage to be read by the student and used for language practice without necessarily being understood or appreciated in its own right as reading material. The main reading tasks in *Reward* encourage the students to gain access to what might be described as the heart of the matter, or the reason why the passage was written in the first place. It is very easy for pedagogical tasks to distract the student from the pleasure of the passage. So, it is only after the heart of the matter has been reached that the passage is used for more obviously linguistic purposes. In *The Day of the Dead*, activities 2 and 3 create an opportunity for the students to think about the ideas suggested by the passage. The subsequent tasks are more concerned with language learning.

Photos: possible questions

What can you see in the photos?
Where do you think the festival takes place?
What do you think the festival celebrates?
Do you think the festival is a happy or a sad one?
Do you think it could take place in your country?

READING AND VOCABULARY

1 Aim: to prepare for reading, to present the words in the box; to pre-teach some important items from the passage.

● Ask the students what they know about the Day of the Dead in Mexico. If anyone has any suggestions, write any key vocabulary items on the board.

● Ask the students to look at the words in the box and to predict which ones they are likely to see in the passage.

2 Aim: to read for main ideas; to react to the passage.

● The activity is intended to help the students read and react to the text as a whole. For many cultures, the most surprising thing is that the festival is a celebration of death. It is not meant to be a sad occasion. Ask the students to read and find the strangest or most interesting piece of information in the passage.

● Ask the students if they have experienced a religious festival like this.

3 Aim: to read for specific information.

● Ask the students to re-read the passage and to decide what the photos show. They can write a short caption for each photo.

GRAMMAR

1 Aim: to practise asking questions in the passive.

● Ask the students to read the information about the passive in the grammar box.

● Ask the students to do the exercise. Suggest that they will find the 'answers' in the passage; somewhere around these answers will be the information needed to write questions.

> **Possible questions**
> 1 What is November 1 called?
> 2 What is prepared for the dead?
> 3 What is baked by the men?
> 4 What is put on the table?
> 5 What is felt by the family?
> 6 What is given to the community?
> 7 What is arranged around a wooden frame?
> 8 What is lit up by hundreds of torches.

● It may be easier to do this activity orally with the whole class.

● Ask the students to write down the questions when you have finished checking orally.

2 Aim: to practise using the passive.

● You may need to point out that the agent *by* may be useful in many of the sentences if you are particularly interested in *who* or *what* performed the action.

> **Answers**
> 1 All Souls' Day is celebrated on November 1.
> 2 It is called the Day of the Dead by the Mexicans.
> 3 A feast is prepared.
> 4 Bread is often bought in markets.
> 5 The spirit is felt by the family.
> 6 The gifts are arranged around a wooden frame.
> 7 The families are carried to the island by boats.
> 8 The souls of the dead are attracted by the lights of the candles.

● Check the students' answers and help them decide why the agent is or isn't necessary in each sentence.

3 Aim: to decide if a sentence should be active or passive.

● This activity is designed to help students be more aware of whether an active or a passive sentence is better. Do the activity orally.

> **Answers**
> 1 b We are more interested in Fiat cars than who makes them.
> 2 b We are more interested in what is sold than who sells them.
> 3 a We are more interested in the subject, ie *my mother* than what she does.
> 4 a We are more interested in *my brother* than the object of the sentence.
> 5 a We are more interested in what is drunk than in who drinks it.
> 6 a We are more interested in cats and one of their characteristics, than in warm beds and the fact that they attract cats.

SPEAKING AND WRITING

1 Aim: to practise using the passive in descriptions of events and processes.

● Ask the students to work on their own and to make notes on an important ritual or festival. Ask them to write a few sentences using passives.

2 Aim: to practise speaking and writing notes.

● Ask the students to go round and find out which festival or ritual other people have chosen. They should choose one festival or ritual, ask questions and make notes about it because they will have to write a passage about it in activity 3.

3 Aim: to practise writing.

● Ask the students to write a passage about their partner's festival or ritual using the notes they made in activity 2.

● You may like to ask the students to do this activity for homework.

READING AND VOCABULARY

1 You are going to read about Mexico's Day of the Dead celebrations. Which of the words in the box do you expect to see?

> belief Buddhist candle Catholic celebration
> cemetery dead highlight mosque Muslim perfume
> saint soul symbol temple torch tourist

2 Read *The Day of the Dead* and decide what is the most surprising or interesting piece of information in the passage.

3 Work in pairs. Read the passage again. Decide what the photos show and write a caption for each one.

GRAMMAR

> **Present simple passive**
> **You form the present simple passive with *am/is/are* + past participle.**
> *A splendid feast **is prepared**.*
> *Shops **are filled** with symbols of death.*
>
> **Questions**
> *What **is** November 1 **called**?*
>
> **You use the passive when you are more interested in the object of the sentence or you don't know who or what does something.**
> *A bowl of water is put on the table.*
> **(You aren't interested in who put it there.)**
> *Shops are filled with symbols of death.*
> **(You don't know who filled the shops.)**
> **If you are more interested in the object, but you know who or what does something, you use *by*.**
> *A special kind of bread is baked **by** the men.*
> **If you are more interested in the subject, you use an active sentence.**
> *The men **bake** a special kind of bread.*

1 Here are some answers to questions about the passage. Write the questions using the passive.

1 The Day of the Dead.
2 A feast.
3 A special kind of bread.
4 A bowl of water and a cloth.
5 The dead person's spirit.
6 The food.
7 The gifts for the dead.
8 The lake surrounding Janitzio.

What is November 1 called ?

2 Rewrite these sentences in the passive.

1 They celebrate All Souls' Day on November 1.
2 The Mexicans call it the Day of the Dead.
3 They prepare a feast.
4 They often buy the bread in markets.
5 The family feels the spirit.
6 They arrange the gifts around a wooden frame.
7 Boats carry the families to the island.
8 The lights of the candles attract the souls of the dead.

All Souls' Day is celebrated on November 1.

3 Look at these pairs of active and passive sentences. In each pair, which do you think is the better sentence?

1 a They make Fiat cars in Italy.
  b Fiat cars are made in Italy.
2 a They sell Macdonalds hamburgers in many countries.
  b Macdonalds hamburgers are sold all over the world.
3 a My mother does a lot of cooking.
  b A lot of cooking is done by my mother.
4 a My brother writes poetry.
  b Poetry is written by my brother.
5 a Coffee is drunk in most countries.
  b They drink coffee in most countries.
6 a Cats like warm beds.
  b Warm beds are liked by cats.

SPEAKING AND WRITING

1 Think about an important ritual or festival in your country. Make notes on:

– what it's called
– what preparations are made
– what is drunk
– what is celebrated
– what food is cooked
– what gifts are given

2 Find someone in your class who has chosen a different ritual or festival. Ask questions about it and make notes, and answer questions about the one you chose.

What is it called? — *The Palio.*
When does it take place? — *On two days in July and August.*

3 Write a passage describing the ritual or festival your partner chose.

The highlight of the year for people from Siena is called the Palio. It takes place on…

Mind your manners!

Making comparisons (3): *but, although, however*

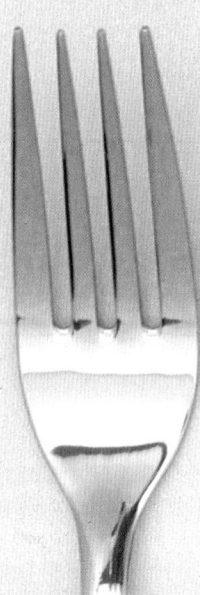

VOCABULARY AND READING

1 Look at the words in the vocabulary box. Group the words under these headings: *things to eat, things on the table, things to cook with, parts of the body* and *things to say when eating or drinking.*

> bowl cheers chips chin cup
> dish elbow fork hand
> ice cream jam knife lap
> melon napkin neck pasta
> plate pot saucepan saucer
> sausage spoon steak
> table cloth teaspoon toast

things to eat: chips...

2 Work in pairs and check your answers. Think of two or three other words which you can add to each group. (You can use a dictionary if necessary.)

3 Read the questions in *Mind your manners!* and think about your answers.

4 Work in pairs and discuss your answers.
In my family we have dinner at ten o'clock.

5 Tell the rest of the class what your answers are. Use these expressions.

Both of us think that...
Neither of us thinks that...
I think ... but Elena thinks...

Mind your manners!

1 What do you say at the start of a meal?

2 What time do you have lunch and dinner?

3 How long does a typical lunch or dinner last?

4 Do you usually use a knife and fork? If so, which hands do you hold them in?

5 Do you use a napkin? If so, where do you put it?

6 At which meals do you eat the following food?
melon pasta fish steak

7 Where do you put your knife and fork when you have finished your meal?

8 Where do you put your hands when you're at the table but not eating?

9 Do you eat cake with a fork or a spoon?

10 What food do you often eat with your fingers at the dining table?

11 When do you usually drink coffee and tea?

12 When can you smoke during a meal?

13 What do you say and do when someone raises their glass?

14 Do you have soup in the summer?

15 Do you eat salad in the winter?

35

GENERAL COMMENTS

Cross-cultural training

Even if your students all come from the same country or have the same cultural background, they are likely to have different answers for some of the questions. Remember that one way of becoming more sensitive to other people's cultures is to become more aware of one's own culture.

Jigsaw listening

There is a further example of jigsaw listening with one tape recorder in this lesson. For a description of the background to this technique, see the teachers' notes for Lesson 15.

Correction

The material in this lesson is likely to generate quite a lot of discussion. Use the opportunity for fluency practice, and don't worry about the students' accuracy. It is very disturbing for students if you correct them while they are involved in trying to put an idea or opinion across. Make a note of any mistakes and discuss them with the students at a later stage.

Main photo

A scene from a Marx brothers' movie, with Harpo Marx.
What can you see in the photo?
Do you know who the man is?
What's he doing?
Do you think this is good manners?
When do you think the photo was taken?
Do you think table manners have changed in recent years?

VOCABULARY AND READING

1 Aim: to present the words in the vocabulary box and to pre-teach some important words.
● Ask the students to group the words under the five headings.

> **Answers**
> **things to eat:** chips, ice cream, jam, melon, pasta, sausage, steak, toast
> **things on the table:** bowl, cup, dish, fork, knife, napkin, plate, saucer, spoon, table cloth, teaspoon
> **things to cook with:** knife, pot, saucepan, fork, spoon
> **parts of the body:** chin, elbow, hand, lap, neck
> **things to say when eating or drinking:** cheers

2 Aim: to personalise and extend the vocabulary list.
● Ask the students to think about other words to add to their lists. Ask for suggestions and write them on the board. If necessary, they can use their dictionaries.

3 Aim: to read and answer the questions.
● Ask the students to read the questions and think about their answers. The questions all concern aspects of behaviour at the dining table or conventions concerning food. In Britain, many children are given explicit training by their parents in table manners. Other social conventions are acquired more implicitly by role modelling.

4 Aim: to practise talking about table manners.
● Ask the students to work in pairs and discuss their answers to the questions.

5 Aim: to practise talking about table manners.
● Lead a class discussion about the questions. The students can use the expressions suggested. Do most people have the same table manners?

LISTENING AND SPEAKING

1 Aim: to listen for specific information.

● 🔊 Ask the students to work in three groups, A, B and C. The members in each group should look at their instructions in the Communication activities. Each group has different questions to answer. Ask them to listen and to write notes in answer to their specific questions.

● Ask the students in each group to discuss their answers, and check that everyone has got the same information.

2 Aim: to discuss what they have heard, to reconstitute the two listening passages and to complete the charts.

● Ask the students to form new groups of three people, with one person from group A, one person from group B and one person from group C. They should work together and talk about the notes they made in activity 1. Remember (from Lesson 15) that the rule is that they can only talk about information which they were specifically asked to listen for and write down. Ask them to answer the questions with as much detail as possible.

● You may like to check their answers orally.

> **Answers**
> 1 Nothing.
> 2 1pm for lunch, and 7pm for dinner.
> 3 About half an hour.
> 4 Fork in left hand, knife in right.
> 5 Yes, on your lap.
> 6 Melon: breakfast, lunch and dinner; pasta: lunch and dinner; fish: breakfast, lunch and dinner; steak: lunch and dinner.
> 7 Together in the centre of the plate handles pointing towards you.
> 8 On your lap.
> 9 A fork.
> 10 Chicken; bread;cheese; fruit.
> 11 Coffee: anytime; tea: in the morning and in the afternoon.
> 12 Before and after.
> 13 'Cheers'.
> 14 Sometimes.
> 15 Sometimes.

● Point out that Stephen's answers are typical of most British people, but that many people may do things differently.

GRAMMAR

1 Aim: to focus on *but, however* and *although*.

● Ask the students to read the information about making comparisons in the grammar box.

● Ask the students to do the exercise. Explain that more than one word can be used to complete the sentences.

> **Answers**
> 1 Many people have dinner quite early, **although/but** we eat quite late, at about nine.
> 2 Most people have milk in their tea, **although/but** I prefer lemon.
> 3 I hold my fork in my left hand to cut food **but** change to my right hand to eat.
> 4 I usually drink coffee in the morning, **although/but** I sometimes have a cup after dinner.
> 5 We eat melon at the start of a meal. **However** some people have it at the end.
> 6 You don't usually smoke while you're eating. **However**, it's OK to smoke after the meal.

2 Aim: to practise using *but, however* and *although*.

● Ask the students to write sentences comparing table customs in their family with those of Stephen.

SOUNDS

Aim: to focus on stress and intonation in sentences with *but, however* and *although*.

● 🔊 Ask the students to listen to the sentences and notice the stress and intonation patterns. When *although* starts a sentence, the voice usually rises at the end of the clause. With however, the voice falls on the word. They should repeat the sentences.

SPEAKING AND WRITING

1 Aim: to practise speaking and writing.

● Divide the students into groups of three or four. Ask each group to think of a social occasion. They should write down three customs, two false and one true. Encourage them to be as creative or amusing as they like. For the false custom, they may like to describe something which you should certainly not do!

● If you have time, ask each group to prepare a little sketch incorporating the customs.

2 Aim: to practise speaking.

● Ask the students to present their customs to the rest of the class; the other students should guess which is the false information.

● If you have time, ask the students to act out the sketches they may have prepared in activity 1.

LISTENING AND SPEAKING

1 Work in groups of three. You're going to hear Stephen, who is English, talking about table manners.

Student A: Turn to Communication activity 6 on page 99.
Student B: Turn to Communication activity 4 on page 98.
Student C: Turn to Communication activity 14 on page 100.

2 Now work together and talk about Stephen's answers to the questions in *Mind your manners!*

▭ Listen again and check.

GRAMMAR

Making comparisons (3): *but, however, although*

You use *but*, *however* and *although* to make a comparison which focuses on a difference. You put *but* at the beginning of a sentence or to join two sentences.
We drink coffee in the morning. **But** *we don't drink it in the afternoon.*
We drink coffee in the morning **but** *we don't drink it in the afternoon.*

You use *although* at the beginning of a subordinate clause. You need to separate the subordinate and the main clause with a comma.
We usually have dinner at six, **although** *some people have dinner later.*

You can put the subordinate clause at the beginning or at the end of the sentence.
Although *we usually have dinner at six, some people have dinner later.*

You use *however* at the beginning of a sentence. It is followed by a comma.
We drink coffee in the morning. **However,** *we don't drink it in the afternoon.*

---

1 Complete these sentences with *however* or *although*.

1 Many people have dinner quite early, ___ we eat quite late, at about nine.
2 Most people have milk in their tea, ___ I prefer lemon.
3 I hold my fork in my left hand to cut food, ___ I change to my right hand to eat.
4 I usually drink coffee in the morning, ___ I sometimes have a cup after dinner.
5 We eat melon at the start of a meal. ___ , some people have it at the end.
6 You don't usually smoke while you're eating. ___ , it's OK to smoke after the meal.

2 Write sentences making comparisons between table customs in your family and Stephen's.

SOUNDS

▭ Listen to the sentences in *Grammar* activity 1. Notice the stress and intonation of sentences with *however* and *although*.

Now say the sentences aloud.

SPEAKING AND WRITING

1 Work in groups of three and write down three examples of table manners – two true and one false.
You should always eat thick soup with a knife and fork.
In a restaurant, you can attract the waiter's attention by whistling.
After a meal, say goodbye and shake hands with all the other people.

2 Present your three examples of table manners to the rest of the class. Try to add as much information and context as possible.

The rest of the class must guess which is the false one.

Progress check 31-35

VOCABULARY

1 Some verbs are followed by a particle (an adverb or a preposition). These are called multi-part verbs.

take out   She took out a pen.
listen to   He's listening to the radio programme.

Use your dictionary to find out how many multi-part verbs you can make with the verbs and the adverbs or prepositions below.

Verbs   come   fall   look   pick   stand   take   turn   throw

Particles   at   away   down   in   off   round   up

2 Now complete these sentences with a suitable particle from the list above.

1 He took ___ his coat.
2 I picked ___ my wallet.
3 She fell ___ the stairs.
4 I threw the wrapping ___ .
5 He turned ___ and looked ___ me.
6 She stood ___ when we came ___ .

3 With some multi-part verbs, you can put the noun object after or before the particle.

take out   She took out a pen. She took a pen out.
put on   He put on his hat. He put his hat on.

But you must put the pronoun object between the verb and the particle.

She took it out.   He put it on.

Rewrite these sentences, replacing the noun with a pronoun.

1 Don't throw the newspaper away.
2 He turned the radio down.
3 He took his hat off.
4 He washed the dishes up.
5 He gave up smoking.
6 I'll find out her name.

4 With other multi-part verbs you always put the object after the particle.

listen to   She's listening to the radio programme. She's listening to it.
look at   He looked at the old woman. He looked at her.

Rewrite these sentences with pronouns.

1 She thought about John.
2 We voted for the socialists.
3 I waited for my mother.
4 He looked at the picture.
5 She listened to the radio.
6 We looked after their cat.

5 Look at the vocabulary boxes in Lessons 31 – 35 again. Choose words which are useful to you and write them in your *Wordbank*.

GRAMMAR

1 Join the two parts of the sentence with *when*.

1 I/play tennis/hurt my leg.
2 He/walk to work/see his friend.
3 They/watch television/fall asleep.
4 She/talk to me/start to cry.
5 We/sit in garden/hear a loud noise.
6 I/look for a pen/find some money.

1 I was playing tennis when I hurt my leg.

2 Rewrite the sentences in 1 using *while*.

1 While I was playing tennis I hurt my leg.

3 Write questions.

1 What/do/8am yesterday?
2 Who/talk to/last night?
3 Why/work so hard/last week?
4 What/talk about/this morning?
5 Where/have dinner/last Sunday?
6 Why/laugh/just now?

1 What were you doing at 8am yesterday?

4 Rewrite the sentences with *too* + adjective or *not* + (adjective) *enough*.

1 It's too noisy in here. (quiet)
2 It wasn't dark enough to sleep. (light)
3 I was too cold. (warm)
4 The region is too industrial. (rural)
5 The hotel room was too dirty. (clean)
6 The jacket wasn't big enough. (small)

1 It isn't quiet enough in here.

Progress check 31–35

GENERAL COMMENTS

You can work through this Progress check in the order shown, or concentrate on areas which may have caused difficulty in Lessons 31 to 35. You can also let the students choose the activities which they would like to or feel the need to do.

VOCABULARY

1 Aim: to present multi-part verbs.

● Multi-part verbs are phrasal (with adverbs) or prepositional verbs, but it is not obvious from looking at an example of a phrasal or prepositional verb which type it might be. In fact, you can often only tell what type it is by checking where you put the noun or pronoun object. So, knowing the name of the type of verb is not going to help the students cope with its structural properties. In fact, in this brief explanation, only transitive phrasal verbs and prepositional verbs are presented; intransitive phrasal verbs and phrasal-prepositional verbs (eg *look out for, take up with*) are covered in intermediate *Reward*.

> **Answers**
> **come:** at, away, down, in, off, round, up
> **fall:** away, down, in, off
> **look:** at, away, down, in, round, up
> **pick:** at, off, up
> **stand:** at, away, down, in, up
> **take:** away, down, in, off, round, up
> **turn:** away, down, in, off, round, up
> **throw:** at, away, down, in, off, up

2 Aim: to practise using multi-part verbs.

> **Answers**
> 1 He took **off** his coat.
> 2 I picked **up** my wallet.
> 3 She fell **down** the stairs.
> 4 I threw the wrapping **away**.
> 5 He turned **round** and looked **at** me.
> 6 She stood **up** when we came **in**.

3 Aim: to focus on word order of transitive phrasal verbs.

● Multi-part verbs present two main difficulties for students. The first is that a verb's meaning is not necessarily obvious from its separate parts (*give up, stand for*). The second is that there are strict rules about the position of the objects.

> **Answers**
> 1 Don't throw it away.　2 He turned it down.
> 3 He took it off.　4 He washed them up.
> 5 He gave it up.　6 I'll find it out.

4 Aim: to focus on word order of prepositional verbs.

> **Answers**
> 1 She thought about him.
> 2 We voted for them.
> 3 I waited for her.
> 4 He looked at it.
> 5 She listened to it.
> 6 We looked after it.

5 Aim: to revise vocabulary.

● It may be useful to go back over Lessons 31 to 35 and categorise the multi-part verbs according to the position of the objects. There are a number of multi-part verbs in Lesson 31.

GRAMMAR

1 Aim: to revise *when* + past continuous.

> **Answers**
> 1 I was playing tennis when I hurt my leg.
> 2 He was walking to work when he saw his friend.
> 3 They were watching television when they fell asleep.
> 4 She was talking to me when she started to cry.
> 5 We were sitting in the garden when we heard a loud noise.
> 6 I was looking for a pen when I found some money.

2 Aim: to revise *while* + past continuous.

> **Answers**
> 1 While I was playing tennis I hurt my leg.
> 2 While he was walking to work he saw his friend.
> 3 While they were watching television they fell asleep.
> 4 While she was talking to me she started to cry.
> 5 While we were sitting in the garden we heard a loud noise.
> 6 While I was looking for a pen I found some money.

3 Aim: to revise asking questions using the past continuous.

This activity is designed to practise the past continuous. However, it is also possible to use the past simple with these questions.

> **Answers**
> 1 What were you doing at 8am yesterday?
> 2 Who were you talking to last night?
> 3 Why were you working so hard last week?
> 4 What were you talking about this morning?
> 5 Where were you having dinner last Sunday?
> 6 Why were you laughing just now?

4 **Aim: to revise *too* + adjective and *not* + (adjective) *enough*.**

Answers
1 It's not quiet enough in here.
2 It was too light to sleep.
3 I wasn't warm enough.
4 The region isn't rural enough.
5 The hotel room wasn't clean enough.
6 The jacket was too small.

5 **Aim: to revise the passive.**
● Ask the students to write their own answers to these questions. Make sure they use a passive.

6 **Aim: to revise the passive.**

Answers
1 A lot of tea is drunk in Britain.
2 A lot of meat is eaten in Argentina.
3 Dinner is eaten by the Spanish at 10pm.
4 Medicine is bought at the chemist's.
5 Your teeth are examined by the dentist.
6 Newspapers are used for wrapping things.

7 **Aim: to revise *although*.**

Answers
1 She likes coffee, although she prefers tea.
2 She doesn't usually have time to eat in the mornings, although at weekends she has a large breakfast.
3 She usually has a sandwich for lunch, although she sometimes has a salad when she goes out with friends.
4 She doesn't smoke, although she doesn't mind other people smoking in her home.

8 **Aim: to revise *however*.**

Answers
1 She likes coffee. However, she prefers tea.
2 She doesn't usually have time to eat in the mornings. However, at weekends she has a large breakfast.
3 She usually has a sandwich for lunch. However, she sometimes has a salad when she goes out with friends.
4 She doesn't smoke. However, she doesn't mind other people smoking in her home.

SOUNDS

1 **Aim: to focus on /v/ and /w/**
● ▭ Some students may have difficulty in distinguishing between the two sounds. Ask them to complete these words with *v* or *w*. Then play the tape and pause after each word. Ask the students to repeat the words.

Answers
wait wallet move visit want work sandwich drive invite women wear shiver walk behave

2 **Aim: to focus on /h/.**
● ▭ Ask the students to listen and tick the words they hear.

Answers
1 ear 2 hair 3 eye 4 hat 5 hate 6 eat 7 art 8 as

● Ask the students to say the words aloud.

3 **Aim: to focus on stress in multi-part verbs.**
● ▭ Ask the students to listen and repeat the sentences. Make sure they stress the verb and the particle.

WRITING

1 **Aim: to focus on rules of punctuation.**
● At this stage, this activity simply confirms what the students already know about English punctuation.

2 **Aim: to focus on word order.**
● Ask the students to read the story again and decide where the words can go in the story.

Answers to activities 1 and 2
An English couple drove to Paris for a **short** holiday. They found their way to the hotel but couldn't find anywhere to park. Finally they found a **parking** space, but by now it was dark. While they were trying to park, a **friendly** Parisian saw their British car and offered to help. 'You have a meter behind you!' he shouted. So the Englishman moved **back** a little. The man shouted again, 'You have a meter behind you!' The driver moved a little more. The Frenchman waved and shouted, 'You have a meter behind you!' The Englishman, who was getting a little angry by now, reversed the car rather quickly. There was a **loud** crash! The Parisian pointed at a **metal** post which was lying under the car. 'I said, you had a meter behind you!'

3 **Aim: to focus on word order.**
● Ask the students to think of six words which could go into the story and write them down.

● Ask them to work in pairs. They should show their words to their partner and ask him/her to decide where the words can go in the story.

● If time is short you may like to set this activity for homework.

5 Write correct answers to these sentences.

1 Is champagne made in Britain?
2 Are shops closed during the weekend?
3 Is leather used for making clothes?
4 Is tea drunk at night?
5 Are grapes eaten at breakfast?
6 Is the housework done by men?

1 No, it's made in France.

6 Rewrite these sentences in the passive. Use *by* if necessary.

1 They drink a lot of tea in Britain.
2 They eat a lot of meat in Argentina.
3 The Spanish eat dinner at 10pm.
4 You buy medicine at the chemist's.
5 The dentist examines your teeth.
6 You use newspapers for wrapping things.

1 A lot of tea is drunk in Britain.

7 Join these sentences about Catherine, an English woman, by rewriting them with *although*.

1 She likes coffee. She prefers tea.
2 She doesn't usually have time to eat in the mornings. At weekends she has a large breakfast.
3 She usually has a sandwich for lunch. She sometimes has a salad when she goes out with friends.
4 She doesn't smoke. She doesn't mind other people smoking in her home.

8 Rewrite the sentences in 7 with *however*.

SOUNDS

1 Complete these words with *v* or *w*.

_ ait _ allet mo _ e _ isit _ ant _ ork sand _ ich
dri _ e in _ ite _ omen _ ear shi _ er _ alk beha _ e

🔊 Listen and check. Say the words aloud.

2 🔊 Listen and tick (✓) the word you hear.

1 ear hear   2 air hair 3 eye high 4 at hat
5 eight hate 6 eat heat 7 art heart 8 as has

Now say the words aloud.

3 🔊 Listen to the answers to *Vocabulary* activity 3. Notice how the speaker stresses the verb and the particle.

1 Don't throw it away.   4 He washed them up.
2 He turned it down.   5 He gave it up.
3 He took it off.   6 I'll find it out.

Now say the sentences aloud.

WRITING

1 Look at some rules for punctuation in English.

You use commas:
– to separate main and subordinate clauses.
While they were waiting, it started to rain.

– to separate signpost words and phrases, such as *however, as a result, in fact, first, next, after that* from the rest of the sentence.
I like coffee. However, I prefer tea.

Rewrite this passage with the correct punctuation. (You may like to look at *Progress check lessons 1 – 5, Reading and writing* before you begin.)

an english couple drove to paris for a holiday they found their way to the hotel but couldnt find anywhere to park finally they found a space but by now it was dark while they were trying to park a parisian saw their british car and offered to help you have a meter behind you he shouted so the englishman moved back a little the man shouted again you have a meter behind you the driver moved a little more the frenchman waved and shouted you have a meter behind you the englishman who was getting a little angry by now reversed the car rather quickly there was a crash the parisian pointed at a post which was lying under the car i said you had a meter behind you

2 Read the story again and decide where these words can go. Some words can go in more than one position.

short back parking friendly loud metal

3 Think of six more words which could go into the story and write them down. Now work in pairs. Show your words to your partner and ask him/her to decide where they can go in the story.

36 | *Lovely weather*

Might and **may** for possibility

READING

1 Read *When's the best time to visit your country?* and make a list of the best times to visit each country.

2 Work in pairs. Is your country mentioned? Do you agree on the best time to visit? If it isn't mentioned, say when the best time to visit is and add it to your list.

VOCABULARY

1 Look at the box and underline the words for seasons. Which of the other words in the box go with each season in your country?

> autumn dry hurricane flood
> sun freezing storm humid
> rain changeable ice lightning
> wind mist spring snow frost
> summer fog thunder wet
> mild winter

winter: freezing, fog, ice

Which words don't you use to talk about your weather?

2 Which of the nouns in the vocabulary box can you turn into adjectives?
fog – foggy

When's the best time to visit your country?

'May or October is best because it's not too humid, although there may be a few showers.' *Wang Wei, Hong Kong*

'We live in New England, so the best time is late August and September when the weather starts to get cooler and the leaves on the trees change colour.' *Norman, USA*

'July and August are fabulous. It might be quite cool but it'll be very pleasant. But don't come in the winter, it'll be dark all day and freezing cold.' *Ingrid, Sweden*

'We say the spring is the best time to visit us. But it may rain and there's quite a lot of wind, so you may miss the cherry blossom. Come in the autumn, the autumn leaves will be beautiful.' *Reiko, Japan*

'Come to Prague in winter and you'll see how beautiful the city is. There may be snow and there are fewer tourists.' *Frantisek, Czech Republic*

'Have you ever had Christmas dinner on the beach? Come in December, and you won't see a cloud in the sky.' *Kevin, Australia*

'In the summer, it's very hot and humid on the coast, although you might find it a little cooler in Mexico City because it's high up.' *Miguel, Mexico*

'Spring and autumn are the best, the climate is perfect on the Aegean and Mediterranean coasts, although it may be cooler in central Anatolia. You won't get much rain between May and October.' *Yildiz, Turkey*

'In December you'll find it a bit like England in July, not too hot. And it'll be rainy and cold in the winter. But it depends where you want to go. It's a big country and very mountainous.' *Maria Sara, Chile*

36

GENERAL COMMENTS

May and *might*

In Lesson 30, students will have come across *may* to request permission, eg *May I call you Esther?* In this lesson, *may* and *might* are used to talk about a specific possibility in the future, eg *I may call round this evening*. You don't usually use *may* and *might* in questions about possibility; the question is usually rephrased with *Do you think...* + *will* The use of *can* to talk about more general possibility (*It can get very hot in the evenings*) will be covered in intermediate *Reward*. This use is quite close to the use of *may* and *might* in this lesson.

Photo: possible questions

What can you see in the photo?
Where do you think it is? What country?
Could it be in your country?
What's the weather like?
What time of day is it?
What time of year is it?
Would you like to go there?

READING

1 Aim: to read for specific information.

- Begin the lesson by asking the students to say what the weather is like today, and to write up some of the vocabulary items used to describe it.

- Ask the students to suggest words which mean the opposite of the words you have written on the board, eg *cloudy – clear, sunny – rainy*.

- Ask the students when the best time is for visitors to come to their country and have pleasant weather.

- Ask them to read the passage and write down the countries and the best seasons or months to visit.

2 Aim: to react to the text.

- Ask the students to talk about the best time to visit their country, and to write a short paragraph, similar to the paragraphs in the passage. It doesn't matter if their country is mentioned; ask them simply to make their own personal contribution.

- Ask the students to read out their paragraphs or stick them on the wall of the classroom for people to read

VOCABULARY

1 Aim: to present the words in the vocabulary box.

- The students may have already suggested some of the words in the box. Check that they understand the other words.

- Ask the students to find the words for seasons.

- Ask the students to group the other words with the seasons. The groupings will vary from country to country. The answer below is how a British person would answer.

> **Answers**
> **autumn:** changeable, fog, mist, rain
> **winter:** freezing, frost, ice, rain, snow, wet, wind
> **spring:** changeable, flood, mild, rain, wet, wind
> **summer:** changeable, dry, humid, lightning, storm, sun, thunder
> In Britain, you wouldn't use *hurricane* to talk about the weather, because they are extremely rare.

- Ask the students to think about words needed to talk about their weather. They may like to use their dictionaries.

2 Aim: to form adjectives from nouns.

- Encourage the students to broaden their vocabulary by making adjectives from nouns. This is particularly common with words to describe the weather.

> **Possible answers**
> autumnal, sunny, stormy, rainy, icy, windy, misty, snowy, frosty, summery, foggy, thundery, wintry

GRAMMAR

1 Aim: to practise using *because* + might and to give advice.

● Ask the students to read the information about *may* and *might* in the grammar box.

● Ask the students to give advice and explain why. Remind them that they saw *should* for giving advice in Lesson 29.

> **Possible answers**
> 1 You should bring an umbrella because it might rain.
> 2 You should wear warm clothes because it might be cold.
> 3 You should bring your camera because you might want to take some photos.
> 4 You should get some sun-tan lotion because it might be sunny.
> 5 You should buy a good map because you may want to explore the countryside.
> 6 You should bring a swimsuit because you may want to go swimming.

2 Aim: to focus on the difference between possibility and certainty.

● The verb forms or the contexts in the sentences suggest an element of certainty about the arrangement or future plan.

> **Answers**
> 1 **I'm going** to Kenya next week. I've got my ticket.
> 2 **I've booked** into a hotel. I've got a room at the Ambassador.
> 3 **I might go** to Lake Victoria or perhaps to the beach near Mombasa.
> 4 **I might not** spend more than a week on the coast, because I want to see the National Park.
> 5 **I might go** on safari if there's room for me.
> 6 **I'm coming** home on the sixteenth. It's the day before I get married.

3 Aim: to practice using *might* to talk about the weather.

● Ask the students to make their own predictions about what the weather may be like in the future, and to write sentences.

4 Aim: to practise using *might* to talk about possible actions in the future.

● Ask the students to write about what they might do in the situations mentioned. To help them you may like to write on the board some suggestions, eg *stay at home, play tennis, go skiing, light a fire, watch TV, go to the cinema, go to the library, go to bed.*

WRITING

1 Aim: to provide a reason for reading the model letter, and to provide prompts for the letter writing.

● The questions are designed to provide a reason for reading the model letter which is to be used in activity 4 for letter writing. The questions will be useful prompts at that stage.

2 Aim: to focus on the order and organisation of the information in the model letter.

● The order of the questions was not necessarily the order in which you would expect to see the information in a letter. Ask the students to number the questions in the order in which they were answered.

> **Answers to 1 and 2**
> **What type of accommodation might Delphine like?**
> [4] 'You may like to stay in a bed and breakfast, which is a private house with a guest bedroom. You won't have to pay much for this type of accommodation.'
> **What type of clothes should Delphine bring and why?**
> [3] '...it might be a bit cold in the evening, so bring a sweater.'
> **When might be the best time to come?**
> [1] 'I think October is the best time to come.'
> **What might Delphine like to do?**
> [5] 'When you arrive in London you may like to buy a railcard to travel around the rest of the country.'
> **What might the weather be like?**
> [2] 'It's usually quite warm, although there may be some rain, and it might be a bit cold in the evening.'

3 Aim: to make notes in preparation for writing a letter.

● This stage of the activity sequence focuses on collecting information and preparing to write the letter. You may like to ask the students to do this in pairs if they come from the same country, so that they collect as many ideas as possible. This preparation stage is very important, so give them about ten minutes to do this.

4 Aim: to write a letter.

● The final stage of this activity sequence is to use the model letter, the prompt questions and the information and ideas collected in the preceding stages to write a letter.

● The students may like to complete this activity for homework.

GRAMMAR

> *Might* and *may* for possibility
>
> *Might* and *may* are modal verbs. Remember that they have the same form for all persons and you don't use the auxiliary *do* in questions and negatives. You can use *might* or *might not* + infinitive to talk about possible future events.
> It **might** rain tomorrow.   There **might not** be much sun.
>
> You can also use *may* or *may not*.
> **May** is a little more sure than *might*.
> It **may** rain tomorrow.   There **may not** be much sun.
>
> Remember you use *will* or *won't* to make predictions.
> It**'ll** be dry and sunny tomorrow over the whole country.
> There **won't** be much wind.

1 Imagine you're talking to a friend who wants to visit your country. Give the following advice and explain why. Use *because + might*.

1 bring an umbrella   4 get some sun-tan lotion
2 wear warm clothes   5 buy a good map
3 bring your camera   6 bring a swimsuit

You should bring an umbrella because it might rain.

2 Choose the correct form of the verb.

1 *I'm going/I might go* to Kenya next week. I've got my ticket.
2 *I've booked/I may book* into a hotel. I've got a room at the Ambassador.
3 *I'm going /I might go* to Lake Victoria or perhaps to the beach near Mombasa.
4 *I won't/I might not* spend more than a week on the coast, because I want to see the National Park.
5 *I'm going/I might go* on safari if there's room for me.
6 *I'm coming/I might come* home on the sixteenth. It's the day before I get married.

3 Say what the weather might or might not be like:

– tomorrow   – next week
– next month   – in ten years' time

4 Think about next weekend. Say what you might or might not do if...

1 ...it's raining   4 ...it's sunny
2 ...it's snowing   5 ...you have work to do
3 ...it's cold   6 ...you feel tired

1 I might stay at home and watch TV.

WRITING

1 Delphine wrote to her friend Benita asking for some advice. Read Delphine's letter and Benita's reply, and find the answers to these questions.

a What type of accommodation might Delphine like?
b What type of clothes should Delphine bring and why?
c When might be the best time to come?
d What might Delphine like to do?
e What might the weather be like?

> Dear Benita,
> At last I'm thinking of coming to Britain. When's the best time to come? Thanks for your help!
> Love, Delphine

> Dear Delphine,
>
> It's good to hear you may want to visit Britain. I think October is the best time to come. It's usually quite warm, although there may be some rain, and it might be a bit cold in the evening, so bring a sweater. Best of all, there won't be so many tourists. Make sure you bring an umbrella. You may like to stay in a bed and breakfast, which is a private house with a guest bedroom. You won't have to pay much for this type of accommodation. When you arrive in London you may like to buy a railcard to travel around the rest of the country.
>
> I hope this is useful.
>
> Best Wishes, *Benita*

2 Number the questions in the order that Benita answers them.

3 Imagine an English friend wants to visit your country and has asked you when the best time is to visit. Make notes answering the questions in 1.
Accommodation: stay in a bed and breakfast...

4 Write a letter to an English friend who wants to visit your country. Use the letter in 1 to help you and your notes. Make sure you answer the questions in full sentences, and in the order you numbered them in 2.

37 | *Help!*

First conditional

VOCABULARY AND SPEAKING

1 Look at the words in the box. Which can you use to describe the incident in the picture?

accident electricity steal break victim kill button catch fire consulate dangerous passport shock ambulance wallet fire flood injured package gas ground floor gun burn mugger plug plug in press rescue drown driving licence bomb switch on switch off explode witness unplug burglar

2 Put the words in the box with the following incidents. Many can go with more than one situation.

a an emergency at home

b a road accident

c a bomb alert

d a mugging

e a flood

Check your answers with another student.

3 Look at these sentences. Decide in which incident you might hear them. Can you guess what the underlined pronouns refer to?

1 'If you touch <u>it</u>, you'll get a shock.'

2 'If you don't give <u>it</u> back, I'll call the police!'

3 'If <u>it</u> doesn't stop rising, we won't be able to escape.'

4 'If you go to the consulate, they'll give you another <u>one</u>.'

5 'If you lose <u>them</u>, <u>they</u>'ll replace them in a couple of days.'

6 'If you light a match, <u>it</u>'ll cause an explosion.'

7 'If we call an ambulance, <u>they</u>'ll take her to hospital.'

8 'If no one picks <u>it</u> up, I'll call the guard.'

GRAMMAR

First conditional
You use the first conditional to talk about a likely situation and to describe its result. You talk about the likely situation with *if* + present simple. You describe the result with *will* or *won't*.
You separate the two clauses with a comma.
*If you **give** it to me, I'll let you go.*
*If you **don't give** it back, I'll call the police.*
*If it **doesn't stop** rising, we **won't be able to** escape.*
You often use the first conditional for promises, threats or warnings.

1 Match the two parts of the sentences.

1 If you unplug the machine,

2 If you give me the money,

3 If we don't escape,

4 If they give you another passport,

5 If they replace your traveller's cheques,

6 If there's an explosion,

7 If she goes to hospital,

8 If the guard thinks it's a bomb,

a ...there won't be any trouble when you leave the country.

b ...you'll be able to carry on with your holiday.

c ...I'll let you go.

d ...we'll drown.

e ...you won't get a shock.

f ...they'll mend her broken leg.

g ...he'll ask everyone to leave the train.

h ...there'll be a lot of damage.

2 Complete the sentences.

1 If I have enough money next summer,...

2 If I have time this week,...

3 If my friends are doing nothing tonight,...

4 If I work hard next week,...

5 If I finish this book,...

6 If I learn English,...

37

GENERAL COMMENTS

First conditional

There are many possible ways of making a conditional sentence with present and future tenses, eg *If you've got his number, why don't you give him a ring? If you see John, tell him I'm OK. If you leave butter in the sun, it melts. If you will come this way, I'll tell her you're here.* But the conditional which causes many students problems is the one for likely situations and their results, where the verb in the *if* clause goes in the present simple, and the verb in the main clause in the future. The other common error in writing is missing out the comma which separates the two clauses. This is obviously not serious, but it makes longer sentences more difficult to read. Lesson 39 covers the second conditional for unreal or hypothetical events and situations.

Help!

The theme of the lesson is emergency situations. If you ask the students to share their own experiences of emergency situations, it is important not to cause them embarrassment or distress. In communicative methodology, in which the student is invited to contribute actively both to the learning process and to the activities in class, it is very easy to transgress social conventions concerning personal feelings and privacy. The risk is similar to a cross-cultural situation in which someone causes or perceives offence where none was intended.

Illustration: possible questions

What can you see in the picture?
Where do you think it is? In a town or in the country? In Britain or in your country?
Who do you think the woman is?
What can you see just behind her, from where we're looking?
Who do you think the person in the shadow is?
What is he or she doing there?

VOCABULARY AND SPEAKING

1 Aim: to focus on the words in the vocabulary box.

● Ask the students if they have ever been in an emergency situation. Ask them to talk about what happened, if you think this can be done without causing distress.

● Ask the students to look at the picture and say which words might be used to describe it.

> **Possible answers**
> mugger steal wallet witness victim

2 Aim: to group the vocabulary items under the headings to help the process of acquisition.

● The students don't need to use all of these words productively but they may find them useful for receptive purposes if they start reading English language newspapers.

> **Possible answers**
> **an emergency at home:** accident, ambulance, break, burglar, burn, button, catch fire, dangerous, drown, electricity, explode, fire, flood, injured, gas, ground floor, kill, plug, plug in, press, shock, steal, switch on, switch off, unplug, victim
> **a road accident:** accident, ambulance, burn, catch fire, dangerous, explode, fire, injured, kill, rescue, shock, witness, victim
> **a bomb alert:** bomb, dangerous, explode, injured, kill, package, rescue, victim
> **a mugging:** consulate, dangerous, driving licence, injured, gun, kill, mugger, passport, shock, steal, wallet, witness, victim
> **a flood:** dangerous, drown, flood, injured, rescue, victim

3 Aim: to present the first conditional.

● The first conditional is presented in each of these sentences. Ask the students to decide in which emergency situation they might hear the sentences. You can do this orally with the whole class.

> **Probable answers**
> 1 an emergency at home 2 a mugging 3 a flood
> 4 a mugging 5 a mugging 6 an emergency at home
> 7 a road accident 8 a bomb alert
> The pronouns probably refer to: 1 an electric cable
> 2 a bag or wallet 3 water 4 a passport 5 traveller's cheques, the bank 6 the flame 7 the ambulance officers 8 a bag or suspect package

GRAMMAR

1 Aim: to focus on the first conditional.

● Ask the students to read the information about the first conditional in the grammar box.

● Ask the students to do the exercise.

> **Answers**
> 1e c 3d 4a 5b 6h 7f 8g

2 Aim: to practise using the first conditional.

● Ask the students to write their own sentences saying what they will do in the circumstances mentioned. Check that they are using the present simple in the *if* clause, and are separating the two clauses with a comma.

SOUNDS

1 Aim: to focus on the sound /ł/.

● 📼 Many students find the /ł/ sound (the 'dark l') that comes at the end of a word rather difficult. If they find this activity difficult, don't insist on perfect pronunciation. Their intelligibility will not be affected. Ask them to listen and repeat the phrases.

2 Aim: to focus on merged sounds.

● 📼 The students will already know that certain words merge with other words in connected speech. Ask them to listen and mark the words the speaker links.

> **Answers**
>
> If you touch that, it'll explode.
>
> If no one picks it up, I'll do it.
>
> If you don't stop eating, you'll be ill.
>
> If you don't let go, I'll scream.
>
> If you don't give it to me, I'll tell your uncle.
>
> If you ask her, she'll answer.

● Play the tape again and pause after each sentence. Ask the students to say each sentence aloud.

LISTENING AND SPEAKING

1 Aim: to listen for main ideas.

● 📼 The students have already discussed the topic of emergency situations, so to a certain extent they will be ready for the listening activity. Explain that the incident is in three parts, and that you're going to play the first part. Ask them to listen and decide what the emergency incident is.

> **Answer**
> A mugging.

2 Aim: to check everyone has understood what happened.

● Check that everyone has understood the main ideas of the passage by asking one or two students to re-tell the story so far.

3 Aim: to prepare for listening to the second part.

● Ask the students to look at the extracts and to discuss what happens next.

● 📼 Explain to the students that they should listen and number the extracts in the order they hear them. Play the tape.

> **Answers**
> ...they had got some good news [1]
> ...now it was me who was feeling sorry [8]
> ...the young man wasn't Australian [3]
> ...admitted he was guilty [5]
> ...the bank clerk called the police [4]
> ...was unemployed and had a family to look after [7]
> ...a young man was trying to change some Australian money [2]
> ... 'I'm sorry, I'm really sorry.' [6]

4 Aim: to prepare for listening to the third part and to practise using the first conditional.

● Ask the students to discuss the situation and to complete the first conditional sentences.

5 Aim: to prepare for listening.

● Ask the students to predict how the story finishes.

● 📼 Play the tape and ask them to listen and find out if they guessed correctly.

● Ask the students if they think Kate made the right decision.

6 Aim: to practise speaking.

● Ask the students to talk about emergency situations they might have been in. You can also ask them to write about it for homework.

SOUNDS

1 🔊 Listen and repeat these phrases.

I'll	I'll do that	she'll	she'll show you
you'll	you'll like it	we'll	we'll love it
he'll	he'll call you	they'll	they'll learn it

2 🔊 Listen and mark the words the speaker links.

1 If you touch that, it'll explode.
2 If no one picks it up, I'll do it.
3 If you don't stop eating, you'll be ill.
4 If you don't let go, I'll scream.
5 If you don't give it to me, I'll tell your uncle.
6 If you ask her, she'll answer.

Now say the sentences aloud.

LISTENING AND SPEAKING

1 🔊 You're going to hear Kate, an Australian woman, describing an incident in three parts. Listen to the first part and decide what the incident is. Choose from the list in *Vocabulary and speaking* activity 2.

2 Work in pairs and check you understood what happened.

3 Work in pairs. Look at these extracts from the second part and try and guess what happens next. (The extracts are in the wrong order.)

...they had got some good news ☐
...now it was me who was feeling sorry ☐
...the young man wasn't Australian ☐
...admitted he was guilty ☐
...the bank clerk called the police ☐
...was unemployed and had a family to look after ☐
...a young man was trying to change some Australian money ☐
...'I'm sorry, I'm really sorry.' ☐

🔊 Now listen to the second part of the story and number the extracts as you hear them.

4 Work in pairs. Predict the possible results of these situations. There may be more than one result.

1 If the police let the man go,...
2 If the man gets a fine,...
3 If the magistrate sends the man to prison,...

5 Work in pairs. How do you think the story finishes?

🔊 Listen to the third part and find out. Do you think Kate made the right decision?

6 Have you ever been involved in an incident like Kate? Tell your partner about it.

My perfect weekend

Would for imaginary situations

My perfect weekend

We asked Stephen from Leeds, and Paula from Nottingham about their perfect weekend. Here are the questions and their answers.

1 Where would you go?
2 How would you travel?
3 Where would you stay?
4 Would you take a companion?
5 What essential piece of clothing or kit would you take?
6 What would you have to eat and drink?
7 Would you take anything to read?
8 Who would be your least welcome guest?
9 What luxury would you take?
10 What three things would you most like to do?

Stephen

a My lap-top computer.
b My swimming trunks, because I love the sea.
c Well, I'd like to take my dog, but I wouldn't be allowed to bring him back to England. So I'd go on my own.
d Sit on the beach, have a good lunch and read a good book.
e To a small town I know in France, by the sea. But I won't say where!
f Yes, 'The Kingdom by the Sea', by Paul Theroux.
g In a friend's house right by the beach. It's very quiet, a great place to relax.
h The local wine, bread, cheese and those enormous tomatoes you can get there.
i The Prime Minister. He would only make me angry.
j By train. I wouldn't go by plane – I hate it.

Paula

a Good shoes. I expect to be on my feet most of the day.
b At the Waldorf Hotel. It's quite an old-fashioned place, but the service is very good.
c Go shopping, have something to eat ... and then go shopping again.
d Probably not. Maybe I'd pick up a book at the airport, something by Jackie Collins.
e Pastrami on rye, bagels, in fact anything from those amazing delicatessens they have in New York. Oh, and champagne, of course.
f New York. I love the place. The shops are just wonderful.
g My ex-husband.
h By Concorde. More time for shopping!
i Yes, my best friend Shirley. We do everything together. She's the only person who likes shopping more than I do.
j My credit cards. But they're more of a necessity, actually.

38

GENERAL COMMENTS

Would

Would is a modal verb. The students have already come across *would* + infinitive in *What would you like? I'd like a Coca Cola.*

My perfect weekend

The topic is a lighthearted one, and the lesson is designed to give the students some straightforward language work and some motivating skills practice before the more significant lesson on the second conditional in Lesson 39.

Photos: possible questions

Who do you think these two people are? What do they do?
What are they doing?
How old are they?
Where are they?
Can you think of things they like doing?
Can you think of any details about their lifestyles?

VOCABULARY AND SPEAKING

1 Aim: to focus on the words in the vocabulary box; to pre-teach some difficult words from the reading passage.

● Ask the students to say which is their most important single possession. Do they think it is a luxury or a necessity?

● Ask them to think about the words in the box and decide which are luxuries and which are necessities.

2 Aim: to practise talking about luxuries and necessities.

● Ask the students to discuss their answers to activity 1. You may like to do this activity orally.

● The items have been chosen to generate cross-cultural comparison in a multi-cultural group, but in mono-cultural groups, you may like to ask the students to guess which cultures or countries might consider the items to be necessities.

● Ask the students to think of other items which they consider to be necessities.

● Then ask them to talk about their necessities with the rest of the class. This is likely to generate a lot of discussion, so give them plenty of time.

READING

1 Aim: to read and react to the passage.

● Ask the students to read the passage and to match the questions and the answers.

Answers		
	Stephen	**Paula**
1	e	f
2	j	h
3	g	b
4	c	i
5	b	a
6	h	e
7	f	d
8	i	g
9	a	j
10	d	c

2 Aim: to prepare to use *would* for imaginary situations.

● Ask the students to prepare for *Grammar* activity 2 by thinking about how they would spend their perfect weekend.

GRAMMAR

1 Aim: to practise writing questions with *would*.

● Ask the students to read the information about *would* in the grammar box.

● Explain that the sentences contain answers to questions which were not used in the passage. Ask the students to write the questions.

> **Possible answers**
> 1 What sports would you play?
> 2 What film would you watch?
> 3 What music would you listen to?
> 4 Would you go out to eat?
> 5 Would you buy a newspaper?
> 6 When would you go back home?

2 Aim: to practise writing sentences with *would*.

● Ask the students to use their answers to *Reading* activity 2 and to write full sentences in answer to the questions in the passage.

● For students who finish sooner than others, ask them to write full answers to the questions they wrote in activity 1.

SOUNDS

Aim: to focus on merged sounds in connected speech.

● 🔲 Explain that some sounds in clusters of consonants sound different or even disappear in connected speech. Ask the students to listen and notice how the *'d* is pronounced.

● Ask the students to say the sentences aloud. It isn't essential for them to pronounce these sentences accurately, but they should realise that if they do so, they will sound much closer to a native speaker accent, and may be easier for other people to understand.

LISTENING

1 Aim: to listen for main ideas.

● 🔲 In this activity, the students are asked to listen to the six speakers and decide which question they are answering. The students don't need to understand every word the speakers are saying.

> **Answers**
> Alex: question 6
> Barbara: question 10
> Alan: question 4
> Daniel: question 8
> Emma: question 1
> Jane: question 9

2 Aim: to check comprehension, and to practise speaking and using *would*.

● Ask the students to discuss their answers to activity 1 in pairs.

● 🔲 Play the tape again for the students to check their answers.

SPEAKING

1 Aim: to practise speaking and using *would*.

● Ask the students to discuss their perfect weekend. They can use their answers to *Grammar* activity 2. Make sure they ask for extra information.

2 Aim: to practise speaking and using *would*.

● Ask the students to imagine how other people in the class would answer the questions. You can do this orally.

3 Aim: to practise speaking and using *would*.

● Ask the students to find out if they guessed correctly.

● The students may like to use these answers to write about the other students' perfect weekends for homework.

VOCABULARY AND SPEAKING

1 Look at the words in the box. Are the things *luxuries* or *necessities*?

cassette recorder cat central heating champagne comb
companion computer credit card dog insurance jewellery
novel orange juice pen religion rice salt seafood soap
swimming trunks telephone washing machine water

2 Work in pairs and discuss your answers to 1. Now talk about other things that are necessities and luxuries for you.

READING

1 Read *My perfect weekend* and match the questions with Stephen and Paula's answers.

2 Think about your answers to the questions. Would you do the same things as Stephen and Paula?

GRAMMAR

Would for imaginary situations
Would **is a modal verb. Remember that modal verbs have the same form for all persons and you don't use the auxiliary** *do* **in questions and negatives.**
You use *would* **to talk about the consequence of an imaginary situation. You often use the contracted form** *'d* **or** *wouldn't.*

Where **would** you go? *I'd go to a small town in France.*
How **would** you travel? *I wouldn't go by plane.*

Questions	**Short answers**
Would *you take a companion?*	*Yes, I would.*
Would *you take anything to read?*	*No, I wouldn't.*

1 Here are some more answers from *My perfect weekend*. Write the questions.

1 'I'd play tennis and maybe go for a swim.'
2 '*Gone with the Wind*. I've seen it five times already.'
3 'Verdi's *Requiem*. It makes me cry.'
4 'Yes, I'd have dinner in a small restaurant in the countryside.'
5 'No, I don't like to read a newspaper when I'm on holiday.'
6 'Sunday night, or if possible, Monday morning.'

2 Write your answers to the questions in *My perfect weekend*. Write full sentences.

SOUNDS

Listen to these sentences. Notice how you don't always hear the /d/ in *'d* before a verb beginning with /t/ or /d/.

1 I'd take my friend.   4 She'd talk to her friends.
2 He'd do some work.   5 We'd drink champagne.
3 I'd tell you a story.   6 They'd travel by plane.

Now say the sentences aloud.

LISTENING

1 Listen to six people answering a question about their perfect weekend. Put the number of the question they answer by the name of the speaker.

Alex ☐   Barbara ☐   Alan ☐
Daniel ☐   Emma ☐   Jane ☐

2 Work in pairs. Say what their answers were.
Alex would drink coconut milk.

Listen again and check.

SPEAKING

1 Work in pairs. Ask and answer the questions about your perfect weekend. Ask for extra information about each question, using *would*.
Where would you go?
I'd go to my grandmother's house.
Why would you like to go there?
Because she would be pleased to see me.

2 Choose two or three of the people in your class and imagine how they would answer the questions.
I think Marco might go to the beach.
Yes, and he might go by motorbike.

3 Now ask the people you chose in 2 and find out if you guessed correctly.

Second conditional

GRAMMAR

> **Second conditional**
>
> **You use the second conditional to talk about an imaginary or unlikely situation and to describe its result. You talk about the imaginary or unlikely situation with *if* + past simple. You describe the result with *would* or *wouldn't*.**
>
> *If I **had** a lot of money, I **would** give some away.*
> *If a stranger **asked** me for money, I **wouldn't** give him any.*
>
> **You form the second conditional with *if* + past tense, *would* + infinitive. You separate the two clauses with a comma.**
>
> *If I found some money in the street, I'd keep it.*

1 Match the two parts of the sentences.

1 If I was extremely rich,   a I'd go to Jamaica.
2 If I won a free holiday,   b I'd give up my job.
3 If I lost my wallet,   c I'd live in New York.
4 If I spoke English fluently,   d I'd go to the police.
5 If I changed my job,   e I'd get a better job.
6 If I could live anywhere,   f I'd be much happier.

2 Write sentences saying what you would do if...

– a stranger asked you for money
– you didn't have any money
– someone lied to you
– someone offered you an expensive gift

READING AND LISTENING

1 Work in pairs. You're going to read a story called *The umbrella man* by Roald Dahl. Here are some words from part 1. What do you think is going to happen?

mother London dentist raining taxi passengers small man seventy umbrella favour suspicious help

Now read part 1 and check.

The umbrella man

Part 1

I'm going to tell you about a funny thing that happened to my mother and me yesterday evening. Yesterday afternoon, my mother took me to London to see the dentist. After that, we went to a cafe. When we came out of the cafe it was raining. 'We must get a taxi,' my mother said. Lots of them came by, but they all had passengers inside them.

Just then a man came up to us. He was a small man and he was probably seventy or more. He said to my mother politely, 'Excuse me.' He was under an umbrella which he held high over his head.

'Yes?' my mother said, very cool and distant.

'I wonder if I could ask a small favour of you,' he said. I saw my mother looking at him suspiciously. She is a suspicious person, my mother. The little man was saying, 'I need some help.'

39

GENERAL COMMENTS

Second conditional

The second conditional was presented implicitly in Lesson 38. In this lesson it is presented more explicitly. Make sure the students realise that the second conditional is for unreal or hypothetical situations and their results, something that is imaginary or unlikely to happen. The past tense used in the main clause is not used to express past time. *If I were you,...* is a common set phrase, and *were* can also be used for the third person singular, eg *If he were rich, he'd...* although it is less common than for the first person singular. The *if* clause can either come at the beginning or at the end of the sentence. In this lesson, the verb in the *if* clause is in the past simple, although there are other possibilities, eg, *If I was sitting there, I'd close the window. If you were feeling ill, you'd go home, wouldn't you? Would* can only be used in the *if* clause in polite requests. *If you would come with me, I'll take you to the manager*; this is not a second (unreal) conditional.

The umbrella man

The story in this lesson was written by Roald Dahl, who died in 1991 and who was famous for his children's books and for his collections of slightly surreal short stories such as *Kiss Kiss* and *Someone like you*. His *Tales of the Unexpected* were turned into a series of short films for television, which have been seen in many countries.

Reading and vocabulary

Try not to explain too many items of unfamiliar vocabulary. Although the story is quite long and may contain a number of new words, you are likely to spoil the students' enjoyment if you interrupt and exploit each part for vocabulary. If you do want to talk about the vocabulary, wait until the end of the story. The aim is to give fluency practice in reading, at least until everyone has finished the story.

Grammar

Exceptionally in *Reward*, this lesson begins with the grammar focus, because *would* has already been introduced, and to present and practise the second conditional at a later stage would spoil the enjoyment of the story.

Illustration: possible questions

Who do you think these people are?
Do they know each other? What's the relationship between them all?
Where are they?
What's the situation?
What's the man holding?
Have the mother and daughter got an umbrella?
Do you think they'd like one?

GRAMMAR

1 Aim: to present the second conditional.
● Ask the students to read the explanation in the grammar box and then do the exercises. Ask them to match the two clauses and make second conditional sentences. You can do this activity orally.

Answers	
1 If I was extremely rich,	b I'd give up my job.
2 If I won a free holiday,	a I'd go to Jamaica.
3 If I lost my wallet,	d I'd go to the police.
4 If I spoke English fluently,	e I'd get a better job.
5 If I changed my job,	f I'd be much happier.
6 If I could live anywhere,	c I'd live in New York.

2 Aim: to practise writing second conditional sentences; to prepare for reading.
● Ask the students to write sentences saying what they would do in the circumstances mentioned. The questions reflect dilemmas which confront the character in the story.

READING AND LISTENING

1 Aim: to pre-teach some difficult vocabulary from the story; to predict what is going to happen; to prepare for reading.
● Explain any words in the list that you think the students may have difficulty with, and then ask the students to work in groups of two or three to predict what is going to happen. The amount of time you spend on this activity depends on how motivated your students appear at the time. The story is quite long, and you may need to create an artificial motivation to start reading. If the students are slow getting down to work, it may be better to spend more time predicting the story so they are carefully prepared for reading. However, if the students appear interested in reading the story, it's best to let them get on with it.

2 Aim: to check comprehension and to practise using the second conditional.

● Ask the students to react to the story and say what they would do in these circumstances. Once again, if the motivation to read on is strong, don't spend too much time on this activity. At this stage, the focus of the work is fluency in reading rather than the accurate manipulation of the target structure.

3 Aim: to predict what is going to happen next in the story.

● Ask the students to work together and predict what is going to happen next by looking at the phrases and by numbering them in the right order. They should also guess who is speaking.

● 🔲 Ask them to listen and check their answers. Play the tape.

> **Answers**
> a 'Why don't you walk home?'
>   [5] (mother)
> b 'He's in some sort of trouble.'
>   [1] (writer)
> c 'Thank you, madam, thank you.'
>   [8] (old man)
> d 'I've never forgotten it before,'
>   [2] (old man)
> e 'I think I'd better just give you the taxi-fare.'
>   [6] (mother)
> f 'I would never accept money from you like that!'
>   [7] (old man)
> g 'I'm offering you this umbrella to protect you.'
>   [4] (old man)
> h 'Are you asking me to give you money?'
>   [3] (old man)

4 Aim: to practise using the second conditional.

● Ask the students to say what they would do in the circumstances if they were one of the three characters. Do this activity orally and elicit suitable answers using the second conditional.

● Ask the students to predict what is going to happen next.

● Ask the students to read on and check their predictions.

5 Aim: to predict what has happened and what will happen next.

● 🔲 By this stage, the students should be involved in the story and it is important not to interrupt this authentic motivation by obvious language practice. For this reason, there is no need for an artificial, pedagogical task to make them listen to the end of the story. They should be genuinely motivated to do this on their own.

VOCABULARY

Aim: to focus on the words in the vocabulary box.

● You may think there are other items you wish to draw the students' attention to. These words were selected, however, because they are probably the most useful and relevant for students at this level. The list is also a manageable number of words to learn.

WRITING

1 Aim: to write the story from another point of view.

● This activity involves a lot of creativity on the part of the students, but within a well-established context. Ask them to think about the story from the old man's point of view and to discuss the reasons for his behaviour.

● Ask the students to write a sentence starting the story from his point of view.

2 Aim: to practise writing in collaboration with other people.

● If you organise the writing activity in the way suggested, the students will have the opportunity of genuinely working together to produce their pieces of writing. This kind of collaborative writing can be very supportive to students who have difficulty with writing English. Each time they receive their neighbour's sentences, they should read them and add another sentence continuing the story on that piece of paper (not the story they started writing in 1).

● If time is short, they may like to finish this activity for homework, working alone. It would, nevertheless, be useful to start the activity in class.

2 Work in pairs. What would you say and do in these circumstances? Would you be suspicious?

3 Work in pairs. Here are some phrases from part 2 of the story. Who do you think is speaking? What do you think is going to happen? Put the phrases in the right order.

a 'Why don't you walk home ?'
b 'He's in some sort of trouble.'
c 'Thank you, madam, thank you.'
d 'I've never forgotten it before,'
e 'I think I'd better just give you the taxi-fare.'
f 'I would never accept money from you like that!'
g 'I'm offering you this umbrella to protect you.'
h 'Are you asking me to give you money?'

🔊 Now listen to part 2 of the story and check.

4 Work in pairs. Talk about what you would do in the circumstances if you were:

– the narrator   – the narrator's mother   – the old man

What do you think happens next? Read part 3 of the story and find out.

'Come under here and keep dry, darling,' my mother said. 'Aren't we lucky! I've never had a silk umbrella before.'
'Why were you so unpleasant to him?' I asked.
'I wanted to be sure he was a gentleman. I'm very pleased I was able to help him.'
'There he goes,' I said. 'Over there. He's crossing the street. He's in a hurry.'
We watched the little man. When he reached the other side of the street, he turned left, walking very fast.
'He doesn't look very tired, does he, mummy? He doesn't look as if he's trying to get a taxi, either.'
My mother was standing very still. 'He's up to something. Come with me.' We crossed the street together. It was raining very hard now, but we were under the silk umbrella.
'He said he was too tired to walk and now he's almost running.'
'He's disappeared!' I cried. 'Where's he gone?'

5 Work in pairs and answer these questions.

1 Where has the old man gone?
2 Do you think they will see him again?
3 How do you think the narrator and her mother feel?
4 How do you think the story finishes?

🔊 Now listen to the last part of the story.

VOCABULARY

Here are some useful words and expressions from the story. Check you know what they mean.

cafe dentist exchange excuse me expect favour gentleman happen pleased pound put on run silk stand still suspicious take out taxi trouble umbrella unpleasant

Are there any other words or expressions you would like to add to this list?

WRITING

1 Think about the old man's behaviour. Why do you think he behaved like this? Write a sentence describing what happened from his point of view. Think about what happened before he met the narrator and her mother.
It was a fine day when I went out, but soon it started raining, which was good, because I was feeling rather thirsty.

2 When you are ready, give your sentence to another student and you will receive an opening sentence from someone else. Read it, then write another sentence to continue the story, with as many details as possible. Change stories like this every time you write a sentence, until you finish the story. Write at least six sentences.

40 *How unlucky can you get?*

Past perfect: *after,* *when* and *because*

VOCABULARY AND LISTENING

1 You are going to listen to a story about an unlucky traveller. Here are some words from the story. What do you think is going to happen?

accident	arrest	break down	capital	cost	crash	exhaust pipe	fall off	guard
journey	mend	musician	nightmare	pack	police	queue	rusty	set off	spend

2 🔲 Now listen to the story. Did you guess correctly in 1?

3 Work in pairs. Look at these extracts from the story and number them in the correct order.

- ☐ He didn't have very much money left.
- ☐ He decided to go to Britain.
- ☐ He bought his ferry ticket.
- ☐ The police stopped him.
- ☐ He drove for a hundred miles.
- ☐ They let him through.
- ☐ He packed his bags and set off in his car.
- ☐ He had to queue all day at the border.
- ☐ At Calais the car broke down.
- ☐ He drove all night through Hungary.
- ☐ He drove off the boat.
- ☐ In France he crashed the car.
- ☐ The exhaust pipe fell off.
- ☐ The border guards looked carefully at his passport.

4 🔲 Listen again and check.

GRAMMAR

Past perfect: *after,* *when* and *because*

You use the past perfect to talk about one past action that happened before another past action. You often use *after,* *when* and *because*, and you use the past simple for the second action.

After he **had decided** to go to Britain, he **packed** his bags.
When he **had packed** his bags into his car, he **set off**.
Because he **had spent** so much, he **had** very little money.

You form the past perfect with *had* + past participle.
The car's exhaust pipe **had fallen** *off.*

You often use the contracted form *'d*.
He'd wanted to be a rock musician for years.

40

GENERAL COMMENTS

Past perfect

This tense may appear to be formed in a similar way in some students' native languages but may not operate in the same way. This is a brief introduction to the tense, and if your students have difficulty using it, they may need extra practice using material from the Practice Book and the Resource Packs. It will also be covered more fully in intermediate *Reward*. You only use the past perfect when you are already talking about the past and need to refer to an action or event which happened even earlier than the actions and events in which you are primarily interested. As soon as you become more interested in this earlier action or event, you use the simple past.

Contractions

You may need to point out that the contracted form *'d* stands for *had* in this lesson. In Lessons 38 and 39 it stands for *would*. The context usually makes this clear.

How unlucky can you get?

The story is known as an *urban legend*, which is a story common to many cultures and countries but told as if it had actually happened to a friend of someone the speaker knows. Apparently, this story has been told with other characters and nationalities involved. This version takes place in 1989 before the Berlin Wall came down but after the Hungarian borders had been opened. The M25 is the motorway which goes all the way round London and is about 100 miles long.

Photo

A queue of Trabant cars at the German Democratic Republic/Hungarian border in August 1989.
What can you see in the photo?
What's the situation?
Where does it take place?
Who are the people in the cars?
How do you think they feel?

VOCABULARY AND LISTENING

1 Aim: to present the words in the box, to predict what the story is going to be about; to prepare for listening.
● Write the words in the box on the board. Ask the students to look at the words and discuss how they might appear in a story called *How unlucky can you get?* Spend five or ten minutes on this important preparatory activity. The students should do this activity in groups of four or five.

● Ask the groups to tell the class their predictions.

2 Aim: to listen for main ideas.
● 🔊 Ask the students to listen to the story and see if their predictions were correct.

3 Aim: to listen for specific information.
● 🔊 Ask the students to number the events in the order they heard them. Play the tape again.

> **Answers**
> [10] He didn't have very much money left.
> [1] He decided to go to Britain.
> [9] He bought his ferry ticket.
> [14] The police stopped him.
> [13] He drove for a hundred miles.
> [6] They let him through.
> [2] He packed his bags and set off in his car.
> [4] He had to queue all day at the border.
> [8] At Calais the car broke down.
> [3] He drove all night through Hungary.
> [11] He drove off the boat.
> [7] In France he crashed the car.
> [12] The exhaust pipe fell off.
> [5] The border guards looked carefully at his passport

● You may like to check these answers orally. It is important to check that everyone has got the sequence of events correctly.

GRAMMAR

1 Aim: to focus on the sequence of events with *when*.
● Ask the students to read the information about the past perfect in the grammar box.

● Ask the students to decide which sentence is true according to the story.

> **Answer**
> *He'd left the boat when he heard a loud noise* is the true sentence. First he left the boat, then he heard a noise. *He left the boat when he heard a loud noise* would mean that he heard the noise and then left the boat, possibly even because he'd heard it.

2 Aim: to focus on the use of the past perfect.
● Ask the students to do the exercise.

> **Answers**
> 1 The people at the border **had decided** to leave their homes.
> 2 He felt nervous but the guards **let** him through.
> 3 He entered the West and he **drove** across Southern Europe.
> 4 At Calais the car broke down and it **cost** a lot of money to mend it.
> 5 He **had spent** most of his money by the time he got to England.
> 6 He heard a loud noise. He saw that the exhaust pipe **had fallen** off.

3 Aim: to focus on the sequence of tenses.
● Ask the students to do the exercise.

> **Answers**
> 1 He set off.
> 2 He waited all day at the border.
> 3 He drove across Southern Europe.
> 4 He heard a loud noise.

4 Aim: to focus on the sequence of tenses.
● Ask the students to do the exercise.

> **Answers**
> 1 After he'd packed his bags.
> 2 Because he had wanted to be a rock musician for many years.
> 3 After he had driven all night through Hungary.
> 4 Because thousands of other people had decided to leave their homes and go to the West.
> 5 After he had arrived in France.
> 6 Because the crash had damaged the exhaust pipe.
> 7 Because he had heard a loud noise.
> 8 After he had driven a hundred miles.

SOUNDS

Aim: to focus on merged sounds in connected speech.
● ▭ Explain that it is sometimes difficult to distinguish between a past simple and a past perfect in connected speech, because the contracted *'d* often merges into the following sounds. Play the tape and ask the students to tick the sentences they hear.

> **Answers**
> He decided to go to Britain.
> He'd packed his bags.
> He'd got to the Austrian border.
> He'd entered the West.
> He'd arrived in France.
> He had very little money.

● ▭ Ask the students to repeat the sentences. They should try to merge the 'd sound into the sound which follows. You can play the tape again and pause after each sentence so that the students can say it aloud.

SPEAKING AND WRITING

1 Aim: to predict the end of the story.
● Ask the students to work together and predict the end of the story. Ask them to think of other bits of bad luck that the man might have had.

2 Aim: to practise writing using the past perfect and *after, when* and *because*.
● Ask the students to write the story using the past perfect with *after, when,* and *because* and the events in *Vocabulary and listening* activity 3.

3 Aim: to read the end of the story.
● By now the students should be motivated enough to read the end of the story without having to perform an accompanying task.

4 Aim: to write notes about a series of unlucky events.
● Ask the students to think about times when they have been very unlucky. They can invent any details they would like to include.

5 Aim: to practise speaking and to use the past perfect and *after, when* and *because*.
● Ask the students to work in pairs and to ask and answer questions about each other's unlucky events.

6 Aim: to write a paragraph about a series of unlucky events.
● Ask the students to write a paragraph about their partners' series of unlucky events.

● You may like to ask the students to do this for homework.

1 Look at these sentences and explain the difference between them.

  a He left the boat when he heard a loud noise.

  b He had left the boat when he heard a loud noise.

2 Choose the best tense.

  1 The people at the border *decided/had decided* to leave their homes.

  2 He felt nervous but the guards *let/had let* him through.

  3 He entered the West and he *drove/had driven* across Southern Europe.

  4 At Calais the car broke down and it *cost/had cost* a lot of money to mend it.

  5 He *spent/had spent* most of his money by the time he got to England.

  6 He heard a loud noise. He saw that the exhaust pipe *fell/had fallen* off.

3 Answer these questions using the past simple.

  1 What did the man do after he'd packed his bags?

  2 What did he do after he'd driven all night through Hungary?

  3 What did he do after he'd entered the West?

  4 What happened after he had left the boat?

4 Answer these questions using the past perfect. Start your answer with the word in brackets.

  1 When did he set off for Liverpool?   (After...)

  2 Why did he decide to go to Britain?   (Because...)

  3 When did he have to queue all day at the border?   (After...)

  4 Why did he have to queue?   (Because...)

  5 When did he have a crash ?   (After...)

  6 Why did the car make more noise?   (Because...)

  7 Why did he stop the car?   (Because...)

  8 When did the police see him?   (After...)

SOUNDS

Listen and tick (✓) the sentences you hear.

1 He decided to go to Britain.   He'd decided to go to Britain.

2 He packed his bags.   He'd packed his bags.

3 He got to the Austrian border.   He'd got to the Austrian border.

4 He entered the West.   He'd entered the West.

5 He arrived in France.   He'd arrived in France.

6 He had very little money.   He'd had very little money.

Now say the sentences aloud.

SPEAKING AND WRITING

1 Work in pairs. How do you think the story finishes?

2 Rewrite the story joining the events in *Vocabulary and listening* activity 3 with *after, when* or *because* + past participle.

After he'd decided to go to Britain, he packed his bags and set off in his car.

3 Turn to Communication activity 17 on page 100 to find out how the story finishes.

4 Have you ever had an unlucky experience? Make a list of everything that happened, but don't write the events in the order they happened.

  – *missed my bus*

  – *had an argument with my boss*

  – *got to work late*

  – *forgot my briefcase*

  – *went home*

5 Work in pairs. Show each other the list of events. Ask and say the order in which they happened.

Did you miss your bus after you had forgotten your briefcase? Yes, I did.

6 Take your partner's notes and write a paragraph describing what happened to him/her. Use the past perfect and *after, when* and *because*.

Progress check  36-40

VOCABULARY

1 Look at these phrases with *make* and *do*.

do   do well   do harm   do business   do the washing up
do the housework   do your homework   do the shopping

make   make the bed   make the coffee   make a phone call
make a mistake   make a noise

You can put many words and phrases with *make* and *do*. *Make* often suggests creating something, and *do* often suggests work, but there are many exceptions.

Use your dictionary and find out if these words and phrases go with *make* or *do*.

an appointment   an arrangement   a cake   notes   friends   the cleaning
damage   your best   the ironing   a decision   a cup of tea   someone a favour

2 You use adverbs to describe verbs. You form an adverb by adding *-ly* to the adjective.

peaceful – peacefully   even – evenly   appropriate – appropriately

Adjectives ending in *-y* drop the *-y* and add *-ily*.
noisy – noisily   tidy – tidily

Adjectives ending in *-le* drop the *-e* and add *-y*.
comfortable – comfortably   sensible – sensibly

The adverb of *good* is *well*.
She speaks good Italian. She speaks Italian well.

Fast, hard, late, and *early* are both adjectives and adverbs.

Make adverbs from these adjectives.

cultural   local   slow   quick   heavy   quiet   expensive   beautiful
interesting   formal   fashionable   tidy   healthy   nervous   polite   happy

3 Complete these sentences with an adverb from 2.

1 The rain fell very ___ .
2 She was singing very ___ .
3 He dressed very ___ .
4 He spoke rather ___ .
5 It was raining so we drove very ___ .
6 We live ___ , not far from here.

4 Look at the vocabulary boxes for lessons 36 – 40 again. Choose words which are useful to you and write them in your *Wordbank*.

GRAMMAR

1 Rewrite these sentences with *might (not)*.

1 It will possibly rain today.
2 It is possible that I will stay a week.
3 She possibly won't ring me tonight.
4 It is possible that the plane won't leave on time.
5 He will possibly arrive soon.
6 It will possibly be less noisy.

1 It might rain today.

2 Write six things you may or may not do this weekend.

3 Make sentences with *if*.

1 leave now/catch your train
2 stay in bed/feel better
3 work hard/get a good job
4 eat carrots/be able to see in the dark
5 go shopping/spend a lot of money
6 ride a bike/save energy

1 If you leave now, you'll catch your train.

4 Write questions about your perfect day.

1 Where/go?
2 What/do in the morning?
3 Where/go in the afternoon?
4 How/get there?
5 What/do in the evening?
6 When/go home?

1 Where would you go?

Progress check 36–40

GENERAL COMMENTS

You can work through this Progress check in the order shown, or concentrate on areas which may have caused difficulty in Lessons 36 to 40. You can also let the students choose the activities which they would like to or feel the need to do.

VOCABULARY

1 Aim: to focus on idiomatic phrases with *do* and *make*.

● The students will have realised that *do* and *make* are very common words in English. Their native language may not make a distinction between the two; in this case, they may find this activity useful. Ask the students to use a dictionary and decide whether you use *make* or *do* with the words and phrases.

Answers
make: an appointment, an arrangement, a cake, notes, friends, a decision, a cup of tea
do: the cleaning, damage, your best, the ironing, someone a favour

2 Aim: to focus on the formation of adverbs.

● The students have already come across certain adverbs in pre-intermediate *Reward.* In Lesson 2 adverbs of frequency (*always, sometimes, occasionally,* etc.) were presented. This activity focuses on how to form adverbs from adjectives.

Answers
culturally, locally, slowly, quickly, heavily, quietly, expensively, beautifully, interestingly, formally, fashionably, tidily, healthily, nervously, politely, happily

3 Aim: to practise using adverbs.

● The students should use the adverbs they formed in activity 2 for this activity. Remind them that adverbs go with a verb, and adjectives go with a noun.

Possible answers
1 The rain fell very heavily.
2 She was singing very beautifully/nervously.
3 He dressed very fashionably.
4 He spoke rather quietly.
5 It was raining so we drove very slowly.
6 We live locally, not far from here.

4 Aim: to review the vocabulary presented in *Reward.*

● In this last lesson of pre-intermediate *Reward*, ask the students to spend ten or fifteen minutes looking through all the vocabulary boxes in the book, looking for ten or twelve words which they have forgotten. They should write them down.

● Ask the students for the words they have written down and write them on the board. Ask other students to explain the words.

GRAMMAR

1 Aim: to revise *might (not).*

Answers
1 It might rain today.
2 I might stay a week.
3 She might not ring me tonight.
4 The plane might not leave on time.
5 He might arrive soon.
6 It might be less noisy.

2 Aim: to revise *may (not).*

● Ask the students to write sentences saying what they may or may not do at the weekend.

3 Aim: to revise the first conditional.

Answers
1 If you leave now, you'll catch your train.
2 If you stay in bed, you'll feel better.
3 If you work hard, you'll get a good job.
4 If you eat carrots, you'll be able to see in the dark.
5 If you go shopping, you'll spend a lot of money.
6 If you ride a bike, you'll save energy.

4 Aim: to revise asking questions with *would*.

Answers
1 Where would you go?
2 What would you do in the morning?
3 Where would you go in the afternoon?
4 How would you get there?
5 What would you do in the evening?
6 When would you go home?

5 Aim: to revise writing sentences with *would*.
● Ask the students to write answers to the questions they wrote in activity 4.

6 Aim: to revise the second conditional.

Answers
1 f 2 b 3 d 4 e 5 a 6 c

7 Aim: to revise the second conditional.
● Ask the students to write sentences saying what they would do in the circumstances mentioned.

8 Aim: to revise the past simple and the past perfect.

Answers
1 After I had left the office, I went straight home.
2 When they arrived at the station, they had missed the train.
3 She walked slowly because she had hurt her ankle.
4 He wrote to me after he had had the accident.
5 She came downstairs when she had changed her clothes.
6 She was late because she had got lost.

9 Aim: to revise because and *after* + past perfect.

Answers
1 Because he had eaten too much.
2 After she had paid the bill.
3 After he had read the reviews.
4 After she had learnt English.
5 Because he had forgotten the way.
6 After she had read it.

SOUNDS

1 Aim: to focus on /w/ and /r/.
● ▭ Some students may have problems in pronouncing /w/ and /r/. Ask the students to say the words aloud. Then play the tape and ask them to repeat the words and put them in two columns.

Answers
/w/: world weeks wash water washing wear
/r/: return religion relative repair wrapping remind

2 Aim: to focus on /ɔː/ and /aʊ/.
● These diphthongs can be difficult to pronounce. Ask the students to say the words and to put them in two groups.

● ▭ Play the tape and pause after each word. Ask the students to repeat the words.

Answers
/ɔː/: poor law more moor roar saw bore sure pour four war
/aʊ/: power tower sour hour flower

● Point out how the same pronunciation can be represented by different spellings of the words in this activity.

3 Aim: to focus on stress timing.
● Remind the students that the stressed words in connected speech are the words which the speaker considers to be important. Ask the students to predict which words are likely to be stressed.

● ▭ Play the tape and ask the students to check their predictions.

Answers
A man from Leipzig in Germany, who had wanted to be a rock musician for many years, decided to go to England. When he had packed his bags into his old Trabant car, he set off for Liverpool, home of the Beatles, to make his name as a musician.

● Ask the students to write the important words on a separate piece of paper.

● They should turn to the instructions in the Communication activity, and reconstitute the passage as accurately as possible from the important words.

SPEAKING

1 Aim: to practise speaking and using the second conditional.
● Ask the students to work in pairs and to discuss the situations in which they would do the things.

2 Aim: to practise speaking.
● Ask the students to work in groups and discuss their answers to activity 1. They should choose a situation which amuses, surprises or shocks them and write a dialogue. Give them fifteen or twenty minutes, if possible, for this stage of the activity.

3 Aim: to practise speaking.
● Ask the groups in turn to act out their dialogues for the rest of the class.

5 Write answers to the questions in 4 about your perfect day.

6 Match the two parts of the sentence.

1 If you took the train,
2 If he left now,
3 If they bought a house in France,
4 If I learned English,
5 If you visited my parents,
6 If we moved house,

a ...they'd be pleased to see you.
b ...he'd be home by seven.
c ...we wouldn't be able to buy a new car.
d ...we'd spend our holidays with them there.
e ...I'd go round the world.
f ...you'd get there in three hours.

1 If you took the train, you'd get there in three hours.

7 Write sentences saying what you would do if...

1 you found £50 in the street.
2 someone asked you to lend them £50.
3 you spoke several languages.
4 your car broke down on the motorway.
5 you met an old friend on the way to work/school.
6 you had an important exam to take.

8 Rewrite these sentences with the past simple or the past perfect form of the verbs in brackets.

1 After I (leave) the office, I (go) straight home.
2 When they (arrive) at the station, they (miss) the train.
3 She (walk) slowly because she (hurt) her ankle.
4 He (write) to me after he (have) the accident.
5 She (come) downstairs when she (change) her clothes.
6 She (be) late because she (get) lost.

9 Answer the questions with *because* or *after* and the words in brackets.

1 Why did he feel ill?   (eat too much)
2 When did she leave?   (pay the bill)
3 When did he buy the book?   (read the reviews)
4 When did she learn French?   (learn English)
5 Why did he get lost?   (forget the way)
6 When did she give the newspaper back?   (read it)

1 Because he had eaten too much.

SOUNDS

1 Say these words aloud. Is the underlined sound /w/ or /r/? Put the words in two columns.

world   return   weeks   religion   wash   relative
water   repair   washing   wrapping   wear   remind

[cassette] Listen and check.

2 Say these words aloud. Is the underlined sound /ɔː/ or /aʊ/? Put the words in two columns.

poor   power   law   tower   sour   hour   more   moor
roar   saw   bore   flower   sure   pour   four   war

[cassette] Listen and check.

3 [cassette] Listen and underline the stressed words in this passage.

> A man from Leipzig in Germany, who had wanted to be a rock musician for many years, decided to go to England. When he had packed his bags into his old Trabant car, he set off for Liverpool, home of the Beatles, to make his name as a musician.

4 Write the stressed words on a separate piece of paper. Now turn to Communication activity 8 on page 99.

SPEAKING

1 Work in groups of two or three. In what situations would you...

– ask someone how long they are going to stay
– tell a friend the truth if it would hurt them
– ask someone to leave
– tell someone a lie
– lose your temper with a friend
– ask someone to be quiet

2 Work in groups and discuss your answers to 1. Choose a situation which amuses, surprises or shocks you and write a dialogue.

A: Hello, lovely to see you! When are you leaving?
B: Well, if it's all right with you, I'll stay for about six months.
A: What? Er, I mean, how nice...!

3 Act out your dialogues to the rest of the class.

Communication activities

1 *Lesson 15*
Listening and speaking, activity 2

Student A: 📼 Listen and find out what Karen has for breakfast and what Pat has for lunch.

When the recording stops, turn back to page 35.

2 *Lesson 16*
Listening and speaking, activity 1

Student A: 📼 Listen and find out what type of entertainment karaoke is, and what type of music they play. Find out where they perform tango and the reasons why people enjoy it.

When the recording stops, turn back to page 38.

3 *Progress check 16 – 20*
Speaking and writing

Student B: Dictate these sentences to *Student A* in turn. Write down the sentences *Student A* dictates.

1 _____
_____

2 So when a man finally got tickets he was surprised to find an empty seat between him and the next person, a woman dressed in black.

3 _____
_____

4 The woman replied, 'Yes, we bought them some months ago but then my husband died.'

5 _____
_____

6 The woman said, 'Well, they're all at the funeral.'

4 *Lesson 35*
Listening and speaking, activity 1

Student B: 📼 Listen and put a tick (✓) by Stephen's answers to these questions.

– What time does he have lunch and dinner?
   midday and 5pm ☐   12.30pm and 6pm ☐
   1pm and 7pm   ☐

– Does he use a napkin? If so, where does he put it?
   tucked under his chin ☐   tied round his neck ☐
   on his lap   ☐

– Where does he put his hands when he's at the table but not eating?
   on the table ☐   he puts his elbows on the table ☐
   on his lap ☐

– When does he usually drink coffee and tea?

	morning	afternoon	any time
coffee			
tea			

– Does he have soup in the summer?
   often ☐   sometimes ☐   never ☐

When the recording stops, turn back to page 83.

5 *Lesson 23*
Vocabulary and listening, activity 3

Group A: 📼 Listen to Barry and find the answers to these questions.

1 When is Australia Day?
2 What does the Melbourne Cup celebrate?
3 In what month is the Melbourne Cup?
4 What do people do before the race?
5 Who takes the day off work?

When the recording stops, check you have all got the same answers. Then turn back to page 54.

6 *Lesson 35*
Listening and speaking, activity 1

Student A: 📼 Listen and put a tick (✓) by Stephen's answers to these questions.

– What does he say at the start of a meal?
'Enjoy your meal!' ☐   'Cheers!' ☐   nothing ☐

– Does he usually use a knife and fork? If so, which hands does he hold them in?
fork in the left hand, knife in the right ☐
he cuts with the fork in his left hand and the knife in his right, then puts the fork in his right hand to eat ☐
he doesn't use a knife, he holds the fork in his left hand ☐

– Where does he put his knife and fork when he has finished a meal?
together in the centre of plate, with the handles pointing towards him ☐
together on the plate, slightly sideways ☐
on the plate in a V-shape ☐

– What food does he often eat with his fingers at the dining table?
chicken ☐   bread ☐   cheese ☐   cake ☐
chocolate ☐   fruit ☐   pie ☐   lettuce ☐   chips ☐

– What does he say when someone raises their glass?
'Cheers!' ☐   'Health!' ☐   nothing ☐

When the recording stops, turn back to page 83.

7 *Lesson 14*
Speaking, activity 2

Student A: Give *Student B* directions from places in column 1 to places in column 2. Tell them where you start from but not where you're going to.

1	2
Trinity College	Abbey Theatre
the Custom House	the Bank of Ireland
the Irish Parliament	Pearse Station

Change round when you're ready.

8 *Progress check 36 – 40*
Sounds, activity 4

Without looking at the passage, write it out in full. Use the stressed words to help you.

Now look back at the passage and check your version.

9 *Lesson 11*
Reading, activity 3

Mostly 'a'
You're extremely ambitious. You're never satisfied with your life and you're always trying to improve things. Try to relax and take things easy!

Mostly 'b'
You're fairly ambitious. You are very aware that life has much to offer, but you don't feel you can achieve very much. Keep trying but don't make yourself unhappy.

Mostly 'c'
You're so unambitious, you don't even know the meaning of the word. Look it up in a dictionary, if you can be bothered.

10 *Progress check 6 – 10*
Writing, activity 2

Read the passage and complete your version of the story.

> **The least successful annual conference was in 1985 when the Association of British Travel Agents went to Sorrento. The flight from Gatwick Airport to Naples was delayed because of fog. Many people were ill because of something they ate. The organisers asked the Minister of Development to give a speech to the delegates in the forum at Pompeii, so a local travel agent decided to drop 3,500 roses as a gesture of friendship. As the Minister began his speech, a plane flew low over the audience and dropped the roses, but the flowers landed outside the forum. No one heard the speech because the engine noise was so loud. Minutes later it returned with some more roses, and missed again. Five times it passed over Pompeii, and each time the roses landed on Mount Vesuvius. The last time it flew over the delegates so low that it forced them to lie on the ground.**
>
> Adapted from *The return of heroic failures*, by Stephen Pile

11 *Lesson 16*
Listening and speaking, activity 1

Student C: Listen and find out who performs karaoke. Find out what type of entertainment tango is and what type of music they play.

When the recording stops, turn back to page 38.

12 *Lesson 15*
Listening and speaking, activity 2

Student B: Listen and find out what Karen has for lunch and what Pat has for dinner.

When the recording stops, turn back to page 35.

13 *Lesson 18*
Vocabulary, activity 2

The words you chose to describe the first person show the kind of person you'd like to be. The words you chose for the second person show how you think other people see you. The words you use to describe the third person show the real you, your true character!

14 *Lesson 35*
Listening and speaking, activity 1

Student C: Listen and put a tick (✓) by Stephen's answers to these questions.

– How long does a typical lunch or dinner last?
15 minutes ☐   30 minutes ☐   45 minutes ☐

– At which meals does he eat the following food?

	melon	pasta	fish	steak
breakfast				
lunch or dinner				
never				

– Does he eat cake with a fork or a spoon?
fork ☐   spoon ☐   fork or spoon ☐   neither ☐

– When can he smoke during a meal?
before, during and after ☐
before and after ☐   never ☐

– Does he eat salad in the winter?
often ☐   sometimes ☐   never ☐

When the recording stops, turn back to page 83.

15 *Lesson 14*
Speaking, activity 2

Student B: Give *Student A* directions from places in column 1 to places in column 2. Tell them where you start from but not where you're going to.

1	2
St Stephen's Green	the National Library
Merrion Square	the Custom House
Lower Baggott Street	Trinity College

Change round when you're ready.

16 *Progress check 16 – 20*
Speaking and writing

Student A: Dictate these sentences to *Student B* in turn. Write down the sentences *Student B* dictates.

1 *The Phantom of the Opera* was one of London's most popular musicals and it was difficult to reserve seats.

2 _____

_____

3 He said, 'It took me a long time to get tickets for this show.'

4 _____

_____

5 The man said, 'I'm so sorry. But why didn't you ask a friend or a relative to come with you?'

6 _____

_____

17 *Lesson 40*
Speaking and writing, activity 3

The police stopped him and asked him where he had come from and what was wrong with the car. Unfortunately, he had learnt his English from Beatles songs, so he couldn't understand the police.

The police couldn't understand him either, and when they discovered that he had no money, they arrested him and sent him back to Germany. You can still see the rusty Trabant in the bushes, next to the M25.

18 *Lesson 23*
Vocabulary and listening, activity 3

Group B: 📼 Listen to Barry and find the answers to these questions.

1 What does Australia Day celebrate?
2 On what day of the week is the Melbourne Cup and at what time of day?
3 How old is the race?
4 How long does it last?
5 Who is interested in the race?

When the recording stops, check you have all got the same answers. Then turn back to page 54.

19 *Lesson 16*
Listening and speaking, activity 1

Student B: 📼 Listen and find out where the perform karaoke and the reasons why people enjoy it. Find out who performs tango.

When the recording stops, turn back to page 38.

20 *Lesson 15*
Listening and speaking, activity 2

Student C: 📼 Listen and find out what Pat has for breakfast and what Karen has for dinner.

When the recording stops, turn back to page 35.

21 *Lesson 13*
Listening, activity 2

Caracas
GUYANA
VENEZUELA
SURINAM
FRENCH GUIANA
Cali
Bogotá
COLOMBIA
Quito
Amazon
EQUADOR
PERU
Recife
Lima
L. Titicaca
BRAZIL
Machu Picchu
La Paz
Brasília
BOLIVIA
Belo Horizonte
PARAGUAY
Rio de Janeiro
CHILE
São Paulo
Asunción
Mendoza
Córdoba
Valparaíso
URUGUAY
Santiago
Buenos Aires
Montevideo
ARGENTINA
Bahía Blanca
Patagonia
PACIFIC OCEAN
SOUTH ATLANTIC OCEAN

Grammar review

CONTENTS

Present simple

Form

You use the contracted form in spoken and informal written English.

Be

Affirmative	Negative
I'm (I am)	I'm not (am not)
you	you
we 're (are)	we aren't (are not)
they	they
he	he
she 's (is)	she isn't (is not)
it	it

Questions	Short answers
Am I?	Yes, I am.
	No, I'm not.
Are you/we/they?	Yes, you/we/they are.
	No, you/we/they're not.
Is he/she/it?	Yes, he/she/it is.
	No, he/she/it isn't.

Have

Affirmative	Negative
I	I
you have	you haven't (have not)
we	we
they	they
he	he
she has	she hasn't (has not)
it	it

Questions	Short answers
Have I/you/we/they?	Yes, I/you/we/they have.
	No, I/you/we/they haven't.
Has he/she/it?	Yes, he/she/it has.
	No, he/she/it hasn't.

Regular verbs

Affirmative		Negative	
I		I	
you	work	you	don't (do not) work
we		we	
they		they	
he		he	
she	works	she	doesn't (does not) work
it		it	

Questions	Short answers
Do I/you/we/they work?	Yes, I/you/we/they do.
	No, I/you/we/they don't.
Does he/she/it work?	Yes, he/she/it does.
	No, he/she/it doesn't.

Question words with *is/are*
What's your name?   Where are your parents?

Question words with *does/do*
What do you do?   Where does he live?

Present simple: third person singular
(See Lesson 2)

You add *-s* to most verbs.
takes, gets

You add *-es* to *do, go* and verbs ending in
-ch, -ss, -sh and *-x*.
goes, does, watches, finishes

You add *-ies* to verbs ending in *-y*.
carries, tries

Use
You use the present simple:

● to talk about customs and habits. (See Lesson 1)
  In my country men go to restaurants on their own.

● to talk about routine activities. (See Lesson 2)
  He gets up at 6.30.

● to talk about a habit. (See Lesson 5)
  He smokes twenty cigarettes a day.

● to talk about a personal characteristic. (See Lesson 5)
  She plays the piano.

● to talk about a general truth. (See Lesson 5)
  You change money in a bank.

Present continuous

Form
You form the present continuous with *be* + present participle
(*-ing*). You use the contracted form in spoken and informal
written English.

Affirmative		Negative	
I'm (am) working		I'm not (am not) working	
you		you	
we	're (are) working	we	aren't (are not) working
they		they	
he		he	
she	's (is) working	she	isn't (is not) working
it		it	

Questions	Short answers
Am I working?	Yes, I am.
	No, I'm not.
Are you/we/they working?	Yes, you/we/they are.
	No, you/we/they aren't.
Is he/she/it working?	Yes, he/she/it is.
	No, he/she/it isn't.

Question words
What are you doing?   Why are you laughing?

Present participle (*-ing*) endings

You form the present participle of most verbs
by adding *-ing*.
go – going, visit – visiting

You add *-ing* to verbs ending in *-e*.
make – making, have – having

You double the final consonant of verbs of one syllable
ending in a vowel and a consonant, and add *-ing*.
get – getting, shop – shopping

You add *-ing* to verbs ending in a vowel and *-y* or *-w*.
draw – drawing, play – playing

You don't usually use these verbs in the continuous form.
*believe feel hear know like see smell sound
taste think understand want*

Use
You use the present continuous to say what is happening
now or around now. There is an idea that the action or state
is temporary. (See Lesson 5)
It's raining.   I'm learning English.

Past simple

Form

You use the contracted form in spoken and informal written English.

Be

Affirmative	Negative
I	I
he was	he wasn't (was not)
she	she
it	it
you	you
we were	we weren't (were not)
they	they

Have

Affirmative	Negative
I	I
you	you
we	we
they had	they didn't (did not) have
he	he
she	she
it	it

Regular verbs

Affirmative	Negative
I	I
you	you
we	we
they worked	they didn't work
he	he
she	she
it	it

Questions	Short answers
Did I/you/we/they work?	Yes, I/you/we/they did.
he/she/it	he/she/it
	No, I/you/we/they didn't.
	he/she/it

Question words

What did you do?   Why did you leave?

Past simple endings

You add *-ed* to most regular verbs.
walk – walked, watch – watched

You add *-d* to verbs ending in *-e*.
close – closed, continue – continued

You double the consonant and add *-ed* to verbs ending in a vowel and a consonant.
stop – stopped, plan – planned

You drop the *-y* and add *-ied* to verbs ending in *-y*.
study – studied, try – tried

You add *-ed* to verbs ending in a vowel + *-y*.
play – played, annoy – annoyed

Irregular verbs

There are many verbs which have an irregular past simple. For a list of the irregular verbs which appear in **Reward Pre-intermediate**, see page 112.

Pronunciation of past simple endings

/t/ *finished, liked, walked*
/d/ *continued, lived, stayed*
/ɪd/ *decided, started, visited*

Expressions of past time

(See Lesson 8)

yesterday, the day before yesterday, last weekend, last night, last month, last year

Use

You use the past simple:

● to talk about a past action or event that is finished. (See Lessons 6, 7 and 8)
  We walked towards each other.

Future simple (*will*)

Form

You form the future simple with *will* + infinitive. You use the contracted form in spoken and informal written English.

Affirmative	Negative
I	I
you	you
we	you
they 'll (will) work	they won't (will not) work
he	he
she	she
it	it

Questions	Short answers
Will I/you/we/they work?	Yes, I/you/we/they will.
he/she/it/	he/she/it/
	No, I/you/we/they won't.
	he/she/it/

Question words

What will you do?   Where will you go?

Expressions of future time

tomorrow, tomorrow morning, tomorrow afternoon,
next week, next month, next year, in two days,
in three months, in five years

Use

You use the future simple:

● to make a prediction or express an opinion about the future. (See Lesson 12)
I think most people will need English for their jobs.

● to talk about decisions you make at the moment of speaking. (See Lesson 13)
I'll give you the money right now.

● to talk about things you are not sure will happen with probably and perhaps. (See Lesson 13)
He'll probably spend three weeks there.
Perhaps he'll stay two days in Rio.

● to offer to do something. (See Lessons 13 and 30)
OK, I'll buy some food.

Present perfect simple

Form

You form the present perfect with *has/have* + past participle. You use the contracted form in spoken and informal written English.

Affirmative	Negative
I	I
you 've (have) worked	you haven't (have not) worked
we	we
they	they
he	he
she 's (has) worked	she hasn't (has not) worked
it	it

Questions	Short answers
Have I/you/we/they worked?	Yes, I/you/we/they have.
	No, I/you/we/they haven't.
Has he/she/it worked?	Yes, he/she/it have.
	No, he/she/it hasn't.

Past participles

All regular and some irregular verbs have past participles which are the same as their past simple form.
Regular: *move – moved, finish – finished, visit – visited*
Irregular: *leave – left, find – found, buy – bought*

Some irregular verbs have past participles which are not the same as the past simple form.
go – went – gone   be – was/were – been
drink – drank – drunk   ring – rang – rung

For a list of the past participles of the irregular verbs which appear in **Reward Pre-intermediate**, see page 112.

Been and gone

He's been to America = He's been there and he's back here now.
He's gone to America = He's still there.

Use

You use the present perfect:

● to talk about past experiences. You often use it with *ever* and *never*. (See Lesson 21)
Have you ever stayed in hospital?
I've had food poisoning several times.
I've never broken my leg.

● to talk about a past action which has a result in the present. It is not important when the action happened. You often use it to describe changes. (See Lesson 22)
She's got married.
I've moved to a new flat.
Have you found a new job?

You often use *just* to emphasise that something has happened very recently.
She's just had a baby.

● to talk about an action or state which began in the past and continues to the present. You use *for* to talk about the length of time. (See Lesson 23)
I've been here for two hours.

You use *since* to say when the action or state began.
I've been here since 8 o'clock.

8 o'clock	Now – 10 o'clock
I arrived.	*I am still here.*

Remember that if you ask for and give more information about these experiences, actions or states, such as *when*, *how*, *why* and *how long*, you use the past simple.

When did you stay in hospital? In 1975.
How long did you stay there? A week.

Past continuous

Form

You form the past continuous with *was/were* + present participle. You use the contracted form in spoken and informal written English.

Affirmative	Negative
I	I
he was working	he wasn't (was not) working
she	she
it	it
you	you
we were working	we weren't (were not) working
they	they

Questions	Short answers
Was I/he/she/it working?	Yes, I/he/she/it was.
	No, I/he/she/it wasn't.
Were you/we/they working?	Yes, you/we/they were.
	No, you/we/they weren't.

You don't usually use these verbs in the continuous form.
*believe feel hear know like see smell sound
taste think understand want*

Use

You use the past continuous:

● to talk about something that was in progress at a specific time in the past. (See Lesson 31)

What were you doing at nine o'clock yesterday morning? I was going to work.

8.30am	**9am**	**9.30am**
I left home. —————————➤		*I arrived at work.*

● to talk about something that was in progress at the time something else happened or interrupted it. You join the parts of the sentences with *when* and *while*. The verb in the *when* clause is usually in the past simple.
(See Lessons 31 and 32)
I was watching television when there was a knock at the door.

The verb in the *while* clause is usually in the past continuous.
While they were visiting the Palace, one of the women had an idea.

Remember that you use *when* + past simple to describe two things which happened one after the other.
When the two women got closer, they saw some people in eighteenth-century clothes.

Past perfect

Form

You form the past perfect with *had* + past participle. You use the contracted form in spoken and informal written English.

Affirmative	Negative
I	I
you	you
we	we
they 'd (had) worked	they hadn't (had not) worked
he	he
she	she
it	it

Questions	Short answers
Had I/you/we/they worked?	Yes, I/you/we/they had.
he/she/it/	he/she/it/
	No, I/you/we/they hadn't.
	he/she/it/

Use

You use the past perfect:

● to talk about one past action that happened before another past action. You often use *after, when* and *because*, and you use the past simple for the second action. (See Lesson 40)

After he had decided to go to Britain, he packed his bags.

Earlier past	Past	Now
He decided to go to Britain.	*He packed his bags.*	➤

When he had packed his bags into his car, he set off. Because he had spent so much, he had very little money.

Verb patterns

There are several possible patterns after certain verbs which involve *-ing* form verbs and infinitive constructions with or without *to*.

-ing form verbs

You can put an *-ing* form verb after certain verbs.
(See Lesson 4)
*I love walking.   She likes swimming.
They hate lying on a beach.*

Remember that *would like to do something* refers to an activity at a specific time in the future.
I'd like to go to the cinema next Saturday.

Try not to confuse it with *like doing something* which refers to an activity you enjoy all the time.
I like going to the cinema. I go most weekends.

Use

You use an *-ing* form verb:

- to describe the purpose of something after *to be for*. (See Lesson 25)
  A cassette player is for playing cassettes.

- to ask people to do things after *would you mind*. (See Lesson 30)
  Would you mind lending it to me?

To + infinitive

You can put *to* + infinitive after many verbs. Here are some of them:

agree decide go have hope learn leave like need offer start try want

Use

You use *to* + infinitive:

- with *(be) going to* and *would like to*. (See below)

- to describe the purpose of something. (See Lesson 25)

You use a cassette player to play cassettes.

Going to

You use *(be) going to*:

- to talk about future intentions or plans. (See Lessons 11 and 25)
  I'm going to be a doctor. (I'm studying medicine.)

- to talk about things which are arranged and sure to happen with *(be) going to*. (See Lesson 13)
  I'm going to visit South America. I've got my ticket.

You often use the present continuous and not *going to* with *come* and *go*.
Are you coming tonight?
NOT ~~Are you going to come tonight?~~
He's going to South America.
NOT ~~He's going to go to South America.~~

Would like to

You use *would like to*:

- to talk about ambitions, hopes or preferences. (See Lesson 11)
  I'd like to speak English fluently.

Modal verbs

The following verbs are modal verbs.
can could may might must should will would

Form
Modal verbs:

- have the same form for all persons.
  I must leave.   He must be quiet.

- don't take the auxiliary *do* in questions and negatives.
  Can you drive?   You mustn't lean out of the window.

- take an infinitve without *to*.
  I can type.   You should see a doctor.

Use

You use *can*:

- to express general ability, something you are able to do on most occasions. (See Lesson 27)
  I can swim a hundred metres.

- to say what you're allowed to do or what it is possible to do. (See Lesson 28)
  You can cross when the light is green.

- to ask for permission. (See Lesson 30)
  Can I smoke?

- to ask people to do things. (See Lesson 30)
  Can you speak louder, please?

Can is a little less formal than *could*.

You use *can't*:

- to say what you're not allowed to do or what it is not possible to do. (See Lesson 28)
  You can't cross when the light is red.

  You can also use *mustn't*. (See Lesson 26)
  You mustn't cross when the light is red.

You use *could*:

- to express general ability in the past. (See Lesson 27)
  When I was five, I could swim but I couldn't write my own name.

- to ask for permission. (See Lesson 30)
  Could I leave now?

- to ask people to do things. (See Lesson 30)
  Could you help me?

  Could is a little more formal than *can*.

You use *may*:

- to ask for permission. (See Lesson 30)
  May I call you Esther?

- to talk about possible future events. (See Lesson 36)
  It may rain tomorrow.

  May has almost the same meaning as *might*.

You use *mght*:

- to talk about possible future events. (See Lesson 36)
  It might rain tomorrow.

  You don't usually use *might* in questions.

You use *must*:

- to talk about something you are obliged to do. The obligation usually comes from the speaker and it can express a moral obligation, necessity, strong advice or a strong suggestion. (See Lesson 26)
  It's late. I must go now. You really must stop smoking.

You often use *must* to talk about safety instructions.
You must fasten your seatbelt.

Have to has almost the same meaning as *must* but the obligation comes from a third person. You often use it to talk about rules.
The government says you have to do military service.

You use *mustn't*:

● To talk about something you're not allowed to do. (See Lesson 26)
  You mustn't smoke here.

  You can also use *can't*.
  You can't smoke here.

You use *should*:

● to give advice. It can also express a mild obligation or the opinion of the speaker. (See Lesson 29)
  You should take some exercise. You shouldn't smoke.

● *Ought(n't) to* has the same meaning as *should(n't)*.
  You ought to take some exercise.

For the uses of *will* see *Future simple (will)*

You use *would*:

● to ask for permission with *mind if*. (See Lesson 30)
  Would you mind if I borrowed it?

● to ask people to do things with *mind + -ing* form verbs. (See Lesson 30)
  Would you mind lending it to me?

● to talk about the consequences of an imaginary situation. (See Lesson 38)
  I'd go to a small town in France.

Conditionals

First conditional

Form

You form the first conditional with *if* + present simple, *will* + infinitive.
If you touch it, you'll get a shock.

Use

You use the first conditional to talk about a likely situation and to describe its result. You talk about the likely situation with *if* + present simple. You describe the result with *will* or *won't*. (See Lesson 37)
If you give it to me, I'll let you go.
If it doesn't stop rising, we won't be able to escape.

You often use the contracted form in speech and informal writing. The *if* clause can come at the beginning or at the end of the sentence.

Second conditional

Form

You form the second conditional with *if* + past simple, *would* + infinitive.
If I found some money in the street, I'd take it to the police.

You can also use the past continuous in the *if* clause.
If it was raining, I'd take an umbrella.

Use

You use the second conditional to talk about an imaginary or unlikely situation and its result. You talk about the imaginary or unlikely situation with *if* + past tense. You describe the result with *would* or *wouldn't*. (See Lesson 39)
If I had a lot of money, I would give some away.
If a stranger asked me for money, I wouldn't give him any.

You often use the contracted form in speech and informal writing. The *if* clause can come at the beginning or at the end of the sentence. It is still common to see *were* and not *was* in the *if* clause.
If I were you, I'd go home

Present simple passive

Form

You form the present simple passive with *am/is/are* + past participle.
A splendid feast is prepared.
Shops are filled with symbols of death.

Use

You use the passive to focus on the object of the sentence. You can use it when you don't know who or what does something. The object of an active sentence becomes the subject of a passive sentence. (See Lesson 33)
A bowl of water is placed on the table.

If you are more interested in the object but you know who or what does something, you use *by*.
A special kind of bread is baked by the men.

Have got

Form

You use the contracted form in spoken and informal written English.

Affirmative	Negative
I	I
you 've (have) got	you haven't (have not) got
we	we
they	they
he	he
she 's (has) got	she hasn't (has not) got
it	it

Questions	Short answers
Have I/you/we/they got?	Yes, I/you/we/they have.
	No, I/you/we/they haven't.
Has he/she/it got?	Yes, he/she/it has.
	No, he/she/it hasn't.

Use

You use *have got* to talk about facilities, possession or relationship. (See Lesson 10)
I've got a new car.

You don't use *have got* to talk about a habit or routine.
I often have lunch out. NOT ~~I often have got lunch out.~~

You don't usually use *have got* in the past. You use the past simple of *have*.
I had a headache yesterday. NOT ~~I had got a headache.~~

Questions

You can form questions in two ways:

- with a question word such as *who, what, which, where, how, why.*
  What's your name?

- without a question word.
  Are you English?

You can put a noun after *what* and *which*. (See Lesson 1)
What time is it?   Which road will you take?

You often say *what* to give the idea that there is more choice.
What books have you read lately?

You can put an adjective or an adverb after *how*.
(See Lessons 15 and 20)
How much is it?   How long does it take by car?
How fast can you drive?

You can use *who, what* or *which* as pronouns to ask about the subject of the sentence. You don't use *do* or *did*.
(See Lessons 1 and 7)
What's your first name?
Who organised the first package trip?

You can use *who, what* or *which* and other question words to ask about the object of the sentence. You use *do* or *did*.
(See Lessons 1 and 7)
Who did he take on the first package trip?

Defining relative clauses

- You use a defining relative clause to define people, things and places. The information in the defining relative clause is important for the sense of the sentence. (See Lesson 24)

You use *who* for people.
A druggist is someone who sells medicine in a shop.

You use *which* for things.
A subway is a railway which runs under the ground.

You often use *that* instead of *who* or *which*.
A druggist is someone that sells medicine in a shop.
A subway is a railway that runs under the ground.

You use *where* for places.
A parking lot is a place where you park your car.

You can use *that* after a superlative adjective instead of *who, which* or *where*. (See Lesson 19)
Who is the nicest person that you know?
What is the most expensive thing that you own?

Remember that there is no comma before a defining relative clause. In speech there is no pause.

Articles

You can find the main uses of articles in Lesson 3. Here are some extra details.

You use *an* for nouns which begin with a vowel.
an armchair

You use *one* if you want to emphasise the number.
One hundred and twenty-two.

Before vowels, you pronounce *the* /ðiː/

You do not use the definite article with parts of the body.
You use a possessive adjective.
I'm washing my hair.

Plurals

You can find the main rules for forming plurals in Lesson 3.

Possessives

You can find the main uses of the possessive *'s* in Lesson 9.

You can find a list of possessive adjectives in Lesson 9.

Expressions of quantity

Countable and uncountable nouns

Countable nouns have both a singular and a plural form.
(See Lesson 15)
an apple – some apples, a melon – some melons,
a potato – some potatoes, a cup – (not) many cups,
a biscuit – a few biscuits

Uncountable nouns do not usually have a plural form.
some wine, some cheese, some fruit, (not) much meat,
a little coffee

If you talk about different kinds of uncountable nouns, they become countable.
Beaujolais and Bordeaux are both French wines.

Expressions with countable or uncountable nouns

You can put countable or uncountable nouns with these expressions of quantity.
lots of apples, lots of cheese, hardly any apples, harldy any cheese, quite a lot of fruit, quite a lot of potatoes

Some and any

(See Lesson 15)

Affirmative   *There's some milk in the fridge.*
   There are some apples on the table.

Negative   *I haven't got any brothers.*
   There isn't any cheese.

Questions

You usually use *any* for questions.
Is there any sugar?

You can use *some* in questions when you are making an offer or a request, and you expect the answer to be *yes*.
Would you like some tea?
Can I have some sugar, please?

Much and many

You use *many* with countable nouns and *much* with uncountable nouns. (See Lesson 15)
How many eggs would you like?
How much butter do you need?

Too much/many, not enough, fewer, less and more

You can put a countable noun in the plural after *too many*, *not enough* and *fewer*.
There are too many people.
There aren't enough clean rivers.
In Britain there are fewer men than women.

You can put an uncountable noun after *too much, not enough, more* and *less*.
There's too much noise.   *There isn't enough farmland.*
There's more pollution.

You can put an adjective after *too* or between *not* and *enough*.
The sea is too polluted.   *The air isn't clean enough.*

Making comparisons
Comparative and superlative adjectives

Form
You add *-er* to most adjectives for the comparative form, and *-est* for the superlative form. (See Lesson 18)
cold colder coldest   cheap cheaper cheapest

You add *-r* to adjectives ending in *-e* for the comparative form, and *-st* for the superlative form.
large larger largest   fine finer finest

You add *-ier* to adjectives ending in *-y* for the comparative form, and *-iest* for the superlative form.
happy happier happiest
friendly friendlier friendliest

You double the last letter of adjectives ending in *-g, -t,* or *-n*.
hot hotter hottest   thin thinner thinnest

You use *more* for the comparative form and *most* for the superlative form of longer adjectives.
expensive more expensive most expensive
important more important most important

Some adjectives have irregular comparative and superlative forms.
good better best   bad worse worst

More than, less than, as ... as

● You put *than* before the object of the comparison. (See Lesson 19)
Children wear more casual clothes than their parents.

● You use *less ... than* to change the focus of the comparison.
Parents wear less casual clothes than their children.

● You can put *much* before the comparative adjective, *more* or *less* to emphasise it.
They're much less formal than they were.

● You use *as ... as* to show something is the same.
They're as casual as teenagers are all over the world.

● You use *not as ... as* to show something is different.
Dresses are not as popular as in Western countries.

But, however, although

You use *but, however,* and *although* to make a comparison which focuses on a difference. (See Lesson 35)

You put *but* at the beginning of a sentence or to join two sentences.

We drink coffee in the morning. But we don't drink it in the afternoon.
We drink coffee in the morning but we don't drink it in the afternoon.

You use *although* at the beginning of a subordinate clause. You need to separate the subordinate and the main clause with a comma.
We usually have dinner at six, although some people have dinner later.

You can put the subordinate clause at the beginning or at the end of the sentence.
Although we usually have dinner at six, some people have dinner later.

You put *however* at the beginning of a sentence. You put a comma after it.
We drink coffee in the morning. However, we don't drink it in the afternoon.

So, because

- You can join two sentences with *so* to describe a consequence.
  She often took the plane so she didn't look at the safety instructions.

- You can join the same two sentences with *because* to describe a reason.
  She didn't look at the safety instructions because she often took the plane.

Prepositions of place

(See Lesson 14)

Prepositions of time and place:
in, at, on, to

(See Lesson 16)

Use

You use *in*:

- with times of the day: *in the morning, in the afternoon*
- with months of the year: *in March, in September*
- with years: *in 1996, in 1872*
- with places: *in London, in France, in the cinema*

You use *at*:

- with times of the day: *at night, at seven o'clock*
- with certain expressions of time: *at the weekend*
- with places: *at the theatre, at the stadium*

You use *on*:

- with days, dates: *on Friday, on 15th July*

You use *to*:

- with places: *Let's go to London.*

Adverbs of frequency

Use

You use an adverb of frequency to say how often things happen. (See Lesson 1)
They always take their shoes off.
We usually take chocolates or flowers.
We often wear jeans and sweaters.
We sometimes arrive about fifteen minutes before.
We never ask personal questions.

Pronunciation guide

/ɑː/	park	/b/	buy
/æ/	hat	/d/	day
/aɪ/	my	/f/	free
/aʊ/	how	/g/	give
/e/	ten	/h/	house
/eɪ/	bay	/j/	you
/eə/	there	/k/	cat
/ɪ/	sit	/l/	look
/iː/	me	/m/	mean
/ɪə/	beer	/n/	nice
/ɒ/	what	/p/	paper
/əʊ/	no	/r/	rain
/ɔː/	more	/s/	sad
/ɔɪ/	toy	/t/	time
/ʊ/	took	/v/	verb
/uː/	soon	/w/	wine
/ʊə/	tour	/z/	zoo
/ɜː/	sir	/ʃ/	shirt
/ʌ/	sun	/ʒ/	leisure
/ə/	better	/ŋ/	sing
		/tʃ/	church
		/θ/	thank
		/ð/	then
		/dʒ/	jacket

Irregular Verbs

Verbs with the same infinitive, past simple and past participle

cost	cost	cost
cut	cut	cut
hit	hit	hit
let	let	let
put	put	put
read /ri:d/	read /red/	read /red/
set	set	set
shut	shut	shut

Verbs with the same past simple and past participle, but a different infinitive

bring	brought	brought
build	built	built
burn	burnt/burned	burnt/burned
buy	bought	bought
catch	caught	caught
feel	felt	felt
find	found	found
get	got	got
have	had	had
hear	heard	heard
hold	held	held
keep	kept	kept
learn	learnt/learned	learnt/learned
leave	left	left
lend	lent	lent
light	lit/lighted	lit/lighted
lose	lost	lost
make	made	made
mean	meant	meant
meet	met	met
pay	paid	paid
say	said	said
sell	sold	sold
send	sent	sent
sit	sat	sat
sleep	slept	slept
smell	smelt/smelled	smelt/smelled
spell	spelt/spelled	spelt/spelled
spend	spent	spent
stand	stood	stood
teach	taught	taught
understand	understood	understood
win	won	won

Verbs with same infinitive and past participle but a different past simple

become	became	become
come	came	come
run	ran	run

Verbs with a different infinitive, past simple and past participle

be	was/were	been
begin	began	begun
break	broke	broken
choose	chose	chosen
do	did	done
draw	drew	drawn
drink	drank	drunk
drive	drove	driven
eat	ate	eaten
fall	fell	fallen
fly	flew	flown
forget	forgot	forgotten
give	gave	given
go	went	gone
grow	grew	grown
know	knew	known
lie	lay	lain
ring	rang	rung
rise	rose	risen
see	saw	seen
show	showed	shown
sing	sang	sung
speak	spoke	spoken
swim	swam	swum
take	took	taken
throw	threw	thrown
wake	woke	woken
wear	wore	worn
write	wrote	written

Tapescripts

Lesson 1 — Speaking and listening, activity 2

Situation 1

PATRICIA Thierry! How lovely to see you. You do look well.
THIERRY How are you Patricia?
PATRICIA We're all very well, really well. Come in, come in.
THIERRY Patricia, this is my friend Rosario Rodriguez.
PATRICIA Hello, Rosario, how do you do. I'm very pleased to meet you.
ROSARIO How do you do.
PATRICIA Do come in. We're in the back room, come on through. Now, when was the last time we got together?...

Situation 2

WOMAN Excuse me!
WAITER Yes, can I help you?
WOMAN Yes, I'd like a Coke, please.
WAITER Certainly. With ice and lemon?
WOMAN Sorry, I don't understand.
WAITER Would you like ice and lemon in your Coke?
WOMAN Oh, yes, please.
WAITER Here you are.
WOMAN Thank you. How much is this?
WAITER Ninety pence.

Lesson 1 — Speaking and listening, activity 3

1 What's your first name?
2 How old are you?
3 How much do you earn?
4 Where do you live?
5 Are you married?
6 What do you do?
7 Do you have any brothers and sisters?
8 Where do you come from?

Lesson 2 — Listening, activity 1

Speaker 1

JO-ANN I get up around seven o'clock and have breakfast. Then I have to leave home at about eight, I guess. It takes me about fifteen minutes to ride the subway downtown and start work.
Q Do you like your job?
JO-ANN It's OK. Some days it's fine, some days you get some real tricky customers.
Q Do you get time off for lunch?
JO-ANN Sure. I stop around twelve-thirty and then at one o'clock I take a walk in the park – just to get some fresh air, you know. If the weather's OK, I'll have a sandwich or something there.
Q Mmm. Does it get busy in the afternoon?
JO-ANN No more than in the morning, I guess. When the bank's open, people call at all times.
Q So when do you leave work?
JO-ANN I leave work at about six-fifteen in the evening, and at six-thirty I take the subway back home. I don't do much when I get home, except watch TV. I guess I go to bed at around eleven-thirty.

Speaker 2

GEORGE I get up very early these days, much earlier than when I was at work, at about six o'clock, I guess.
Q Why so early?
GEORGE Oh, I like to see the sun rise, have a walk on the beach, you know. If it's hot, I go for a swim in the sea before breakfast.
Q And what time do you have breakfast?
GEORGE I get breakfast ready for about eight-fifteen and take it up to Hilary who's still in bed. Then we go shopping in the local mall, meet some friends, have lunch, that sort of thing.
Q What time do you have lunch, then?
GEORGE At about half-past twelve, I guess. And in the afternoon, I go for another walk, maybe play a little golf, have a swim, go down to the community centre to join Hilary.
Q And what about in the evening?

GEORGE Well, I meet my friends at about five-thirty and have a drink or two at the golf club. We talk for about an hour or so, then I go home for dinner at seven o'clock. Maybe we have friends round, maybe we go for dinner, it just depends. But most nights we're in bed by ten-thirty. It's a great life!

Lesson 2 — Sounds, activity 2

/s/ takes sits asks talks
/z/ goes sings arrives offers has serves does
/ɪz/ finishes refuses washes watches

Lesson 5 — Listening and vocabulary, activity 2

Conversation 1

WOMAN Good morning.
CASHIER Good morning. Can I help you?
WOMAN I'd like two hundred pounds worth of Italian lira, please.
CASHIER Traveller's cheques or cash?
WOMAN Traveller's cheques.
CASHIER Could I have your passport, please? Thank you. Would you sign and date them, please?
WOMAN Here you are.
CASHIER That makes four hundred and twenty one thousand lira. If you'd like to go over to the cash desk, the cashier will give you the money.
WOMAN Thank you.
CASHIER Thank you. Goodbye.

Conversation 2

WAITRESS Good evening sir. A table for two?
MAN Yes, but could we sit by the window, please?
WAITRESS Certainly sir. Come this way. Here you are sir, madam. Can I take your coats?
WOMAN Thank you.
WAITRESS Here are the menus. Can I get you an aperitif?
MAN Not for me, thank you.
WOMAN No, I think we'll just have some wine with the meal, please.
WAITRESS Could I suggest the lamb this evening? It's done in spices, it's a Lebanese dish, a speciality of the chef. I'm sure you'll like it. It comes with mixed vegetables.
MAN Sounds lovely.
WOMAN Well, I'd like to have a look at the menu anyway.
WAITRESS Certainly, madam. I'll come back when you're ready to order.

Conversation 3

WOMAN A return ticket to London, please.
OFFICIAL Are you travelling after nine twenty?
WOMAN Yes.
OFFICIAL That's eleven pounds fifty. Would you like a travelcard?
WOMAN What's that?
OFFICIAL It gives you unlimited travel on the underground and the buses all day.
WOMAN OK, I'll have a travelcard as well.
OFFICIAL That'll be thirteen fifty. Thank you.
WOMAN Thank you.

Conversation 4

MAN What would you recommend for a cough?
CHEMIST Well, I can give you some cough medicine, but if it's very bad, you ought to see a doctor.
MAN No, it's not too bad. I thought I'd try something from the chemist's first.
CHEMIST Have you got a headache and a temperature?
MAN Yes.
CHEMIST Well, I should try this medicine for your cough and these tablets for the headache and the temperature. They're a kind of aspirin which you can dissolve in a cup of hot water.
MAN Thank you. How much is that?
CHEMIST That'll be three pounds twenty, please.
MAN Thank you.

Progress check 1 – 5 — Sounds, activity 1

/s/ gets looks smokes wants
/z/ does goes leaves sings
/ɪz/ finishes refuses

Lesson 6   Sounds, activity 2

/t/   finished   walked   danced
/d/   continued   enjoyed   called
/ɪd/   started   wanted   expected

Lesson 8   Listening and speaking, activity 2

A few years ago, I was going from London to Paris to join my husband and children. I checked in early and the assistant offered to change my ticket from tourist class to business class, which was very nice, and as she gave me my boarding pass, she told me there was no delay, because there was little air traffic. So all in all, I was in a very good mood. I went to the business class lounge and had some coffee, then they called the flight, right on time. I had a row of seats to myself because there weren't many other passengers and we all sat down, put on the seat belts and waited for take-off. The cabin crew went through the safety instructions and I thought, 'Why do they do this on every flight? It's only forty-five minutes to Paris.'

Then we took off, it was lovely morning so we had a fabulous view of the south coast as we flew towards France. But then the pilot made an announcement. 'I am sorry but we appear to have some technical difficulties, nothing to worry about, but we will land at an airport just outside Paris while we get things sorted out. 'He sounded very calm, but I knew that something was the matter. The flight attendants went through the cabin telling us to put on our safety belts. Then the captain came on again. 'Well, I'm afraid that the situation is more serious than we thought and we will be landing as soon as we reach the French coast.' So for the first time in my life I looked at the safety instructions and checked the emergency exits. In a very short time we landed, and as soon as the plane came to a stop, the attendants told us to get out as quickly as possible using the emergency chutes, and to get away from the plane. Nobody panicked, but as soon as we were all out of the plane, we ran as fast as we could. They told us later that they thought there was a bomb on board. It was only afterwards that I realised how frightened I was.

Lesson 8   Listening and speaking, activity 5

I arrived in this town quite late, about ten o'clock, I suppose, and looked for somewhere to stay. I didn't have a reservation anywhere, because I didn't expect to stay there. The main hotel was full, but the woman on the reception desk told me there was a small guest house down the road. I left the car where it was and I went down the road and in fact I walked past the guest house, because I saw there were no lights on. I knocked on the door – there was no bell, or anything – and a man opened the door. He was unshaven and wearing a very dirty vest, and the television was on. I asked if he had a room and without saying a word, he picked up a key from behind the desk and pushed a registration form at me. I wrote my name, but not my address. I don't know why, I felt there was something wrong. I saw that all the keys were on their hooks, so either all the other guests were out or I was the only person to stay there. I followed him up some very dirty stairs – there was no lift, of course – and he showed me a very dusty room. He gave me the key and closed the door. I sat down on the tiny single bed and wondered what to do. It was a horrible room with a washbasin, but no shower or toilet or anything. I decided I wanted to check out but I was frightened by the man downstairs. I ran downstairs and walked quickly past him, but he didn't look up. I quickly said, 'I left my suitcase in my car', and ran out of the door, up the street, and locked myself in the car – I was so scared. I slept in the car that night!

Lesson 9   Vocabulary and listening, activity 2

I have quite a large family, actually. My grandmother lives with my mother and father, and she's called Jacqueline. Then I have a brother who's called Raymond, who lives in Cassis, that's quite near Marseille, and a sister, called Chantal who lives in Marseille. My aunt is called Christine and my uncle is called Tony. Then, of course, there's my husband Vincent. His family comes from Montpellier, and that's where we live now. Oh, I nearly forgot, my mother is called Marie and my father is Georges. We're very close.

Lesson 10   Vocabulary and listening, activity 2

Q	What's Mestre like?
CATHERINE	Oh, it's not bad. I mean, people don't really visit Mestre at all, because they're on their way through to Venice. But, in a way, it's easier to live in Mestre, rather than Venice.
Q	Why's that?

CATHERINE	Well, it's cheaper, for one and it's a lot more convenient being on the mainland.
Q	Oh, right.
CATHERINE	But it's quite industrial, really, it's got a port and quite a few factories. I mean, it's not a pretty place to look at.
Q	What's the rest of the town like?
CATHERINE	Well, as far as culture goes, it hasn't got a cathedral or an art gallery. They're all in Venice. But the shops are very good, I think, and it's got a nice park as well.
Q	Any sports facilities?
CATHERINE	Unfortunately, nothing very big, no.
Q	Would you say the night-life is good?
CATHERINE	No, not really, I mean, it's got restaurants – the food can be very good – and a cinema, which we go to quite often. But if you want theatres or discos you have to go to Venice.

Lesson 10   Vocabulary and listening, activity 5

SPEAKER 1	It's quite cheap to live here, although in certain parts of town the accommodation is really very expensive. Everyday things in the shop are very cheap compared with England, although clothes can cost a lot.
SPEAKER 2	Oh, it's incredible. I mean, there are so many cars and lorries. You see, the railways system is not very extensive so everything has to go by road. But at rush hours in the centre of town, it's so busy.
SPEAKER 3	It's a very interesting place to visit, although I wouldn't call the buildings beautiful. But the palace is quite fascinating and well worth spending a day on. And then there are some wonderful churches and mosques.
SPEAKER 4	It's quite small, really, about one hundred thousand people, and the city centre is really very compact, although the suburbs stretch for about five kilometres in each direction.
SPEAKER 5	It rains all the time, it's quite famous for that. Even in the middle of summer you are never certain that the sun will shine all day, so it's difficult to make plans for any outdoor activities.
SPEAKER 6	Well, you can still walk the streets at night and not feel nervous that someone is going to attack you. But I think it's a good idea to be sensible and keep your money and valuables safe in your jacket. You never know, it can happen anywhere, can't it?

Lesson 12   Listening, activity 2

Q	Lynne, um... I'd like to ask you first, um... at what age do people start learning English these days?
LYNNE	Um... well, in many countries children start learning English when they go to school and then they complete their formal training later on but, I think in some countries they're starting to teach English to much younger children and I think this will become more and more common around the world.
GREG	Yeah, um... that's certainly true 'cause I know that er... in some countries they're even having English lessons for six-year-old children, um... so, er... they'll certainly be learning as soon as they start school, if not before.
Q	I see. And do you think that, um... English will soon be the universal language?
LYNNE	Oh, I think most adults already speak some English, um... even if it's only a word or two here and there, because, um... well, English is very common and very useful.
GREG	Mmm... I...
Q	What about you Greg?
GREG	Well, I was just going to say that, er... I think that's right. Because, if you think about it, already there are so many words, er... for example to do with computers, um... er... that are in English and that are used internationally, er... for example, um...'radio, television, football', these are all international words – English words though. So I think in years to come, um... there'll be very few people who don't speak English, not just a few words but, you know, whole sentences, even.
Q	And, er... do you think, Lynne, that teachers will start um... using English to teach other subjects, you know, for instance, geography or science, and that they'll be used in schools all over the world?
LYNNE	Yes, I do. I think that teachers will start experimenting with that. I think it is likely and I think in many ways it's the best way of learning English.

GREG	Mmm...
Q	Greg?
GREG	Um... I'm not sure about that actually, I don't think that's right. I think some will be in English certainly, um... for example, lessons in science, say. But no, I think quite a lot won't be in English – other lessons. There's no reason why every single subject should be in English.
Q	Right. Now, what about, um... British and American life and habits, institutions, do you think that it's important to know about those?
LYNNE	I don't, not at all. I mean, I don't think that English as a language has anything to do with, you know, double-decker buses, and bowler hats, and hamburgers and yellow taxis. I mean, it's an international language and, um... it can be used for communication between, you know, people who don't know each other's language, um... as a tool really. So, I don't think that the cultural roots of English are important at all.
GREG	Oh, sorry. Can I just come in there. I think that's... I really do disagree there, because I think you have to understand, er... the culture of a country, simply because there are some words that mean different things to different people depending on what country they're in, for example, er... the word 'tea', er... can be a drink to some people in one country and in another country it means an entire meal. Um... the word 'police' means different things to different people. You always have to know a little bit about the background and the culture of a country before you can fully understand the language.
Q	Mmm. What about in the work, er... situation. How important is English there, what's its role?
LYNNE	Well, I think it's really important and I think more and more people will use it at work – it's, er... easily understood wherever you come from and I think, well, actually, everyone will need to use more English for their work.
Q	Mmm. Greg?
GREG	Um... I think some people will need to use more English, particularly people working in big companies who have to travel a lot and do a lot of business between countries, but I think for the majority of the population in any country, um... who don't... who aren't involved in international business or moving around or travelling, then I think they'll be very happy sticking to their own language.
Q	And the traditional language class as we know it – do you think that that will continue or will there be other forms of teaching, such as, you know, teaching involving television and computers, using those sort of technologies?
LYNNE	Well, I think that the traditional language class will still exist. Er... I think that personal contact is very important with the language teacher and, er... of course, there is more than one person in a class, you can interact with the other students and I think that that's much more valuable often than just relating to a computer screen or, you know, listening to cassettes.
Q	Mmm. Do you agree with that Greg?
GREG	Not entirely. I think that we live in a computer age now and, um... it's highly likely that computers and other, er... videos for example – all those interactive programmes that you use with videos – will allow people to learn foreign languages in a different way on their own, um... so that you aren't dependent on teachers and other students. I'm not sure, but I think that's how it'll be.
Q	And finally, can I ask you Lynne, do you think that, um... English will ever become more important than, um... the language of the native speaker.
LYNNE	Well, no. I think obviously English is important, but I think your own language and your own culture and traditions are more important to you and I think it's good to respect those and to hold on to them.
GREG	Yes, I agree. I think it would be very arrogant to think that English would be more important than your own language, I mean, 'cause your own culture and your personal identity and your national identity are, after all, far more important aren't they?
LYNNE	Mmm. I think so.
Q	Thank you very much.
GREG	Thank you.

Lesson 12   Sounds, activity 1

First syllable	actor banker chemistry secretary
Second syllable	accountant arithmetic biology computer geography
Third syllable	economics education engineer politician

Lesson 13   Listening, activity 3

CATHY	So, what are you going to do this summer?
DUNCAN	I'm going to visit South America.
CATHY	Great!
DUNCAN	I'm going to be travelling around quite a lot and I'll probably spend three weeks there.
CATHY	So, where are you going?
DUNCAN	To start off, I'm going to fly to Rio and perhaps I'll stay there for a couple of days.
CATHY	Do you know anyone there?
DUNCAN	No, but I've always wanted to go to Rio. It's meant to be quite spectacular. And I'll probably go up the Sugar Loaf mountain, with the cable car, like all the tourists.
CATHY	And where are you going after that?
DUNCAN	I'm going to fly to Santiago in Chile, where I've got some friends. We'll probably spend some time doing some sightseeing and then we're going to lie on the beach for a few days in Valparaiso, which is on the coast, not far away from Santiago. Then I'm flying to Lima where I'm going to meet my girlfriend and then we're going to visit Machu Picchu in the mountains.
CATHY	And visit the ruins? Ah, fabulous.
DUNCAN	Yes, it should be really interesting. And then we'll probably go somewhere on the Amazon, I don't know where yet, but I'd like to spend a week in the jungle. Then we'll probably fly home.
CATHY	Well, have a nice time!

Lesson 14   Listening, activity 1

So, we're here in the very heart of Dublin now, with St Stephen's Green behind you, just over there. If you look to your left, the road over there is Dawson Street, and the street on the right-hand side from where we are now is Merrion Row. And if you look straight ahead, straight down this street, which is Kildare Street, you'll see the Irish Parliament building on your right, and at the end of the street, you can see Trinity College and College Park.

Lesson 14   Listening, activity 2

I think a good short tour of Dublin would start here in St Stephen's Green. Go along Merrion Row and turn left into Merrion Street and go straight ahead as far as Merrion Square. I suggest you walk all the way round Merrion Square, so turn right into Merrion Square South, left into Merrion Square East and left again into Merrion Square North. Go straight ahead into Clare Street and Nassau Street, and then have a walk round Trinity College and College Park. Then walk back along Grafton Street, and you'll find yourself back in St Stephen's Green.

Lesson 14   Sounds, activity 1

/ɑ:/	article charm far bar
/æ/	bank map actor national
/ʌ/	pub Dublin number others bus

Lesson 15   Listening and speaking, activity 2

Q	So, you live in The States. Where exactly do you live?
PAT	In San Francisco.
Q	Uh huh. An, um... we were wondering what, er... what typical meals were like in The States. I mean, what do you have for a typical breakfast?
PAT	Well, a lot of people in The States eat a great deal, but I tend to eat a fairly small amount. Um... I'm quite attached to cereals still, and... so, I'll eat grape nuts or muesli, something like that. But, on the whole, I tend to just have, er... fruit and, um... juicing is very big in The States.
Q	Uh, huh.
PAT	I'll juice a lot of fruit in a blender.
Q	Is there any special fruit that's really good?
PAT	Um... apricots are very good. Apricots, orange, banana.
Q	And, when you juice things, you make a mixture of these fruits, do you?
PAT	Yes, you basically just bung everything in, and, um, juice it and then drink and then that keeps you going for most of the morning.
Q	Lovely. How about lunch? Do you have lunch?
PAT	Well, if I'm going to eat lunch, I tend to go more for, um... either light pasta dishes, or vegetable platter where you have, you know, beans, zucchini, egg plant, that sort of thing.

Q Zucchini is... we call...

PAT Zucchini is courgette.

Q Courgette, and egg plant is...

PAT That's aubergines.

Q Aubergine, yeah.

PAT I've obviously lived there too long, because I use those terms so naturally! But, um... more often than not, probably just a sandwich, which is usually like a French stick. I mean, they call them 'submarines' – they're big French sticks. Full of meat...

Q And would that be your main meal?

PAT No, no, no. The main meal would be in the evening.

Q And what would you have then?

PAT Um... well, San Francisco is... you can eat very well, um... if you go to seafood restaurants, so, and... I like a lot of seafood, so, um... perhaps a favourite is soft shell crabs.

Q Lovely.

PAT ...with steamed vegetables. Perhaps have some sort of seafood chowder before that, and a mixed salad, and finish it all with key-lime pie, perhaps.

Q Key-lime pie. What's that?

PAT I've never quite worked out what it is. But it's got limes in it, and it's a bit like a cheesecake, with lime.

Q Lovely.

Q Karen, how long have you been living in Hong Kong?

KAREN Just over a year now.

Q And... and you've got used to the way of life there, have you?

KAREN Yes, I think so. It's a very racy place, so you have to adapt.

Q Yes, what... what sort of things do you eat? What's a typical breakfast for you?

KAREN Um... well, during the week I'm very busy. I have to get off to, um... the school where I work quite early, so I have quite a quick breakfast – but a substantial one. Um... so usually cereal, toast, um... orange juice, a cup of tea.

Q A bit like an English breakfast, really?

KAREN Yes, yes. I live with two other English people, so we all eat an English breakfast.

Q Yes, yes... And then, er... do you have lunch?

KAREN If I have time, yes. I rush out and get a sandwich, or a baked potato from a local, um... fast-food place, if I can.

Q Yes, yes. No Chinese delicacies?

KAREN Not at lunchtime, no. No time!

Q Yeah, that's not your main meal, then? The main meal would be dinner time?

KAREN In the evening.

Q Yes. yes. And then, what would you eat normally?

KAREN Well, um... we get together – several teachers – and we go out to a Chinese restaurant, or um... there are plenty of other restaurants represented in Hong Kong, – um... Japanese, Indonesian – and we eat there. Um... a favourite Chinese meal would be 'dim sum', probably.

Q Uh, huh. What's that?

KAREN Well, it's like a dumpling, um... which they steam, and inside will be different types of meat or vegetables. It's very nice.

Q Yes. And, are there any, um... deserts, any sweet things to eat?

KAREN The Chinese aren't so good on their deserts, they're very sticky, and a bit too sweet for me, so I try to avoid those.

Q Yeah. Thank you.

Progress check 11 – 15   **Sounds, activity 2**

/tʃ/   charm   chicken   cheese   peach
/ʃ/   politician   traditional   she   old-fashioned   fish

Lesson 16   **Listening and speaking, activity 1**

Q So, Ken, tell me. What is karaoke?

KEN Well, basically, it's singing along to some recorded music. You have a microphone and there's some music playing and you sing the words – in tune, if you can.

Q Where does this take place?

KEN Well, all over Tokyo there are karaoke bars where you can go with friends, have a drink and sing karaoke. It's very common.

Q Who performs it, then?

KEN Anyone. Anyone who feels brave enough to sing in public, it could be you or me or anyone.

Q It sounds as if you need a drink to sing in public. And what kind of music do you sing?

KEN Well, it's traditional Japanese music for older people, but for young people it's mostly well-known Western pop songs, you know, Frank Sinatra, Phil Collins, Madonna, that sort of thing.

Q Why do people enjoy it?

KEN I don't know, really. It's a chance to show that you could be a pop singer as well, I suppose. It's also a way of showing how close you are to your friends. If you can make a fool of yourself in front of these people, then you really are good friends, that kind of thing.

Q What about tango then, Philippa?

PHILIPPA Tango is a very exotic kind of dance in Latin America, and it's especially popular in Argentina, where it came from originally.

Q And where is it performed?

PHILIPPA In concert halls or theatres, or maybe small bars.

Q And who performs tango?

PHILIPPA Well, in the theatre they're mostly professional dancers, although in dance halls and bars, everyone tries to dance the tango if the music is right.

Q And what is the music they use to dance to?

PHILIPPA Well, tango is both the dance and the music. You dance the tango to music specially written for it. They use the violin and the accordion quite a lot for it.

Q And why do you think people enjoy it?

PHILIPPA Well, it's a very passionate dance. It's full of life, it's great fun.

Q Can you dance it?

PHILIPPA Yes, well, not very well, but I try!

Lesson 19   **Sounds, activity 1**

/ə/   smaller   bigger   closer   happier   funnier
/ɪ/   smallest   biggest   closest   happiest   funniest

Lesson 19   **Sounds, activity 3**

1   He's less casual than she is.
    No, he's **more** casual than she is.
2   It's more noisy now than it was.
    No, it's **less** noisy now than it was.
3   Clothes are cheaper here than at home.
    No, clothes are **less** cheap here than at home.
4   He's less optimistic than she is.
    No, he's **more** optimistic than she is.
5   It's easier to get good clothes here.
    No, it's **less** easy to get good clothes here.
6   He's less confident than she is.
    No, he's **more** confident than she is.

Lesson 19   **Listening and speaking, activity 2**

Q Oh, the one thing I don't like about Britain is the weather.

GRAHAM Well, I agree. I've lived in a number of countries around the world, and I always like to come home to Britain, but I don't usually stay for long because I really enjoy the sunshine, and you just can't be sure of getting the sunshine at all in Britain.

Q Even in the summer.

GRAHAM Even in the summer. No, of course, sometimes we have a warm summer's day, but it can be really quite cold most of the time. I'd say the average temperature during the summer was about 18 or 19 degrees.

Q Is it easy to get good clothes in Britain?

GRAHAM I would say yes, it is quite easy, although obviously for really good clothes you have to pay quite a lot of money. But there's the tradition of the cloth industry, um... especially in the Midlands and the North, so at least good quality clothes are available, even if people don't wear them all the time.

Q So you can get good clothes but you have to pay a lot for them?

GRAHAM That's right, in proportion to what most people earn, anyway.

Q Do people dress formally in Britain?

GRAHAM In comparison with other countries, I'd say no, certainly not in social situations. I mean, it depends what you mean by formal, but apart from the office, most men don't wear suits, and many people come to dinner parties dressed very casually. Um... of course there are some social circles where it's still very formal, but I'd say for the average Briton it's fairly informal.

Q With your experience of people from around the world, would you say the British are small or large people?

GRAHAM Er... to be quite honest, I think they are about average. Um... they're not particularly small, no.

Q And what is the design of clothes from Britain like?

GRAHAM I think it's quite good, especially for traditional clothes, er... Burberry, Jaeger and the kind of clothes people wear, or at least used to wear in the countryside. For town-wear and high fashion, there are now some very famous designers, whose clothes sell even in places like Paris and Milan. It's a small industry, but it's good quality, I think.

Lesson 20   Vocabulary and listening, activity 2

Three hundred and forty-six.
Six hundred and seventy-eight.
Two thousand, three hundred and forty-five.
Eighteen thousand, six hundred and sixty-four.
Twenty-four thousand, five hundred and eighty-nine.
One hundred and twenty-three thousand, four hundred and fifty-six.
Two hundred and two.
Fifty-four thousand, five hundred and sixty-six.
Three thousand, four hundred and eighty-one.
Four hundred and seven.
Ten thousand and twenty.

Lesson 20   Vocabulary and listening, activity 4

Q So, what was your most memorable journey, Sarah?

SARAH Well, I was a passenger on this incredible trip across America with a friend who was moving home from New Orleans to Laguna Beach on the California coast.

Q Mmm. How long did it take?

SARAH It took nine days in all. It was great fun!

Q What sort of distance is that?

SARAH About 2,500 miles, and that's pretty direct as well.

Q So where did you go?

SARAH Well, we set off from New Orleans and took the Interstate 10 Highway, which runs all the way from Florida to Los Angeles. We picked it up just outside the city, and we drove through Louisiana as far as San Antonio in Texas, where we stopped for the night.

Q How far was that, then?

SARAH Oh, the first day we did 550 miles.

Q What's San Antonio like?

SARAH It's quite interesting. It's a strange kind of mix of skyscrapers and Indian houses and churches. It's about 200 years old.

Q What's the scenery like in the country? Is it desert or mountainous or what?

SARAH Well, it's not really mountainous at that point, um... sort of hills and small trees. It's very remote, though, and you can drive hundreds of miles between gas stations, so you have to make sure you've got plenty of petrol.

Q It's cheaper there, isn't it?

SARAH Yes, the Americans call it 'gas', and it costs about a dollar a gallon, what's that – about 25 cents a litre, I suppose.

Q And, where did you go then?

SARAH Well, after that, we started to hit the desert proper and we drove 350 miles to Fort Stockton, which is a typical desert motel stop, in the middle of nowhere. Then we drove through the Guadeloupe Mountains National Park to El Paso, on the border between Texas and Mexico. Then between El Paso and Las Cruces you start climbing into the Sierra Madre.

Q Is it true that you can only drive at 55 miles an hour?

SARAH Well, that's the law, although a lot of people go a lot faster. It depends if you think there are any police patrols around.

Q Where to then?

SARAH We turned off and took a detour to Nogales on the Mexican border which was great. We had lunch there, and then headed north to Tucson in Arizona on Route 19.

Q And where did you stay en route?

SARAH In motels. They're incredibly cheap. The most expensive was thirty-five dollars, the cheapest was twenty. And from Tucson we turned west

again and crossed into California. From there it's only 300 miles to San Diego on the coast, which was very beautiful – my first sight of the Pacific Ocean. And then we drove on up the Pacific Coast Highway to Laguna Beach, not far from LA, and arrived at my friend's apartment, with a fabulous sea view and only ten minutes from the beach.

Q It sounds very memorable!

SARAH It was, it really was. But although it was great to arrive, it was much better to travel. Great fun!

Lesson 20   Sounds, activity 2

| 21 | 33 | 421 | 645 | 4,591 |
| 3,542 | 77,889 | 12,523 | 101,456 | 987,241 |

Progress check 16 – 20   Sounds, activity 1

/ʊ/ good book pullover took cook
/uː/ blue shoe boot suit

Lesson 22   Vocabulary and listening, activity 2

Q I expect you've seen many changes since 1989.

HEIDE 1989, that was a very important year for us. We were all very excited when the Wall came down of course, but we had no idea how different our lives were going to be from then on. Of course, the most important thing for my generation was the change of government and the fact that we've had an election for the first time in our lives. We now vote for our national representatives like every other country in the West, and of course we also elect our regional government as well. I was born ten years after they built the wall, in 1971, so I have never lived in a democratic country until now.

Q Have the changes all been good ones?

HEIDE Well, no, I don't think so. We had a fairly good standard of living under the old government, and although we have more possibilities now, the fact is that we haven't got the money to take advantage of them. And then unemployment was unthinkable in the old days, but now it's very common. My father, who was a teacher, lost his job. Just imagine a teacher losing his job! And, there's been a kind of inflation as well, with prices going up so that they match prices on the western part of the country. It's quite hard at times.

Q Do you get the chance to meet people from other countries and talk about standards of living and the lifestyle in their countries?

HEIDE Yes, we do. We can travel now of course, although as usual, money is a problem. And we have a lot of tourists who come to visit us now, which never happened before. So, we get the chance to talk to people from other countries.

Lesson 22   Vocabulary and listening, activity 3

Q Has anything happened to you personally in recent months?

HEIDE Well, for me personally, the most important thing is that I've got married and I've just had a baby boy, called Konrad, who is now one month old. He's totally changed my life, but it's worth it. I mean, my life is very different now, because before I got married I lived with my parents in a very small flat in the east of the city, But now, of course, I live with my husband and Konrad and we've just moved to a new flat in Potsdam, in a new building. It's very comfortable for the three of us. What else? Oh, I've found a new job – it's for a computer firm and I start on Monday. I've bought a new car – a Volkswagen. And I've given up smoking. My husband doesn't smoke and I had to look after myself when I was expecting Konrad, so I've given up and I feel much healthier now.

Lesson 23   Vocabulary and listening, activity 3

Q Are there any, um... special events or festivals which everyone in Australia celebrates?

BARRY Well, there's Christmas and Easter, like you have in Britain, I mean. They're pretty important.

Q I was thinking more of national holidays, like Independence Day or events that happen in your town or region, you know, local festivals or something like that. I don't even know if you have an Independence Day in Australia.

BARRY	Well, that's because we're not really independent, are we? At least, not yet. But I suppose there's Australia Day, which we celebrate on, er... the twenty-sixth of January.
Q	Oh, and what does that celebrate?
BARRY	It's the day Captain Cook arrived in Botany Bay in Sydney in 1788, bringing Europeans to Australia. But the trouble is, we don't really celebrate it very much usually, except on special anniversaries, er... like in, er... 1988 which was two hundred years since he arrived. Um... no, there's one day which people really enjoy and that's the Melbourne Cup in November.
Q	Oh, and what does the Melbourne Cup celebrate?
BARRY	It's a horse race.
Q	A horse race?
BARRY	Yeah, the whole country stops during the horse race, and everyone wants to know which horse wins. We love horse racing.
Q	And, er... you say it takes place in November?
BARRY	Yeah, the first, um... the first Tuesday in November. At two-forty in the afternoon, every year.
Q	And, er... when did it first take place?
BARRY	I think it started in 1874.
Q	Right, so, it's over a hundred and twenty years old. And what exactly happens?
BARRY	Well, people from all over Australia come to Melbourne on special planes and trains for the day and dress up and go to Flemington race course. And they take picnics in their cars which they eat before the race. Um... and everyone bets on the horse they think is going to win. And then the race starts.
Q	At exactly two-forty?
BARRY	That's right. And at two-forty-three, it's all over.
Q	So, it only takes three minutes?
BARRY	Yes, and if you're enjoying yourself too much, you miss it. It's great fun. It's a great social occasion, a kind of, er... social ritual at the start of summer.
Q	Is it a public holiday?
BARRY	It is in the city of Melbourne and the whole of the State of Victoria. Everybody takes the day off. But not in the rest of Australia.
Q	But, so, even people who can't go to Melbourne are interested in the race?
BARRY	Oh yeah. The interesting thing is that the whole of Australia wants to know who wins the Melbourne Cup. Everybody listens to the race on the radio or watches it on television. The traffic stops and in Canberra, the politicians stop work in parliament.
Q	So everyone's involved, even people outside Melbourne.
BARRY	That's right. It's a kind of state occasion for the whole of Australia.

Lesson 24 **Vocabulary and listening, activity 2**

WOMAN	'Bill'. A 'bill' is paper money. You have a 'dollar bill', a 'five-dollar bill' and so on.
MAN	Right... we call that a 'bank note'. Um... 'trousers' are... an item of clothing. Yes...
WOMAN	Oh, I know what trousers are! Yes... we call them 'pants'.
MAN	Oh, right.
WOMAN	Oh, the snack food that's round and flat and fried and thin and very crisp, we call them, er... 'chips... potato chips'.
MAN	Oh, right. Er... cold, you mean? Yeah, we call those, um... 'crisps'. You buy them in a packet... 'crisps'. Um... er... 'car park', well... a place where you park cars.
WOMAN	Yeah, we call that a 'parking lot'. Um... if I need some medicine or something like that, I go to see the 'druggist'.
MAN	Right. Oh, is that the person, or... is that a place or a person?
WOMAN	No, the place is the 'drugstore', the 'druggist' is the person who'll give me the medicine.
MAN	Right, we call that a 'chemist' – but that's the name of the shop. Um... a 'state school' is a school which is, um... funded by the state, it's the opposite of a private school, in other words.
WOMAN	Oh yes, we call that 'public school'.
MAN	Oh, right.
WOMAN	Um... water in a sink comes out of a 'faucet'.
MAN	Ah yes, we call that a 'tap'. Um... 'traffic lights' – do you know what those are? When you're driving along the road, and you have to stop because there are lights...

WOMAN	You have to stop, so we call them 'stop lights'.
MAN	'Stop lights'. Right...
WOMAN	Um... when I get a hamburger I also like to get 'french fries', which are the strips of, er... fried potato.
MAN	Oh, right – 'chips', we call those.
WOMAN	Do you?
MAN	Yes, 'chips'. Um... when you travel around, for example in London on, er... on the train under the ground, that's called the 'underground'.
WOMAN	No...
MAN	Yeah.
WOMAN	It's called a 'subway'.
MAN	No. The 'underground', yeah...
WOMAN	Well, that's what we call it. I fill my car with 'gas'.
MAN	Ah yes, we call that 'petrol'. Um... there's another item of clothing – er... a 'waistcoat', um...
WOMAN	Oh, yeah, that men wear. We call that a 'vest.'
MAN	That's right, it doesn't have any sleeves, yeah? 'Vest'... yeah.
WOMAN	Every town in America has a 'main street' where all the shops and things are.
MAN	Oh, right... No, we call that the 'high street'. Same thing, 'high street'.

Lesson 25 **Listening, activity 2**

SPEAKER 1	Well, it's... it's that stuff you need to put two different pieces together. For instance, two pieces of paper. You put that stuff on one bit of paper and stick the other paper on top of it, for example. Or, you can do that with leather as well, if your shoe gets broken, or you can do that with wood, and things like that.
SPEAKER 2	It's a piece of material. Um... it's a square and it's soft and you use it to... after a bath for drying yourself when you are wet.
SPEAKER 3	It looks like little pieces of wood, very thin little pieces all in a box and, er... at the tip there's a... they are either black or red and it's something you use to light a fire or anything like that.
SPEAKER 4	Er... I want... you put it on when it's hot and you buy it in a bottle, a plastic bottle and you put it on your body and it protects you from the sun.
SPEAKER 5	It's a machine for cleaning. You have a tube and it, er... it sucks the dust. It's a machine for cleaning the carpet or the floor.

Progress check 21 – 25 **Sounds, activity 1**

/ɜː/   surfing   fireworks   learnt   heard   university   word
/ə/   primary   economics   lecture   chosen

Progress check 21 – 25 **Sounds, activity 2**

/ɔː/   sport   awful   walking   performance   four
/ʌ/   up   summer   bus   pub   sunny

Lesson 26 **Vocabulary and listening, activity 3**

Conversation 1

POLICE OFFICER	Good morning, sir.
DRIVER	Good morning, officer.
POLICE OFFICER	Do you realise what speed you were driving at, sir?
DRIVER	Er... about a hundred and ten, I suppose?
POLICE OFFICER	No, sir. When we were following you in the fast lane a few minutes ago, you were driving at one hundred and eighty.
DRIVER	Was I really? Oh, goodness me! Dear, oh dear! I didn't know this car could go that fast! Aren't these German cars magnificent!
POLICE OFFICER	Yes, sir. But we're not on a German autobahn now. We're on a British motorway. You really mustn't drive so fast. You mustn't drive at speeds over one hundred and ten in this country.
DRIVER	Ah. Er... my wife's ill.
PASSENGER	No, I'm not.
DRIVER	We're going to the hospital and the traffic delayed us, so I was driving fast. What did you say, dear?
PASSENGER	I'm not ill.
DRIVER	Ah.
POLICE OFFICER	And did I see you using your car telephone?
DRIVER	I was phoning the hospital.
PASSENGER	No, you weren't.

POLICE OFFICER	You have to have both hands on the wheel when you're driving. You mustn't use the car telephone, you know. Can I see your licence, sir?
DRIVER	My brother's a policeman.
POLICE OFFICER	Is he sir? How interesting.
PASSENGER	No, he isn't.
DRIVER	Aargh! Will you be quiet!
POLICE OFFICER	I think you'd better follow me, sir...

Conversation 2

POLICE OFFICER	Oi! You! Stop!
CYCLIST	Who me?
POLICE OFFICER	Yes, you. Did you see that red light?
CYCLIST	What red light?
POLICE OFFICER	The red traffic light. Even cyclists have to stop at red traffic lights. And have you got any lights?
CYCLIST	No.
POLICE OFFICER	Well, you have to have lights when you ride a bicycle at night.
CYCLIST	Is that so?
POLICE OFFICER	And I saw you riding on the pavement a moment ago.
CYCLIST	Did you?
POLICE OFFICER	You mustn't ride on the pavement.
CYCLIST	Officer, do you know who I am?
POLICE OFFICER	No, madam, but I don't care if you're the Queen of England. Could I have your name, please?
CYCLIST	No, you can't. You'll have to catch me first.
POLICE OFFICER	Oi! Come back here!

Conversation 3

TICKET INSPECTOR	Tickets, please. Tickets, please. Good morning, can I have your tickets please?
PASSENGER	Here you are.
TICKET INSPECTOR	Excuse me, but this is a second class ticket.
PASSENGER	I know.
TICKET INSPECTOR	But you mustn't sit in here. This is a first class carriage.
PASSENGER	But there weren't any seats in the second class carriage. It's standing room only.
TICKET INSPECTOR	I'm sorry, madam, but you either have to go back to the second class carriage and stand, or pay for a first class ticket.
PASSENGER	But I've paid for my ticket. You have to give me somewhere to sit.
TICKET INSPECTOR	I'm sorry, but in Britain the law is your ticket is for transport by train only. It does not give you the right to a seat as well.
PASSENGER	Oh, but that's outrageous!
TICKET INSPECTOR	I'm sorry, but I don't make the rules.

Lesson 27 **Listening and speaking, activity 2**

The Skylight, part 1

The heat, as the taxi drove up the hill, became more violent. The woman sat in the back of the car with a five-year-old boy beside her, his thumb in his mouth.

'When are we there?' the boy asked.

'Soon.'

The child's eyes closed. 'Oh no', she thought. 'He mustn't go to sleep'. She could hear herself telling the story in the cold English spring. 'It's so sensible to take this house for the summer. It's in the mountains, ten minutes drive from Golfe-Juan. Philip will drive the girls but I will take Johnny by air. And the Gachets will have everything ready for us.' But now it was real. She was hot and afraid. Will everything be all right when we get there, she wondered. Suddenly, the driver turned off the road, drove up a narrow track, and stopped. The woman could only see stones and grass.

'But – where?'

He pointed and got out. He picked up their suitcases and walked away. She took the child's hand and followed the driver. Above them, on the terrace, was the square grey house. A small skylight in the roof caught the sun. The shutters and doors were all closed.

'Vous avez la clef, madame?'

'The key? But Monsieur and Madame Gachet are expecting us.'

The driver tried the door. It was locked. She knocked. There was no answer.

Lesson 27 **Listening and speaking, activity 4**

The Skylight, part 2

The driver wanted his money. She paid him and he disappeared. She heard the taxi leaving.

'Why can't we go into the house?' asked Johnny.

'Because it's locked.'

She looked up and saw the skylight.

'If there was a ladder, perhaps we could...'

'There's a ladder.' he said. 'Can we lift it?'

It was quite light.

'Are you going to climb up there?' the child asked.

She hesitated. 'Yes, I suppose so.'

She started to climb. At the top she saw the tiny skylight was open. She couldn't get through it, but a child could do it. She could lower Johnny through, and he could run downstairs and unlock a window. She came down the ladder. He was lying on the ground, nearly asleep.

'Johnny,' she said, 'Would you like to climb the ladder?'

'Can I climb it now?' he asked.

'Yes. Yes, you can. When you've got through the skylight, I want you to do something.' She explained, very carefully.

Together they climbed the ladder. She lowered him through the window until he stood on a table.

'Can you get down?' she asked.

'Yes. Shall I go and open the window now?'

'Yes,' she said, 'And hurry.'

She climbed down the ladder, went to the window and waited. It was getting dark.

'Johnny, it's this one. Are you there, Johnny?'

Give him time, she thought. He's only five. He can't hurry. She climbed up the ladder again and shouted, 'Johnny, can you hear me?' Her voice had no volume, no echo.

Lesson 27 **Listening and speaking, activity 6**

The Skylight, part 3

It was now dark. She went down again and ran round the house, shouting his name. Something has happened to him, I must go for help. She ran to the road and when she saw the lights of the car, she waved her arms to stop it. She started to cry. It was a long time before the three men understood.

'But how can we get in? We have no tools,' they said.

'There's a farm back there. Will you take me?' They let her into the car.

'Turn round. It's back there on the left. There it is!'

They turned off the road. She got out of the car and ran to the front door. A small woman in trousers opened the door.

'My dear, what's happened?'

'You're English?' She told her the story. Another woman appeared.

'Yvonne,' Miss Jardine said, 'Get some tools, a hammer and an axe.'

They all got into the car and went back to the house. They drove up the lane and stopped. She ran to the house, calling 'Johnny? Johnny?'

Lesson 27 **Listening and speaking, activity 8**

The Skylight, part 4

One of the men took an axe and smashed the shutters. She was quickly through the window.

'Johnny, where are you?'

She ran up the stairs. A door on the first floor was open. He was lying on the floor, fast asleep. Surrounding him were lots of toys. She shook him gently. He opened his eyes.

'I like the toys,' he said.

His thumb went back into his mouth and his eyes closed again. She sat with her head on her knees, her arms round her body.

'Oh, thank God, She whispered. 'Oh, thank God.'

Lesson 28 **Listening, activity 1**

| Q | I mean, in Britain for instance, if I, sort of... want to cross the street at a pedestrian crossing and, er... the light is red but there are no cars, is it allowed to cross? I mean... |
| JANE | Oh yes... yeah, you don't have to wait for the pedestrian light to turn green before you cross, like you do in some countries. |

Q	Like Japan you do, and I think Germany you do as well.
JANE	Yes. But no, you can cross.
Q	I see... um... and if... when I'm in a park, for instance, you know, with grass everywhere, um... can I walk on that grass?
JANE	Yes, unless there's a sign that says 'Do not walk on the grass', usually you can always walk on the grass.
Q	Ah, because in France it's virtually impossible...
JANE	Really?
Q	I mean, you'd get a fine, except a few areas. But, that's good... Now, I had a problem last time I wanted drink, it was, um... I think, Sunday in the afternoon, er... around half-past three, and, er... I mean... was it possible to get a glass of beer somewhere because I found it very difficult?
JANE	Ah well, not on a Sunday, probably on any other day it would be OK, I'm talking about pubs.
Q	Yes, because everything was shut.
JANE	Exactly, but there is a possibility, um... of getting a beer or an alcoholic drink in a restaurant.
Q	But I'd have to eat then.
JANE	But you usually have to eat something, that's right. That's usually the rule.
Q	What a bore! Mmm. Yes... And, I wanted to ... I mean, I was told that if you had young children you can drink in a bar? Er... I mean, can you take the children with you, or...?
JANE	Mmm... not into the bar. Sometimes they might have a beer garden or something like that. But if you're talking about very young children – no, they can't come in with you, not to a bar.
Q	Oh dear, so they have to go to separate place if there is such a thing then?
JANE	Well, there isn't usually.
Q	There isn't!
JANE	There isn't usually, no.
Q	And, er... in a restaurant, er... perhaps foreign food or whatever, if I don't know the food, can I go into the kitchen and look at the food?
JANE	And have a look?
Q	Yeah, have a look, I mean, sometimes in France we do that... it's allowed – not every restaurant – is it possible here?
JANE	I've never seen it done.
Q	If you ask nicely?
JANE	I think if you asked very nicely, um... they might say yes but I think normally I would say no.
Q	You just wait for a plate? Oh, I see. And, for all my correspondence and letters and postcards, um... is it possible in England to go to a newsagent or a tobacconist shop and get some stamps there?
JANE	To get the stamps? Yes, yes, um... also, all kinds of places now – supermarkets sell them. Yes, there's usually a sign on the door that indicates that they will sell them.
Q	A friend of mine came here about ten years ago, and I think it wasn't the case then.
JANE	That's right.
Q	So you've changed!
JANE	Yeah, we've changed!
Q	Yes, and what about smoking in the cinema, I mean, if there's no sign for instance which says 'no smoking' is that allowed, you can actually smoke if there's no sign saying you can't?
JANE	Well, yes I think so. Usually I think most of the cinemas are non-smoking now, but then, yes, then there would be a sign. So if there's no sign...
Q	You can get away with it?
JANE	I suppose you can smoke and hope nobody says anything. Yeah. Yeah... I think you can.
Q	And, um... in a cafe for instance, if I want a drink and I don't want to wait at a table or something, can I just go to the bar and, er... pay for it there?
JANE	Oh, I think you... Yes.
Q	That's no problem? It's the same.
JANE	Yes, I think you can do what you want in most cafes.
Q	And, for instance, If I feel a bit ill, er... and I haven't got medical insurance, or much money on me, or... er... Can I make an appointment to see a doctor, any doctor?
JANE	A doctor? Oh yes, Yes, you can, er... get an appointment at any doctor's surgery.
Q	No problem?
JANE	As a temporary patient, it shouldn't be a problem. It is possible.

Q	Oh, that's wonderful. Oh, that's good. And lastly, if I want to take a bus, can I buy a single ticket before I get on? Is that possible?
JANE	A single ticket? Er... I don't think you can on the bus.
Q	So I have to buy it on the bus itself?
JANE	Yes. You can get passes that allow you to go on the bus all day, but just for one single ticket, you must buy it as you get on the bus.
Q	I see. Mmm. Thank you very much.
JANE	OK.

Lesson 29 **Listening, activity 2**

Q	So, Doctor Samuels, what do you advise for people who want to avoid jet lag?
DOCTOR	The thing about flying is that it has a dehydrating effect on the body, and this means that you need to replace the missing liquid with lots of water. It's also better to avoid alcohol, as this only makes the dehydrating effect worse. So, lots of water or juice, and adjust your body clock by adopting the time of your destination as soon as possible. So even if your body tells you it's time for bed, try and stay awake until it's bedtime in your new time zone.
Q	And what can you do to avoid stomach upsets?
DOCTOR	Mmm. One of the main causes of stomach upsets when you're in a foreign country is the water supply. So, the most important thing to do is to drink only bottled or boiled water, and don't forget the ice cubes in your drink too – they may not be from a clean source of water. And don't eat uncooked food, like salad, because it may have been washed in dirty water.
Q	I always get badly bitten when I'm away. What do you suggest for this?
DOCTOR	Well, keep your arms and legs covered in the evening when mosquitoes like to bite most of all, and maybe wear a hat as well. Insect repellent is very useful but you'll find the most determined mosquito will always find a patch of skin to bite.
Q	And sunstroke?
DOCTOR	Oh, well, obviously the most important thing is not to spend all day in the sun. If you come from a country where you don't get much sunshine, I suggest you spend only about twenty minutes or half an hour in the sun during the first day or two and gradually increase the time. And you should wear a hat, because your head is where you're most likely to catch the sun.

Lesson 30 **Reading and listening, activity 2**

WOMAN	OK, here we go, um... question number one: 'You're a guest in someone's home, you'd like a cigarette, what do you say?'
MAN	Um... I think it's got to be 'a', don't you? 'Is it all right if I smoke?'.
WOMAN	I think that's most polite, yes. 'Is it all right if I smoke?' I'll tick that.
MAN	OK, number two: 'A friend suggests you have dinner together in a certain restaurant, at the end of the meal the waiter brings the check, what do you say?'
WOMAN	Ooh, um... well...
MAN	Well, it could be 'c', because he did offer...
WOMAN	'Your friend suggested dinner, and you expect him to pay.'
MAN	...but I think 'b' is...
WOMAN	'Shall we share this?'
MAN	It's... yes. Yeah. 'Shall we share this?' He could always persuade you later, couldn't he?
WOMAN	Question number three: 'You're visiting a friend when the phone rings, what do you expect her to say to the caller?' Oh, 'b'.
MAN	I think it has to be 'b'.
WOMAN	It has to be 'b'.
MAN	Everything else is very bizarre.
WOMAN	'Would you mind if I called you back, I've got a visitor here at the moment.' Yes.
MAN	Uh... number four: 'It's late and your neighbours are playing very loud music, what do you say to them?'
WOMAN	'Turn down the music!'
MAN	Yes! No, they're your neighbours, you have to try and get along. I think you'd start with 'b'.
WOMAN	'Could you turn the music down please?' Number five: 'You meet a Ms Esther Craig for the first time, you don't know how to address her, what do you say?'

MAN	This one's a little odd, isn't it? 'What do I call you?'
WOMAN	No.
MAN	'Would you mind telling me what to call you?'
WOMAN	No.
MAN	Do you think it's...
WOMAN	I think it's 'b'.
MAN	'b'? 'Can I call you Esther? ... Can I call you Esther?' It's friendly. It's...
WOMAN	Yeah, 'b'.
MAN	Yeah. Six: 'You meet someone at a party and get on very well, as she leaves she says, "Nice meeting you, we must do lunch sometime."'.
WOMAN	Do lunch?
MAN	Do lunch!
WOMAN	Ooh, um... I like 'c'. 'That's a great idea, bye.'
MAN	But, that sounds like a kiss-off because you don't... she doesn't... you don't know their phone number. It's got to be 'b' if you're serious about it.
WOMAN	'b'. 'Would you mind giving me your phone number?' You're right, 'b'. Number seven: 'You'd like your friend to lend you a book, what do you say?' Oh, 'c'.
MAN	Well 'c' seems the most polite – 'Would you mind lending it to me?'
WOMAN	Yes, 'a'.
MAN	Very direct. Number eight: 'You're at the information office at the railway station and you want to know some train times. What do you say?'
WOMAN	Mmm.
MAN	Well, 'c' seems a little formal, and English!
WOMAN	'I wonder if you could tell me when the next train to New York is?' Yes, I'd say 'a'. 'When's the next train to New York?'
MAN	'a'. 'When's the next train to New York?' As long as you don't say it too rudely! Say it nicely, 'When's the next train to New York?'
WOMAN	Question number nine: 'Your host serves you food you don't like (ugh!), you eat it but then the host offers you more, what do you say?'
MAN	It's got to be 'a'.
WOMAN	It's got to be 'a'.
MAN	'It was very nice, but no thank you, I've had enough.'
WOMAN	Yes.
MAN	I don't think you really need to have eaten all of the first course, even. Uh, number 10: 'As you are leaving a shop the assistant says, "Have a nice day." What do you say?'
WOMAN	I like 'c'. 'No thanks I've made other arrangements.'
MAN	Yes, but that might be a little too ironic for the shop assistant, so maybe 'a', I think.
WOMAN	'Thank you, same to you, bye.' Yeah.
MAN	Nice and direct, simple, polite.
WOMAN	I think we passed.
MAN	I think so.

Progress check 26 – 30   **Sounds, activity 1**

/əʊ/ wrote vote know only telephone home photo
/ɔɪ/ boy noise royal unemployment

Progress check 26 – 30   **Sounds, activity 2**

/ʃ/ shoe station pressure situation
/tʃ/ teacher temperature
/dʒ/ oxygen passengers stranger

Lesson 31   **Listening and speaking, activity 3**

Part 1

I was at home one afternoon sitting in front of the fire and watching the television, when suddenly there was a knock at the door. I wasn't expecting anyone, so I was quite surprised. I went to the door and opened it, and there was the Queen of England standing at the door, wearing her crown and holding a Harrods bag, as if she was coming home from the shops, just like you or me.

'Hello, Queen,' I said. 'Do come in.'

'Thank you,' she said, and came in.

I showed her into our front room, which we only use when we've got company – after all, it's not every day you have the Queen of England come to visit you, is it?

'This is very good of you,' she said in that voice of hers – well, she

couldn't say it anyone else's voice, could she?

'I went shopping this afternoon, and I am so tired that I really must sit down and rest my feet for a moment.'

So I said to her, 'You sit down, Queen and put your feet up. Would you like a nice cup of tea?' And, er... well, she smiled in that lovely royal way she's got and said, 'Oh, thank you.'

I went into the kitchen and got out the special china tea cups and saucers, it's called 'Royal Doulton', just to make her feel at home, and some chocolate biscuits. I was getting the tea ready when she called out to me, 'The shops are so crowded at this time of year.' So I went back into the front room.

'Yes, they are. What have you bought?' I asked, looking at her Harrods bag.

'I was looking for some curtains when I saw this lovely material, and I thought to myself, that would look very nice in St George's Hall.'

So I asked her, 'Where's that?'

'Windsor Castle,' she said. 'We were planning to redecorate the castle when the fire burnt the place down. Bit of luck, really.'

I said, um... 'Ooh, yes, it gives you a chance to start again, doesn't it?'

Lesson 31   **Listening and speaking, activity 5**

Part 2

I went and got the tea and brought it back and poured her a nice cup of tea and gave it to her, and, er... we carried on talking just as if we were old friends, which, in a way, I suppose we are. But, er... when you think about it, it's quite amazing, isn't it? I mean, there was the Queen sitting on our sofa in our front room. Anyway, we were talking about the weather when suddenly the phone rang. I picked it up. It was Prince Philip.

'Is the Queen there?'

So I said, 'Yes, she is' and gave her the telephone.

'Hello, Philip, I won't be long. Just having a cup of tea. I'll be back at about five o'clock if I don't wait too long for a bus.'

She came back to the sofa and was just finishing her tea when my husband arrived home from work.

'This is the Queen, dear,' I said helpfully. 'She's just been shopping and now she's going home.' So my husband said, um...

'Oh, hello, Queen, pleased to meet you,' and shook hands. 'Would you like a lift home?'

Now my husband never gets the car out in the week, so I knew that this was a very special occasion.

'Thank you so much, but I'll get the bus. The 75 is usually quite good at this time of day,' she replied.

And, er... well, then, she stood up and said thank you and goodbye. Oh, and she was going down the garden path when she stopped, turned to me and waved to me like she does on television. And she was gone!

Lesson 35   **Listening and speaking, activity 1**

Q	Now, Stephen, I'd really like to know, what do you say at the start of a meal in England?
STEPHEN	Uh, you don't really say anything, actually. I mean, you can say, 'Oh, this looks delicious', or something like that, but there's nothing formal that you say.
Q	Really? Because in Germany it's 'guten appetit', so is there such a thing as 'good appetite'?
STEPHEN	No, no. Nothing like that.
Q	Oh, that's surprising. Mmm. And, um... what time, roughly, do you have lunch and dinner?
STEPHEN	Um... I'd say that you have lunch at round about one o'clock, um... and dinner at about seven o'clock. I mean, obviously sometimes it can be later, eight o'clock, even nine o'clock, but normally I would say about seven o'clock.
Q	Right, well that's roughly the same in my country, I think. And, um... tell me, how long does a typical lunch or dinner last?
STEPHEN	I would say that lunch and dinner, I would say they last about half-an-hour, thirty minutes...
Q	Is that all?
STEPHEN	Yes, if it's just an informal lunch. I mean, obviously if it's a dinner party or a special occasion, um... it would last longer. But if it's just a...
Q	Like a family meal?
STEPHEN	Yes, but if it's just an everyday family meal, then half an hour.
Q	Ah, right. And, um... now that's just my curiosity, in which hand do you hold your fork?

STEPHEN Ah, that's simple, you hold your fork in your left hand and knife in the right hand, always.

Q Yeah, that's the same in my country, yeah. And, um... tell me, Stephen, do you actually use a napkin, like, generally in Britain? And if you do, tell me, where would you put it?

STEPHEN Uh, some people do, and some people don't – there are no rules really. Um... if you're in a slightly posh or smart place you're probably more likely to have a napkin. If you do use a napkin you put it on your lap.

Q Right, um... so you wouldn't put it round your neck, like the French?

STEPHEN No, well, generally not.

Q Rather on the lap. And um... tell me, at which meal would you eat the following food, um... melon – when would you eat that?

STEPHEN Melon, well, you could really eat melon for breakfast, um... or for lunch or for dinner, any of those three.

Q Oh, right. And how about pasta?

STEPHEN Pasta, I would say lunch certainly and dinner, but not breakfast.

Q Right. And how about fish?

STEPHEN Fish, you can have fish for breakfast or lunch or dinner. Yes, all three.

Q Oh right. And um... how about steak?

STEPHEN Steak, er... lunch or dinner, not breakfast, though apparently in America they do, but not in this country.

Q All right. And, um... tell me, where do you actually put your knife and fork once you have finished your meal? Is there a specific way you would put them in England?

STEPHEN Um... yes, there is really. What most people do is they put the knife and fork together in the middle of the plate so the handle's pointing towards them and the points are facing away from them.

Q Oh, is that right? Because in Germany you couldn't do that, you'd have to put it slightly sideways; both of them parallel but slightly sideways, it couldn't be in the middle. And, um... yeah, and how about your hands? I mean, during a meal, I mean, or... where would you put your hands when you're at the table but not in the process of eating?

STEPHEN Um... I think most people would put them just on their lap.

Q Really, now, that's interesting because, again in Germany you couldn't do that. You'd have to have your hands on the table, sort of just loosely just lying on the table; they couldn't be under the table.

STEPHEN No, I think most people'd have it most certainly underneath, just on the lap.

Q Uh huh. And um... tell me, how do you eat cake? Do you eat cake with a fork or a spoon?

STEPHEN Um... I think informally you eat cake with your hands, with your fingers. Um... in a slightly more formal situation or if the cake is particularly sticky or messy you might use a fork. Um... so it's either a fork or hands really, a spoon you normally eat things like pies or puddings.

Q Right. Would it be a smaller fork like a cake fork or would it be normal size?

STEPHEN Yes, it would be quite a small fork.

Q Yes, that's the same, actually the same, in Germany. And, um... because you were just saying you eat cake, you can eat cake with your hands. What food do you usually eat with your fingers at the dining table?

STEPHEN Um... chicken, chicken bones normally, you know, bones with the chicken still on them. Um... bread, bread and butter, sandwiches, that kind of thing. Cheese, pieces of cheese. Er... fruit certainly, um... cake, yes, those are the main ones I'd say.

Q Yeah, it's funny. Do you have that... in Germany it would be even rude to eat chicken with a knife and fork.

STEPHEN Oh, really?

Q You have to, I mean, you can eat the chicken breast, of course, but the legs and the wings you would have to eat.

STEPHEN You'd almost have to pick it up?

Q Yes, you have to.

STEPHEN Not quite the same here.

Q Right. And, um... tell me, what... are there any times of the day when you usually drink coffee and tea?

STEPHEN Um... well, anytime really. Um... people drink coffee at anytime certainly, breakfast, mid-morning, after lunch, in the evening. Um... tea – people... a lot of people drink tea at teatime obviously, and some people have tea for breakfast.

Q Right, so that would be a little more specific than the coffee?

STEPHEN A little bit more, yes.

Q And tell me, when can you actually smoke during a meal?

STEPHEN Ah, well, you can't smoke during a meal really, or you don't. People smoke before a meal and after, but not during it.

Q And if that meal consists of several different courses and if it lasts, like, for a few hours?

STEPHEN Um... no. Sometimes if there are lots of people then you might have a break between courses when people can smoke, but not always.

Q Ah, right. Because in Germany you could do that. Between courses that would be no problem.

STEPHEN Yes, well that's acceptable here too.

Q And, um... tell me, what do you actually say... what is the word you use when someone raises their glass?

STEPHEN Oh, you mean to toast someone? 'Cheers!'. 'Cheers' is what you say.

Q And, um... tell me, do you actually, do you have soup in the summer?

STEPHEN You can do. Soup is normally drunk in the winter but quite a lot of people have soup in the summer. You can have cold soup, of course, like 'gazpacho' or something like that, but people occasionally drink warm soup as well. It's perfectly possible.

Q Right, so that's an all-year-round thing really, and... how about salad, um... would you eat salad in the winter?

STEPHEN Sometimes, yes. It's like soup really, um... salad is generally eaten in the summer, most often but, er... you can certainly have it in the winter.

Q Right, well, thank you very much Stephen. That was interesting.

Progress check 31 – 35   **Sounds, activity 1**

wait wallet move visit want work sandwich drive invite women wear shiver walk behave

Progress check 31 – 35   **Sounds, activity 2**

1 ear 2 hair 3 eye 4 hat 5 hate 6 eat 7 art 8 as

Lesson 37   **Listening and speaking, activity 1**

Part 1

I was on holiday in the Lake District, and I was visiting a church in Kendal. I was sitting in the churchyard, relaxing for a moment, and my handbag was beside me, although I was holding the strap. Not the sort of place where you expect anything to happen, is it? Suddenly, someone came up from behind, grabbed my bag and pulled it very hard, breaking the strap. I shouted, first in pain, because when he pulled the bag it hurt my wrist, then in anger as I saw him get on a motorcycle and drive away. I felt awful as I watched my passport, my money, credit cards, various documents disappear down the road. The police were very kind and said that this sort of thing happens too often these days. I thought to myself, 'If I ever catch him, I'll kill him!' I told the consulate about the loss of my passport, and I cancelled my credit cards, got some more money and tried to forget about it. But that wasn't the end of the story.

Lesson 37   **Listening and speaking, activity 3**

Part 2

Four days later, the police rang me at my hotel and said they'd got some good news. A young man was trying to change some Australian money at the bank in Windermere. Now, there aren't many Australians in the Lake District at this time of year, and secondly, the young man wasn't Australian. So the bank clerk called the police, who came very quickly and they stopped the man as he was walking away from the bank. When they questioned him, he broke down and admitted he was guilty. They asked me to go to the police station in Kendal to identify him. Well, when he took my bag, I didn't see his face, so I couldn't really say if it was him. But he recognised me, and said, 'I'm sorry, I'm really sorry.' The police showed me the other things from my bag which he had on him. The man started to cry. The police said he came from Manchester. He was unemployed and he had a family to look after. I was the victim all right, but now it was me who was feeling sorry.

Lesson 37   **Listening and speaking, activity 5**

Part 3

The police said 'If we let him go, he'll probably take someone else's bag in some other town. But if we send him to court, he'll get a fine, which he won't be able to pay, so he'll go to prison. If he goes to prison, either he'll never take anyone else's bag again, or he'll learn how to do it more efficiently. So what do we do?' I didn't know what to say, so... I just felt so guilty and I had to keep telling myself, 'He's the criminal, I'm the victim.' Well, in the end, they sent him to court, and he got a fine, which was small enough for him to pay, but now he's got a criminal record, and will probably try to take someone's bag again. It's crazy. He's sorry. I'm sorry. We're all sorry.

Lesson 38   **Listening, activity 1**

ALEX	I love coconut milk and I never get a chance to drink it fresh in Britain, so that's what I'd drink.
BARBARA	For me the nicest things to do would be to sleep late – I never get the chance to do that with the children, to have a long breakfast with all the Sunday papers and then to go to the beach for a swim.
ALAN	Yes, my wife. She's my best friend as well. Besides, we've only just got married.
DANIEL	The postman. I really wouldn't want to get any news from home. In fact, I probably wouldn't tell anyone where I was going.
EMMA	The most beautiful place I have ever been is to Delphi in Greece. So I'd go there again and see if it's still as beautiful as I remember.
JANE	Well, believe it or not, the video. You see, I'd want to go with the children, but they're still quite young and very demanding. So if I took a video player and some cartoons on videos, I would at least have half an hour to relax without having to look after them.

Lesson 39   **Reading and listening, activity 3**

The umbrella man, part 2

My mother was staring down at him along the full length of her nose. I wanted to say to her, 'Oh mummy, he's a very old man, and he's polite, and he's in some sort of trouble, so be nice to him.' But I didn't say anything.

'I've never forgotten it before,' he said.

'You've never forgotten what?' my mother asked.

'My wallet,' he said. 'I must've left it in my other jacket.'

'Are you asking me to give you money?' my mother said.

'No, I'm offering you this umbrella to protect you and to keep, if you would give me a pound for my taxi fare just to get me home.'

'Why don't you walk home?' my mother asked.

'Oh, I don't think I could manage it. I've gone too far already.'

The idea of getting an umbrella to shelter was very attractive.

'It's a lovely silk umbrella,' the little man said. 'Why don't you take it, madam? It cost me over twenty pounds, but that isn't important because I want to get home.'

'I don't think it's quite right that I should take an umbrella from you worth twenty pounds. I think I'd better just give you the taxi-fare.'

'No, no, no!' he cried. 'I would never accept money from you like that! Take the umbrella, dear lady, and keep the rain off your shoulders.'

She took out a pound and gave it to the little man. He took it and gave her the umbrella. He said, 'Thank you, madam, thank you.' Then he was gone.

Lesson 39   **Reading and listening, activity 5**

The umbrella man, part 4

'He went in that door!' It was a pub. The room we were looking into was full of people and cigarette smoke, and our little man was in the middle of it all, without his hat and coat, and moving towards the bar. When he reached it, he spoke to the barman. The barman gave him a drink. The little man gave him a pound. The barman didn't give him any change. The little man drank it in one go.

'That's a very expensive drink,' I said.

He was smiling now. He went to where his hat and coat were. He put on his hat. He put on his coat. Then, very quickly, he took from the rack one of the many wet umbrellas, and left.

'Did you see that!' my mother shouted.

'Sssh!' I whispered. 'He's coming out.'

He didn't see us. He opened his new umbrella and went down the road. We followed him back to the main street where we met him first, and we

watched as he exchanged his new umbrella for another pound. This time it was with a tall, thin man who didn't even have a hat or a coat. When it was over he went off again, this time in the opposite direction.

'He never goes in the same pub twice,' my mother said. 'I expect he's always hoping for rainy days.'

Lesson 40   **Vocabulary and listening, activity 2**

Just before the Berlin Wall had come down, a man from Leipzig in Germany, who had wanted to be a rock musician for many years, decided to go to England. When he had packed his bags into his old Trabant car, he set off for Liverpool, home of the Beatles, to make his name as a musician.

It was a nightmare journey. He drove all night through Hungary and when he got to the Austrian border he had to queue all day with thousands of other people who had decided to leave their homes and go to the West. He grew nervous while he waited, but the border guards looked carefully at his passport and let him through. After he had entered the West he drove happily across Southern Europe.

He arrived in France and had a minor accident. The crash damaged the exhaust pipe, and the car made more noise than it had done before. At Calais the car broke down and it cost a lot of money to mend it. The ferry ticket was expensive too. He had very little money left now because he had spent so much on the journey. But when he saw the white cliffs of Dover ahead, he felt better.

He left the boat at Dover and he heard a loud noise, so he stopped and got out. The car's exhaust pipe had fallen off. But he headed for London on his way to Liverpool. He was driving along south of London when he came onto the M25 motorway around the capital, which wasn't on his out-of-date map at all. By now, clouds of black smoke were coming from the car, and he was driving noisily in the slow lane. He had driven over a hundred miles around the M25 when the police saw him.

Lesson 40   **Sounds**

1   He decided to go to Britain.
2   He'd packed his bags.
3   He'd got to the Austrian border.
4   He'd entered the West.
5   He'd arrived in France.
6   He had very little money.

Progress check 36 – 40   **Sounds, activity 1**

/w/   world   weeks   wash   water   washing   wear
/r/   return   religion   relative   repair   wrapping   remind

Progress check 36 – 40   **Sounds, activity 2**

/ɔː/   poor   law   more   moor   roar   saw   bore   sure   pour   four   war
/aʊ/   power   tower   sour   hour   flower

Wordlist

The first number after each word shows the lesson in the Student's Book in which the word first appears in the vocabulary box. The numbers in *italics* show the later lessons in which the word appears again.

abroad /əˈbrɔːd/ 7, *11*
accept /əkˈsept/ 1
accident /ˈæksɪdənt/ 37, *40*
accountant /əˈkaʊntənt/ 12
ache /eɪk/ 29
actor /ˈæktə(r)/ 12
advertise /ˈædvətaɪz/ 7
aeroplane /ˈeərəpleɪn/ 28
airport /ˈeəpɔːt/ 8
allow /əˈlaʊ/ 7
ambulance /ˈæmbjʊləns/ 37
angry /ˈæŋgrɪ/ 32
ankle /ˈæŋkl/ 21
anniversary /ˌænɪˈvɜːsərɪ/ 32
answer /ˈɑːnsə(r)/ 1
apples /ˈæplz/ 15
architecture /ˈɑːkɪtektʃə(r)/ 10
arithmetic /əˈrɪθmətɪk/ 12
arm /ɑːm/ 21, *29*
arrest /əˈrest/ 40
arrive /əˈraɪv/ 1, *20*
art gallery /ˈɑːt gælərɪ/ 10
ask /ɑːsk/ 1
asleep /əˈsliːp/ 27
aspirin /ˈæsprɪn/ 13
atmosphere /ˈætməsfɪə(r)/ 32
attractive /əˈtræktɪv/ 17
aunt /ɑːnt/ 9
autumn /ˈɔːtəm/ 36
awful /ˈɔːfl/ 4
axe /æks/ 27

back /bæk/ 21
backpack /ˈbækpæk/ 13
bad /bæd/ 10
bag /bæg/ 6, *25*
bald /bɔːld/ 17
ballet /ˈbæleɪ/ 16
bananas /bəˈnɑːnəz/ 15
bank /bæŋk/ 5, *14, 24*
banker /ˈbæŋkə(r)/ 12
bar /bɑː(r)/ 28
barbecue /ˈbɑːbɪkjuː/ 23
basin /ˈbeɪsn/ 3
bath /bɑːθ/ 3
bathroom /ˈbɑːθruːm/ 3
be at /biː æt/ 31
beach /biːtʃ/ 4, *10, 28, 33*
beard /bɪəd/ 17
beautiful /ˈbjuːtɪfl/ 10, *14, 17*

bed and breakfast
    /bed ən ˈbrekfəst/ 8
bed /bed/ 3
bedroom /ˈbedruːm/ 3
beef /biːf/ 15
beer /bɪə(r)/ 15
belief /bɪˈliːf/ 34
bicycle /ˈbaɪsɪkl/ 26
big /bɪg/ 6
bill (US) /bɪl/ 24
biology /baɪˈɒlədʒɪ/ 12
biscuit /ˈbɪskɪt/ 15, *31*
bite /baɪt/ 29
black /blæk/ 17, *19*
blonde /blɒnd/ 17
blouse /blaʊz/ 19
blue /bluː/ 19
boarding pass /ˈbɔːdɪŋ pɑːs/ 8
boat /bəʊt/ 7
body /ˈbɒdɪ/ 21
bomb /bɒm/ 37
book (v) /bʊk/ 8
border /ˈbɔːdə(r)/ 20
boring /ˈbɔːrɪŋ/ 10
borrow /ˈbɒrəʊ/ 30
bottle /ˈbɒtl/ 15
bowl /bəʊl/ 35
boy /bɔɪ/ 9
boyfriend /ˈbɔɪfrend/ 9
bread /bred/ 15
break down (v) /breɪk ˈdaʊn/ 40
break /breɪk/ 37
breakfast /ˈbrekfəst/ 2
bridge /brɪdʒ/ 32
bring /brɪŋ/ 7
broken /ˈbrəʊkən/ 6
brother /ˈbrʌðə(r)/ 9
brown /braʊn/ 17, *19*
Buddhist /ˈbʊdɪst/ 34
bunch /bʌntʃ/ 6
burglar /ˈbɜːglə(r)/ 37
burn /bɜːn/ 37
bus station /ˈbʌs steɪʃn/ 14
bus /bʌs/ 5, *28*
business class /ˈbɪznɪs klɑːs/ 8
busy /ˈbɪzɪ/ 10
butter /ˈbʌtə(r)/ 15
button /ˈbʌtn/ 37
buy /baɪ/ 5, *7*

cabbage /ˈkæbɪdʒ/ 15
cabin /ˈkæbɪn/ 8
cafe /ˈkæfeɪ/ 10, *39*
call out /ˈkɔːl aʊt/ 31
calm /kɑːm/ 18
camera /ˈkæmrə/ 13, *25*
canal /kəˈnæl/ 28
candle /ˈkændl/ 34
capital /ˈkæpɪtl/ 40
careful /ˈkeəfəl/ 18

carpet /ˈkɑːpɪt/ 3
carriage /ˈkærɪdʒ/ 26
carrots /ˈkærəts/ 15
carry /ˈkærɪ/ 7
cash /kæʃ/ 7
cassette recorder /kəˈset
    rɪˈkɔːdə(r)/ 38
casual /ˈkæʒʊəl/ 19
cat /kæt/ 38
catch fire /kætʃ ˈfaɪə(r)/ 37
cathedral /kəˈθiːdrəl/ 10
Catholic /ˈkæθəlɪk/ 34
celebrate /ˈseləbreɪt/ 23
celebration /ˌseləˈbreɪʃn/ 34
cemetery /ˈsemətərɪ/ 34
central heating
    /ˈsentrəl ˈhiːtɪŋ/ 38
centre /ˈsentə(r)/ 33
century /ˈsentʃərɪ/ 32
chair /tʃeə(r)/ 3
champagne /ʃæmˈpeɪn/ 38
change /tʃeɪndʒ/ 5, *7, 11*
changeable /ˈtʃeɪndʒəbl/ 36
charm /tʃɑːm/ 14
cheap /tʃiːp/ 4, *10*
check /tʃek/ 7
check in /tʃek ɪn/ 8
check out /tʃek aʊt/ 8
cheers /tʃɪəz/ 35
cheese /tʃiːz/ 15
chemist /ˈkemɪst/ 5, *24*
chemistry /ˈkemɪstrɪ/ 12
chicken /ˈtʃɪkɪn/ 15
child /tʃaɪld/ 9
chin /tʃɪn/ 35
chips (GB) /tʃɪps/ 24, *35*
chips (US) /tʃɪps/ 24
chocolate /ˈtʃɒklət/ 31
choose /tʃuːz/ 7
church /tʃɜːtʃ/ 28
cinema /ˈsɪnəmə/ 10, *16*
city /ˈsɪtɪ/ 33
clever /ˈklevə(r)/ 18
cliff /klɪf/ 33
climate /ˈklaɪmɪt/ 10
climb /klaɪm/ 27
close /kləʊz/ 5
closing time /ˈkləʊzɪŋ taɪm/ 16
cloth /klɒθ/ 25
cloud /klaʊd/ 6
club /klʌb/ 16
coast /kəʊst/ 33
coat /kəʊt/ 6, *19*
coffee /ˈkɒfɪ/ 15
cold /kəʊld/ (adj) 4, *10, 18*
cold /kəʊld/ (n) 29
college /ˈkɒlɪdʒ/ 14
comb /kəʊm/ 38
come /kʌm/ 2
come from /kʌm frəm/ 1

come in /kʌm ɪn/ 31
companion /kəmˈpænɪən/ 38
computer /kəmˈpjuːtə(r)/ 12, *38*
concert /ˈkɒnsət/ 16
confident /ˈkɒnfɪdənt/ 18
confusing /kənˈfjuːzɪŋ/ 4
connection /kəˈnekʃn/ 8
consulate /ˈkɒnsjʊlət/ 37
conversation /ˌkɒnvəˈseɪʃn/ 14
cooker /ˈkʊkə(r)/ 3
cost /kɒst/ 20, *40*
cottage /ˈkɒtɪdʒ/ 32
cotton /ˈkɒtn/ 25
country /ˈkʌntrɪ/ 7
countryside /ˈkʌntrɪsaɪd/ 4, *33*
cousin /ˈkʌzn/ 9
crash /kræʃ/ 40
create /kriːˈeɪt/ 7
credit card /ˈkredɪt kɑːd/ 38
crisps /krɪsps/ 24
cross /krɒs/ 5
crowd /kraʊd/ 32
crowded /ˈkraʊdɪd/ 4, *10*
cup /kʌp/ 15, *31, 35*
cupboard /ˈkʌbəd/ 3
curly /ˈkɜːlɪ/ 17
currency /ˈkʌrənsɪ/ 13
curtains /ˈkɜːtnz/ 3
curved /kɜːvd/ 25
customs /ˈkʌstəmz/ 26

dancer /ˈdɑːnsə(r)/ 12
dancing /ˈdɑːnsɪŋ/ 23
dangerous /ˈdeɪndʒərəs/ 10, *37*
dark /dɑːk/ 17
date /deɪt/ 30
daughter /ˈdɔːtə(r)/ 9
dead /ded/ 34
delay /dɪˈleɪ/ 8
dentist /ˈdentɪst/ 39
departure /dɪˈpɑːtʃə(r)/ 8
desert (n) /ˈdezət/ 20, *33*
diary /ˈdaɪərɪ/ 30
die /daɪ/ 7
dining room /ˈdaɪnɪŋ ruːm/ 3
dinner /ˈdɪnə(r)/ 2
dirty /ˈdɜːtɪ/ 4, *10*
disappear /ˌdɪsəˈpɪə(r)/ 32
disco /ˈdɪskəʊ/ 16
discover /dɪsˈkʌvə(r)/ 32
disease /dɪˈziːz/ 29
dish /dɪʃ/ 35
dishwasher /ˈdɪʃwɒʃə(r)/ 3
distance /ˈdɪstəns/ 20
doctor /ˈdɒktə(r)/ 12
dog /dɒg/ 38
door /dɔː(r)/ 3, *27, 31*
double room /ˈdʌbl ˈruːm/ 8
draw /drɔː/ 5
dress /dres/ 19
drink /drɪŋk/ 1

drive /draɪv/ 20, *26*
driver /ˈdraɪvə(r)/ 20
driving /ˈdraɪvɪŋ/ 4
driving licence
  /ˈdraɪvɪŋ laɪsns/ 37
drown /draʊn/ 37
druggist (US) /ˈdrʌgɪst/ 24
dry /draɪ/ 36
duty-free /ˌdjuːtɪ ˈfriː/ 26

ear /ɪə(r)/ 21, *29*
earn /ɜːn/ 1, *11*
east /iːst/ 33
economics /ˌiːkəˈnɒmɪks/ 12
education /ˌedʒʊˈkeɪʃn/ 12, *22*
eggs /egz/ 15
elbow /ˈelbəʊ/ 21, 35
election /ɪˈlekʃn/ 22
electricity /ɪˌlekˈtrɪsətɪ/ 37
elegant /ˈelɪgənt/ 14
employment /ɪmˈplɔɪmənt/ 22
engineer /ˌendʒɪˈnɪə(r)/ 12
entertainment
  /ˌentəˈteɪnmənt/ 10
escalator /ˈeskəleɪtə(r)/ 28
evil /ˈiːvl/ 32
excellent /ˈeksələnt/ 10
exchange /ɪksˈtʃeɪndʒ/ 7, *39*
excuse me /ɪkˈskjuːz miː/ 39
exhaust pipe /ɪgˈzɔːst paɪp/ 40
exhibition /ˌeksɪˈbɪʃn/ 16
expect /ɪkˈspekt/ 39
expensive /ɪkˈspensɪv/ 10
explode /ɪkˈspləʊd/ 37
eyes /aɪz/ 21

face /feɪs/ 17, *21*
factory /ˈfæktərɪ/ 10, *33*
fair /feə(r)/ 17, *23*
fall off /fɔːl ɒf/ 40
famous /ˈfeɪməs/ 14
fare /feə(r)/ 8
farm /fɑːm/ 27
farmland /ˈfɑːmlænd/ 33
fast /fɑːst/ 26
fat /fæt/ 17
father /ˈfɑːðə(r)/ 9
faucet (US) /ˈfɔːsɪt/ 24
favour /ˈfeɪvə(r)/ 39
favourite /ˈfeɪvrɪt/ 32
ferry /ˈferɪ/ 8
field /fiːld/ 33
film /fɪlm/ 16
finally /ˈfaɪnəlɪ/ 31
finance /ˈfaɪnæns/ 22
finger /ˈfɪŋgə(r)/ 21
finish /ˈfɪnɪʃ/ 2
fire /ˈfaɪə(r)/ 37
fireworks /ˈfaɪəwɜːks/ 23
first class /fɜːst klɑːs/ 26
fish /fɪʃ/ 15
fit /fɪt/ 18
flag /flæg/ 23
flat /flæt/ (adj) 6, *33*
flat /flæt/ (n) 3

flood /flʌd/ 36, *37*
flower /ˈflaʊə(r)/ 6
flu /fluː/ 29
fog /fɒg/ 36
food /fuːd/ 4, *5, 10, 13*
foot /fʊt/ 21
foreign /ˈfɒrən/ 7, *11*
forest /ˈfɒrɪst/ 33
fork /fɔːk/ 35
formal /ˈfɔːml/ 19, *30*
fortunately /ˈfɔːtʃənətlɪ/ 31
freedom /ˈfriːdəm/ 22
freezing /ˈfriːzɪŋ/ 36
french fries (US)
  /frentʃ ˈfraɪz/ 24
fridge /frɪdʒ/ 3
friend /frend/ 9
friendly /ˈfrendlɪ/ 4
front /frʌnt/ 31
frost /frɒst/ 36
fruit /fruːt/ 15
funny /ˈfʌnɪ/ 18

gallery /ˈgælərɪ/ 16
gallon /ˈgælən/ 20
garden /ˈgɑːdn/ 3, *31*
gas /gæs/ 37
gas (US) /gæs/ 24
gas station (US)
  /gæs steɪʃn/ 20
generous /ˈdʒenərəs/ 4
gentleman /ˈdʒentlmən/ 39
geography /dʒɪˈɒgrəfɪ/ 12
get /get/ 2, *5, 20*
get dressed /get drest/ 2
get on with someone /get ɒn
  wɪð sʌmwʌn/ 30
get out /get aʊt/ 31
get up /get ʌp/ 2
girl /gɜːl/ 9
girlfriend /ˈgɜːlfrend/ 9
give up /gɪv ʌp/ 7
glass /glɑːs/ 15, *25*
glasses /ˈglɑːsɪz/ 17
glue /gluː/ 25
go /gəʊ/ 1
go into /gəʊ ɪntuː/ 31
go shopping /gəʊ ʃɒpɪŋ/ 2
go to sleep /gəʊ tə sliːp/ 2
good /gʊd/ 10
give /gɪv/ 1
good-looking /gʊd lʊkɪŋ/ 17
government /ˈgʌvnmənt/ 22
grandfather /ˈgrændfɑːðə(r)/ 9
grandmother /ˈgrændmʌðə(r)/ 9
grapes /greɪps/ 15
grass /grɑːs/ 4, *27, 32*
great /greɪt/ 4
green /griːn/ 19
grey /greɪ/ 19
ground floor /graʊnd flɔː(r)/ 37
grounds /graʊndz/ 32
guard /gɑːd/ 26, *32, 40*
guide book /gaɪd bʊk/ 13
gun /gʌn/ 4, *37*

hair /heə(r)/ 17, *21*
hairdryer /ˈheə(r)draɪə(r)/ 25
ham /hæm/ 15
hammer /ˈhæmə(r)/ 27
hand /hænd/ 35
handbag /ˈhændbæg/ 13
hangover /ˈhæŋəʊvə(r)/ 29
happen /ˈhæpən/ 39
harbour /ˈhɑːbə(r)/ 8
hard /hɑːd/ 25
hat /hæt/ 19
have /hæv/ 2
have a shower/bath
  /hæv ə ˈʃaʊə(r)/bɑːθ/ 2
head /hed/ 17, *21, 29*
heart /hɑːt/ 29
heavy /ˈhevɪ/ 6, *25*
high /haɪ/ 25
high blood pressure
  /haɪ blʌd preʃə(r)/ 29
high street /ˈhaɪ striːt/ 24
highlight /ˈhaɪlaɪt/ 34
highway /ˈhaɪweɪ/ 20
hill /hɪl/ 20, *27, 33*
hilly /ˈhɪlɪ/ 33
history /ˈhɪstrɪ/ 12
hold /həʊld/ 5
hold on /həʊld ɒn/ 30
holiday /ˈhɒlədeɪ/ 7, *23*
home /həʊm/ 2, *3*
hospital /ˈhɒspɪtl/ 14, *28*
hot /hɒt/ 4
hotel /həʊˈtel/ 7, *28*
house /haʊs/ 3, *11*
housing /ˈhaʊzɪŋ/ 22
humid /ˈhjuːmɪd/ 36
hurricane /ˈhʌrɪkən/ 36
hurry /ˈhʌrɪ/ 27
husband /ˈhʌzbənd/ 9

ice /aɪs/ 36
ice cream /aɪskriːm/ 35
ill /ɪl/ 29
imaginative /ɪˈmædʒɪnətɪv/ 18
industrial /ɪnˈdʌstrɪəl/ 33
industry /ˈɪndəstrɪ/ 33
inflation /ɪnˈfleɪʃn/ 22
informal /ɪnˈfɔːml/ 30
injured /ˈɪndʒəd/ 37
insect /ˈɪnsekt/ 4, *29*
insurance /ɪnˈʃʊərəns/ 38
intelligent /ɪnˈtelɪdʒənt/ 18
interesting /ˈɪntrəstɪŋ/ 10, *18*
interval /ˈɪntəvl/ 16
introduce /ˌɪntrəˈdjuːs/ 7
invade /ɪnˈveɪd/ 32
island /ˈaɪlənd/ 33

jacket /ˈdʒækɪt/ 19
jam /dʒæm/ 35
jeans /dʒiːnz/ 19
jet lag /ˈdʒet læg/ 29
jewellery /ˈdʒuːəlrɪ/ 38
jobs /dʒɒbz/ 11
jokes /dʒəʊks/ 14
journalist /ˈdʒɜːnəlɪst/ 12

journey /ˈdʒɜːnɪ/ 40
juice /dʒuːs/ 15
jungle /ˈdʒʌŋgl/ 33

kill /kɪl/ 37
kilo /ˈkiːləʊ/ 15, *37*
kind /kaɪnd/ 17, *18*
king /kɪŋ/ 23
kitchen /ˈkɪtʃɪn/ 3
knee /niː/ 21
knife /naɪf/ 35
knock /nɒk/ 27
know /nəʊ/ 1

ladder /ˈlædə(r)/ 27
lake /leɪk/ 33
lamb /læm/ 15
lamp /læmp/ 3
land /lænd/ 8
lane /leɪn/ 26
language /ˈlæŋgwɪdʒ/ 11, *12*
lap /læp/ 35
large /lɑːdʒ/ 10
law and order
  /lɔːr ənd ˈɔːdə(r)/ 22
law courts /lɔː kɔːts/ 14
lawn /lɔːn/ 32
lazy /ˈleɪzɪ/ 18
lead (v) /liːd/ 7
learn /lɜːn/ 11
leather /ˈleðə(r)/ 25
leave /liːv/ 2, *7, 20*
left-wing /left wɪŋ/ 14
leg /leg/ 21, *29*
lend /lend/ 30
lettuce /ˈletɪs/ 15
level crossing /levl ˈkrɒsɪŋ/ 28
library /ˈlaɪbrərɪ/ 14, *28*
lift /lɪft/ 8, *28*
light /laɪt/ 25
lightning /ˈlaɪtnɪŋ/ 36
liquid /ˈlɪkwɪd/ 25
literary /ˈlɪtərərɪ/ 14
live /lɪv/ 1
lively /ˈlaɪvlɪ/ 14
living room /ˈlɪvɪŋ ruːm/ 3
loaf /ləʊf/ 15
local /ˈləʊkl/ 14
lock /lɒk/ 27
long /lɒŋ/ 17, *25*
look at /lʊk æt/ 5
look for /lʊk fɔː/ 31
lose one's way
  /luːz wʌnz weɪ/ 32
low /ləʊ/ 6, *25*
lower (v) /ˈləʊə(r)/ 27
luggage /ˈlʌgɪdʒ/ 8
lunch /lʌntʃ/ 2
lung /lʌŋ/ 29
luxury /ˈlʌkʃərɪ/ 7

machine /məˈʃiːn/ 25
main street (US) /meɪn striːt/ 24
man /mæn/ 9, *17*
manners /ˈmænə(r)z/ 30
map /mæp/ 13

Progress Test 1   Lessons 1–10

SECTION 1: VOCABULARY (30 marks)

1a Underline the odd-one-out and leave a group of three related words. (10 marks)

  b Add one other word to the groups of words. (10 marks)

Example: her my our <u>they</u> _their_

1 after always often sometimes _____

2 afternoon morning night yesterday _____

3 basin bath fridge shower _____

4 aunt cousin husband tourist _____

5 cathedral countryside factory park _____

6 crowded friendly generous polite _____

7 children man students women _____

8 did lived see wrote _____

9 British France Germany Japan _____

10 dirty expensive modern safety _____

2 Complete these sentences with ten different verbs. (10 marks)

Example: I ___buy___ food at the supermarket.

1 I never _____ the shopping.

2 She doesn't _____ a musical instrument.

3 I _____ to music at home.

4 Pleased to _____ you.

5 Where do you _____ from?

6 I _____ about ten cigarettes a day.

7 Where do we _____ for the bus?

8 I can't _____ the weather in Britain.

9 Where do you _____ dinner?

10 They don't _____ glasses.

Progress Test 1 Lessons 1–10

SECTION 2: GRAMMAR (30 marks)

3a Choose ten of these words to complete the first ten spaces in the passage. (10 marks)

Example: a) had b) has c) have

1 a) go b) was c) went
2 a) cousin b) cousins c) cousin's
3 a) a b) an c) the
4 a) because b) but c) so
5 a) get b) got c) has
6 a) has b) is c) isn't
7 a) a b) one c) two
8 a) from b) in c) to
9 a) because b) but c) so
10 a) aren't b) weren't c) didn't

b Complete the last ten spaces with ten of your own words. (10 marks)

When did I last __have__ a holiday? Well, six months ago I (1) _____ to the United States because my (2) _____ is out there. He's (3) _____ engineer. He lives and works in Denver (4) _____ I decided to visit him there. Denver's an interesting city and it's (5) _____ lots of theatres and art galleries. There aren't many tourists and it (6) _____ very expensive. After (7) _____ days in Denver, we took a plane (8) _____ Las Vegas. We had single tickets (9) _____ we wanted to drive back to Denver. Our first stop was the Grand Canyon. There is a hotel in Grand Canyon village but we (10) _____ stay in the hotel. The weather (11) _____ awful but we decided to walk to the bottom of the canyon. There is a hostel at the bottom of the canyon (12) _____ we stayed there. There (13) _____ beds for about thirty people in the hostel and there (14) _____ a kitchen and a dining room. You pay about $50 per person and (15) _____ have dinner and breakfast there. I come from the north-east of England and I (16) _____ some people from my home town in the hostel. The hostel was full and (17) _____ were lots of American people there. I (18) _____ Americans because they're so friendly and polite. The next day we had breakfast (19) _____ seven o'clock and checked out at half past seven. Then we walked back to (20) _____ village and had lunch in the hotel.

4 Write your own questions for these answers. (10 marks)

Example: Yes, I am. This is my husband.
Are you married?

1 I'm twenty-three.
_____

2 I'm an engineer.
_____

3 Fine, thanks.
_____

4 At a quarter past seven in the morning.
_____

5 Yes, one brother and two sisters.
_____

6 We usually go abroad.
_____

7 I watched TV, then I went to bed.
_____

8 I'm writing a letter.
_____

9 No, there isn't.
_____

10 I'd like a beer, please.
_____

PHOTOCOPIABLE

Progress Test 1　Lessons 1–10

SECTION 3: READING (20 marks)

5　Read the passage *At home in the Land of the Rising Sun*. Which of these things is the article about? Tick the box. (2 marks)

a　Traditional Japanese homes.　☐

b　The differences between modern and traditional Japanese homes.　☐

c　Traditional features of modern Japanese homes.　☐

6　Are these sentences true (T) or false (F) or doesn't the passage say (DS)? (10 marks)

Example: Japanese bathrooms are quite small.　*DS*

1　People eat and sleep in separate rooms.　☐

2　A typical Japanese home has cushions and curtains.　☐

3　There are often dried grass mats in Japanese homes.　☐

4　All Japanese apartments have gardens.　☐

5　Japanese gardens are always very small.　☐

7　Find four things the passage says about hospitality in Japan. Make notes. (8 marks)

_____

_____

_____

_____

_____

_____

AT HOME IN THE LAND OF THE RISING SUN

MANY PEOPLE IN JAPAN now live in apartments and their homes have all the latest household equipment. But there are still many old and traditional features in modern Japanese homes.

The *genkan* or hall is always an important feature in Japanese homes. It is usually where the family and guests change their shoes. It is a place of welcome and often has flowers or a picture. When guests leave, everyone comes to the *genkan* to say goodbye.

The main room of a Japanese home is used as the sitting room, dining room and bedroom, and there isn't much furniture, just cushions around a low table. Other features are *oshiire* or cupboards where you put bedding and clothes during the day, *fusuma* or sliding doors to make the living areas into a number of small rooms and many homes still have a *tokonoma*, a small area with flowers, a painting and a *tatami* mat. Guests usually sit in front of the *tokonoma*. The kitchen is often very small, and you never eat or entertain people there.

Modern building materials are of course very common, but there is still some wood and paper in modern homes. There are often bamboo curtains *(sudare)* and heavy wooden shutters *(amado)* on the windows. The most common feature is the *tatami* mat, made of dried grass, about 1.8 by 0.9 metres. The Japanese still measure a room by the number of *tatami* mats it can contain.

A garden is very important, if the home has enough space. It is not a place where children play but somewhere beautiful to look at. If there isn't a garden, there are pots of plants or small bonsai trees on shelves.

Progress Test 1 Lessons 1–10

SECTION 4: WRITING (20 marks)

8 Introduce yourself to your teacher. For example, include some information about your home and family, describe your daily routine and / or say what you like doing. Write 10 – 15 sentences. (20 marks)

_____

_____

_____

_____

_____

_____

_____

_____

_____

_____

_____

_____

_____

_____

_____

_____

_____

_____

_____

_____

Progress Test 2 Lessons 11–20

SECTION 1: VOCABULARY (30 marks)

1a Find fifteen words in the word square. The words are in two directions: ↓ and →. Write the words in five groups. (15 marks)

A jobs _____

_____

B subjects _____

_____

C clothes _____

_____

D town features _____

_____

E things to eat _____

_____

b Add one other word to each group of words. (5 marks)

```
J  E  C  O  N  O  M  I  C  S
O  H  S  S  O  N  I  O  N  S
U  I  T  W  S  U  I  T  H  P
R  S  A  E  N  B  T  M  O  O
N  T  D  A  U  R  I  U  S  T
A  O  I  T  R  E  G  S  P  A
L  R  U  E  S  A  H  E  I  T
I  Y  M  R  E  D  T  U  T  O
S  A  C  T  O  R  S  M  A  E
T  S  C  I  E  N  C  E  L  S
```

2 Complete these sentences with ten different nouns. (10 marks)

Example:
There's an _interval_ of 15 minutes during the concert.

1 I'd like to earn a lot of _____ .

2 How many _____ of biscuits do we need?

3 The seats are in the third _____ on the left.

4 He's got long hair, a _____ and a moustache.

5 He's wearing _____ because he's going running.

6 Does _____ cost $2 a gallon?

7 Is there any _____ in your medical kit?

8 I've got a _____ so I don't need scissors.

9 There's an _____ of her work at the gallery.

10 There was a driver and three _____ in the car.

Progress Test 2 Lessons 11–20

SECTION 2: GRAMMAR (30 marks)

3a Choose ten of these words to complete the first ten spaces in the conversation. (10 marks)

Example: a) go b) going c) to go

1 a) at b) in c) on
2 a) can't b) don't c) won't
3 a) at b) in c) on
4 a) last b) next c) this
5 a) at b) in c) on
6 a) see b) seeing c) to see
7 a) best b) better c) good
8 a) does b) is c) will
9 a) many b) much c) often
10 a) I'll phone b) I'm going to phone c) I'm phoning

b Complete the last ten spaces with ten of your own words. (10 marks)

SANDRA: How about ___going___ to the theatre

(1) _____ the weekend?

DENISE: I'm sorry, I (2) _____ . I'm going to my sister's

(3) _____ London tomorrow for the weekend.

What about one evening (4) _____ week?

SANDRA: OK. How about Wednesday?

DENISE: That's fine. What's (5) _____ ?

SANDRA: *A Comedy of Errors* is at the Playhouse. I saw it
a year ago but I'd like (6) _____ it again. I think it's
Shakespeare's (7) _____ play.

DENISE: What time (8) _____ it start?

SANDRA: Half past seven.

DENISE: How (9) _____ are the tickets?

SANDRA: I don't know.

DENISE: (10) _____ the theatre and find out. I'm sure
they (11) _____ be very expensive. I'll reserve two
seats, (12) _____ you like. Where is the Playhouse,
by the way?

SANDRA: It's in Jarrett Street. You know the main square?
Well, go down Turnpike Street, (13) _____ left at
the crossroads and then take the second turning
(14) _____ the right. It's next (15) _____ the
car park.

DENISE: How (16) _____ is it from the main square?

SANDRA: About five hundred metres. It (17) _____
about ten minutes on foot. But, listen, I'm going to
take my car. I don't (18) _____ using public
transport late (19) _____ night. I'll give you a lift.
I'll come round to (20) _____ flat at about seven
o'clock.

DENISE: Thanks very much. That's very kind of you.

4 Rewrite these sentences. Begin with the words in
brackets. (10 marks)

Example: She's as patient as he is. (He isn't)
He isn't more patient than she is.

1 The post office is opposite the library. (The library)

_____

2 I bought a guidebook because I'm going to
Switzerland. (I'm going)

_____

3 Could you tell me the way to the law courts? (How do)

_____

4 How about going to the opera next week? (Let's)

_____

5 She isn't as confident as he is. (He)

_____

6 The T-shirts are cheaper than the shirts. (The shirts)

_____

7 The chemist is behind the supermarket. (The
supermarket)

_____

8 Fruit is better for you than chocolate. (Chocolate isn't)

_____

9 He's completely bald. (He hasn't)

_____

10 She likes being at university. (She enjoys)

_____

Progress Test 2 Lessons 11–20

SECTION 3: READING (20 marks)

5 Read the passage *Highway to the Andes.* Was James Ferguson's journey horrible? (2 marks)

_____

6 Are these sentences true (T) or false (F) or doesn't the passage say (DS)? (8 marks)

Example: James began his journey in Valencia. [F]

1 Most people fly from Caracas to Mérida. [ ]

2 James ate roast beef on his journey. [ ]

3 Caracas is four hundred and fifty kilometres from Barinas. [ ]

4 Barinas is higher than three thousand metres. [ ]

5 People say that the first part of the journey is the best. [ ]

6 James stopped for a drink in the *páramo.* [ ]

7 Mérida is 4,800 metres above sea level. [ ]

8 He tried a garlic-flavoured ice cream. [ ]

7 What did James see out of his car window? Use these five headings and make notes. (10 marks)

a near Caracas

_____

_____

b near Valencia

_____

_____

c near Acarigua

_____

_____

d near Apartaderos

_____

_____

e near Mérida

_____

_____

Highway to the Andes

'You're going to DRIVE to Mérida? But it's so far – it will be horrible.' This is what my friends said when I told them I wanted to go to Mérida by car.

The Venezuelans love taking planes. Domestic flights are reliable and cheap and there are lots of them. So people thought it a little strange that I wanted to hire a car and drive, what is only 700 kilometres, to the Andean city of Mérida.

But the journey wasn't horrible at all. The roads were in good condition, there was motorway for part of the way and regular petrol stations (where petrol cost less than water). There was even a roadside self-service restaurant between Caracas and Valencia.

I drove out of Caracas, and its mixture of modern skyscrapers and old, poor housing, on a mountainous highway. At first, the scenery was Caribbean: bananas, small villages, coconut trees. When you reach Valencia, orange trees mix with sugar-cane. A little further on, past Acarigua, the country changes into the *llanos* or plains. This is real cowboy country, with huge isolated farms and occasional roadside ovens where you can buy roast beef by the kilo.

Everyone agrees that the last four hours to Mérida are the best, so it's a good idea to spend the night in Barinas and drive the remaining 250 kilometres in the morning. An hour or so out of Barinas and the Andes seem to appear suddenly around the corner. From then on, it's an exciting climb up to 3,000 metres and the Venezuelan *páramo.*

The *páramo* is the cold, windy mountain range. Although there are no trees in these mountains, there are lots of plants and flowers. At 3,600 metres you notice that the air is getting thinner. After a strong cup of Venezuelan coffee at the village of Apartaderos, my heart seemed to beat much faster than usual.

My car didn't like the thin air and slowed down. I didn't mind – this gave me time to look at the mountains. Locally, 'crossing the *páramo*' means dying, and the car seemed to know this.

The scenery on the way down into Mérida is spectacular. Some of it is Andean with its low, stone villages and potato fields and some is Alpine with its rivers full of fish, and inviting hotels. Mérida itself is a pleasant town. It has three claims to fame: the University of the Andes, the world's highest cable-car – to the snowy 4,800-metre Pico Bolívar – and an ice cream shop with the world's biggest number of flavours (400), including garlic and spaghetti bolognese. But this wasn't why I had come.

Adapted from *Highway to the Andes* by James Ferguson.

Progress Test 2 Lessons 11–20

SECTION 4: WRITING (20 marks)

8 Describe someone you know quite well. For example, talk about the person's clothes, appearance and character. Write 10 – 15 sentences.

_____

_____

_____

_____

_____

_____

_____

_____

_____

_____

_____

_____

_____

_____

_____

_____

_____

_____

_____

_____

_____

Progress Test 3 Lessons 21–30

SECTION 1: VOCABULARY (30 marks)

1a Match ten of the words in the box with the first ten words in the list. (10 marks)

1 alcoholic 2 bad 3 blood 4 car 5 cassette 6 food

7 guide 8 heart 9 insect 10 military 11 pedestrian

12 prime 13 railway 14 shoe 15 sleeping 16 state

17 police 18 ticket 19 traffic 20 wine

---

attack bite book cold class drink hours house

kit match parade park pass patrol poisoning

player pressure science terminal tour

---

alcoholic drink _____   _____

_____   _____

_____   _____

_____   _____

_____

b Choose words which go with the other ten words in the list above. Do not use words in the box. Write ten pairs of words. (10 marks)

_____   _____

_____   _____

_____   _____

_____   _____

_____   _____

2 Complete these sentences with ten different nouns. (10 marks)

Example:
   I was ill when I went to Egypt on *holiday* last year.

1 The queen always makes a short _____ before she opens a new building.

2 We've got five _____ on each foot.

3 In British English, a _____ is something you turn on and off to control water in a bath or basin.

4 You use _____ to wash yourself with.

5 Pedestrians mustn't step off the _____ before looking both ways.

6 In my country, you can't feed the animals when you go to a _____ .

7 She's got a _____ because she had too much wine last night.

8 You have to use the stairs when the _____ isn't working.

9 The children had a lot of _____ to play with.

10 A typewriter is a _____ for writing letters with.

Progress Test 3 Lessons 21–30

SECTION 2: GRAMMAR
(30 marks)

3a Choose ten of these words to complete the first ten spaces in the passage. (10 marks)

Example:
   a) change b) changed c) have changed

1 a) get b) getting c) to get
2 a) for b) from c) since
3 a) has bought b) has built c) has moved
4 a) has lived b) lived c) lives
5 a) her b) him c) them
6 a) found b) given up c) had
7 a) where b) which c) who
8 a) can b) must c) should
9 a) can b) have c) must
10 a) can b) can't c) mustn't

b Complete the last ten spaces with ten of your own words. (10 marks)

Dear Stacey,   December 8th
        Many thanks for your letter and for your news. I hope you're keeping well.
    Lots of things _have changed_ for me too. My husband and I have decided (1) _____ divorced. We've known each other (2) _____ ten years so it wasn't easy for us. He (3) _____ to a new flat and now (4) _____ on the other side of the city. The children stay with (5) _____ on Friday and Saturday nights. I've also (6) _____ a job in an off-licence. (An off-licence is a shop (7) _____ you can buy alcohol.) My hours are from 11am to 3pm during the week so I (8) _____ take the children to school every day. Unfortunately I (9) _____ to work Friday and Saturday evenings too. This means I (10) _____ go out at weekends. My sister sometimes looks after the children and then (11) _____ I (12) _____ can go out during the week. _____ at work last week because I had food poisoning. I went out for a meal last Monday evening and I (13) _____ sick all night. The next day I felt terrible and I (14) _____ go to work. I've eaten very little since then and I've (15) _____ a lot of weight. I feel much better now so I (16) _____ back to work yesterday. Have you (17) _____ had food poisoning? It's awful!
    Thanks very much for the photos of the baby. Does he look (18) _____ you or your husband? And what's (19) _____ name?
    It's nearly midnight so I (20) _____ finish now. It would be lovely to hear from you again.
                                    Love,
                                       Vicky

4 Rewrite these sentences. Begin with the words in brackets. (10 marks)

Example: You mustn't ride a motorbike before you're sixteen. (You can't)
    You can't ride a motorbike before you're sixteen.

1 You use it to take photos with. (It's for)

_____

2 We must take some sun-tan lotion. (We have)

_____

3 Can you run five kilometres? (Are you)

_____

4 You should go to the dentist. (You ought)

_____

5 My throat hurts. (I've got)

_____

6 Would you mind telling me the time? (Could you)

_____

7 Never walk along a railway line. (You must)

_____

8 I'll carry your suitcase, shall I? (Shall I)

_____

9 I couldn't read when I was five. (I wasn't)

_____

10 Can you lend me your car this evening? (Can I)

_____

Progress Test 3   Lessons 21–30

SECTION 3: READING (20 marks)

5 Read the leaflet *Belt up*. What is the leaflet about? Choose one of these things. (2 marks)

  a Wearing seat belts in the front of cars. ☐

  b Wearing seat belts in the back of cars. ☐

  c Wearing seat belts in the front and back of cars. ☐

6 Are these sentences true (T) or false (F) or doesn't the passage say (DS)? (10 marks)

Example: The seat belt laws have reduced the number of deaths in accidents. ☐ *T*

1 British car drivers have had to wear seat belts since 1983. ☐

2 Other people's lives may be in danger if you don't wear a seat belt. ☐

3 People who don't wear available seat belts have to pay a fine. ☐

4 Some people don't have to wear rear seat belts. ☐

5 You have to wear seat belts in lorries. ☐

6 All British cars must have seat belts in the back. ☐

7 It is safer to sit in the back of a car. ☐

8 Adults, and not children, should wear the available seat belts. ☐

9 Children should never wear adult seat belts. ☐

10 The new law has prevented at least 370 deaths. ☐

7 What are the current laws on using seat belts in British cars? Make notes. Use no more than 40 words. (8 marks)

_____

_____

_____

_____

_____

_____

_____

BELT UP

WEARING A SEAT BELT SAVES LIVES

Wearing a seat belt in the front of a car has been the law since 1983. This has saved people from at least 370 deaths and 7000 serious injuries each year.

In 1989, it became the law for children under 14 to wear seat belts, or to sit in a special child seat in the back seats of cars. In the first year, it is estimated that the new law saved 200 children from death or serious injury.
Rear seat belts save 140 lives and prevent 1,800 serious injuries every year at current wearing rates.

THE LAW OF FEBRUARY 1992

The law of 1 February 1992 means that adult passengers in the rear seats, as well as those in the front, must wear seat belts if there are seat belts in the rear of the car.

- It is illegal to carry a child under three years of age in the front of a car unless there is a special child seat.
- If there is a child seat in the front, but not the back of the car, children under three years of age must use that seat.
- Children between 3 and 11 years of age under 1.5m tall must use a child seat, or wear an adult seat belt.
- children aged 12 or 13, or a younger child more than 1.5m tall must wear an adult seat belt.

It is not only your life which is in danger if you do not wear an available seat belt. You are also risking the lives of the driver and other passengers. Adults sitting in the back seats of cars without wearing a seat belt can be thrown forward in an accident and can hurt people in front of them.

If you do not wear an available seat belt and the police stop you, you may have to pay a fine.

Do you always have to wear seat belts in the back of cars?
You have to wear seat belts in the back of cars which have them. All new cars now have seat belts in the back but older cars don't have to have them. There could be a medical reason why you should not wear a seat belt. Contact your doctor to check.

Who is responsible for making sure back seat passengers wear their seat belts?
Adult passengers or, if the passengers are children under 14, the driver.

What if there aren't enough seat belts available?
Not every passenger can wear a seat belt if there are not enough seat belts available. If there aren't enough seat belts available, it is safer for adults, and not children, to wear them. Heavier passengers will cause greater injury to others in an accident if they are not wearing a seat belt.

Progress Test 3   Lessons 21–30

SECTION 4: WRITING (20 marks)

8 Imagine that you have gone to the doctor's because you are not feeling well. Write the conversation between the doctor and yourself. For example, talk about the medical problems you have had, explain your current complaint, give the doctor's advice. Write 10 to 15 sentences.

_____

_____

_____

_____

_____

_____

_____

_____

_____

_____

_____

_____

_____

_____

_____

_____

_____

_____

_____

PHOTOCOPIABLE

Progress Test 4   Lessons 31–40

SECTION 1: VOCABULARY (30 marks)

1a Underline the odd-one-out and leave a group of three related words. (5 marks)

 b Add one other word to the groups of words. (5 marks)

Example:
 burglar companion <u>mugging</u> musician _victim_

1 east north south toast _____

2 cup bowl saucer spoon _____

3 autumn spring storm summer _____

4 dry fog ice snow _____

5 did drunk eaten written _____

2a Underline five words or phrases you can put with *do*. (5 marks)

a cake a decision a mistake a noise a phone call

an appointment business friends harm

notes someone a favour the bed the cleaning

the tea your best

 b Write five other words or phrases you can put with *do*. (5 marks)

_____

_____

_____

_____

_____

3 Complete the sentences with ten different particles. (10 marks)

Example: He took ___*out*___ a pen and wrote his name.

1 He put his coat _____ because it was cold.

2 He got _____ his car and escaped the rain.

3 Can you turn the radio _____ a bit? It's too loud.

4 The vacuum cleaner isn't plugged _____ .

5 We'll look _____ your cat when you're on holiday.

6 Who are you going to vote _____ in the next elections?

7 I'd read the letter when I threw the envelope

_____ .

8 I've never been able to give _____ smoking.

9 The children waved _____ their grandmother.

10 You can't watch TV. I've switched it _____ .

© Simon Greenall 1994. This sheet may be photocopied for use in class.

Progress Test 4   Lessons 31–40

SECTION 2: GRAMMAR (30 marks)

4a Choose ten of these words to complete the first
ten spaces in the passage. (10 marks)

Example: a) although b) but c) however

1 a) danger b) dangerous c) dangerously

2 a) has b) had c) was

3 a) had laughed b) laughed c) were laughing

4 a) bad b) good c) well

5 a) never b) often c) sometimes

6 a) stay b) stayed c) were staying

7 a) understanding b) understand c) understood

8 a) am b) were c) would

9 a) can b) might b) won't

10 a) he b) I c) they

b Complete the last ten spaces with ten of your own
words. (10 marks)

Foreign drivers aren't always welcome in Britain,
_____*although*_____ most of them are very good drivers.
However, tourist Graziano Montironi is warning British
motorists that he's (1) _____ . Graziano, on a
motoring holiday, with his wife, Lucia, and their three
young children, reached York yesterday afternoon.

Forty-year-old Graziano, a lorry driver from Rome,
(2) _____ just arrived in York in his car when
heads there started to turn. People (3) _____ when
they saw the signs in his car: 'Sorry, I'm Italian and I'm a
(4) _____ driver. Keep a safe distance. I drive
dangerously. We don't want an accident.'

Graziano said yesterday: 'It's the first time I've been to
England, so I've (5) _____ driven on the left. While
we (6) _____ in a bed and breakfast in Dover, I
explained this to one of the other guests. He was
American so he (7) _____ my problem. "If I
(8) _____ you, I'd put some stickers in your
windows. That (9) _____ help," he said and that's
what (10) _____ did. Italians (11) _____
known as bad drivers and I'm not too proud to admit it.

Anyway, it's my wife's car. She (12) _____ drive
in England because she's too nervous. If we
(13) _____ an accident, it will be my fault. So I want
(14) _____ make sure nothing happens.'

A motorist who pulled up behind Graziano's car said,
'I was going down Tower Street (15) _____ it
suddenly pulled out in front (16) _____ me from
the left. Fortunately, I (17) _____ hit him. I read

what the driver (18) _____ written on his stickers
and gave him enough room. I'd probably do the same if I
(19) _____ to Italy. It seems like a (20) _____
idea!'

5 Rewrite these sentences. Begin with the words in
brackets. (10 marks)

Example:
They speak English and French in Canada. (English)
English and French are spoken in Canada.

1 I was drying some shirts when one of them caught
fire. (While)

_____

2 It's been too foggy recently. (There's)

_____

3 If you don't give that pen to me, I'll take it from you.
(I'll)

_____

4 We like pasta but we don't have it every day.
(Although)

_____

5 They set off when they'd listened to the weather
forecast. (After)

_____

6 You use money for buying things. (Money)

_____

7 There isn't enough wind to go hang-gliding. (It)

_____

8 It might not be very sunny tomorrow. (There)

_____

9 When I plugged the computer in, it exploded. (The
computer)

_____

10 I wouldn't tell a stranger the time if he asked me. (If)

_____

141

Progress Test 4 Lessons 31–40

SECTION 3: READING (20 marks)

6 Read *Dreamtime*. There are five paragraphs in the text. Which paragraphs describe these things? Write the number of the paragraph in the box. (6 marks)

a Aboriginal society. ☐

b Aboriginal life since 1778. ☐

c Aboriginal religious ceremonies. ☐

7 Are these sentences true (T) or false (F) or doesn't the passage say (DS)? (9 marks)

Example:
There are fewer than 30,000 Aborigines today. DS

1 The Aboriginal culture is older than Middle Eastern culture. ☐

2 When the Europeans arrived, most Aborigines lived on the coast. ☐

3 Aborigine groups always lived in the same place. ☐

4 The Aboriginal lifestyle today is the same as it has always been. ☐

5 *Didgeridoos* are important religious celebrations. ☐

6 Women take part in the religious ceremonies. ☐

7 Aborigines' bodies are painted before the ceremonies begin. ☐

8 *Corroborees* can mark important events in an Aborigine's life. ☐

9 Aborigines believe in their own connection with the past and the future. ☐

8 Why were animals and plants important to the Aborigines? Make notes. (5 marks)

_____

_____

_____

_____

_____

_____

Dreamtime

1 THE AUSTRALIAN CONTINENT had its own culture, which was rich and complex in its customs, religions and lifestyles, long before the ancient civilisations in the Middle East, Europe and the Americas developed. For more than 40,000 years before European navigators visited their coast, the Australian Aborigines had lived on this continent. It is estimated that before the arrival of the first European settlers in 1778, the Aborigine population was more than 300,000. At that time, 500 different languages were spoken.

2 The Aborigines lived in groups of 10 to 50 people. They lived a mostly peaceful, nomadic life, moving from one area to another in their search for food. They collected fruit, nuts and insects to eat and hunted kangaroos and emus. They adapted to the harsh environment; they were perfectly at home before white settlers arrived.

3 The older men in the group had to build the group's identity through its religious beliefs and traditions. These developed in close harmony with their environment and were very important to the group members. The idea of Dreamtime or Dreaming was very important in their way of life. This is the time of the creation, when the land, sea, sky and all creatures were made. Dreamtime is a time that existed long ago, although for Aborigines it is an experience which links the past, the present and the future. The Aborigines believed that a person's spirit did not die when the person died. Instead, these spirits – of their parents and grandparents – continued to live on, in rocks, animals, plants or other human forms. So each group formed a special relationship with an animal or a plant, which acted as a symbol of their group identity and a protector. Rock paintings were of special importance and record Aboriginal beliefs. These are also passed on from parents to their children by storytelling and through song and dance.

4 Although the Aboriginal population of Australia is greatly reduced today, some of their customs live on. The ceremony of celebrating with song and dance is called *corroboree*. The dancers wear special clothes and body paint. They are accompanied by drums and other instruments, and *didgeridoos*, hollow pieces of wood producing a strange sound which is said to sound like the calling of the spirits of their ancestors. *Corroborees* are sometimes performed to persuade the spirits to bring rain, or to provide successful hunting. Other *corroborees* are held to mark the time when boys reach manhood, to mourn death or to celebrate love.

5 With the arrrival of the European settlers, the way of life of the Aborigines changed dramatically. Thousands died from European diseases and many had to leave their lands. The Aboriginal culture had prepared the people for everything they might expect to face in life – everything, that is, except white settlers.

Progress Test 4 Lessons 31–40

SECTION 4: WRITING (20 marks)

9 Think about an interesting day you have had recently. Describe what happened. Try to use a variety of different structures. Write 10 to 15 sentences. (20 marks)

Answers Progress Test 1 Lessons 1–10

SECTION 1: VOCABULARY [30 marks]

1a (10 marks: 1 mark for each correct answer.)

1	after	6	crowded
2	yesterday	7	man
3	fridge	8	see
4	tourist	9	British
5	countryside	10	safety

b (10 marks: 1 mark for each appropriate answer.)
1 an adverb of frequency, eg *never, usually*
2 evening (expression of time used with *in the*)
3 toilet (equipment found in a bathroom)
4 a family member, eg *brother, grandfather*
5 a feature or facility of towns, eg *restaurant, university*
6 an adjective for describing people, eg *awful, unfriendly*
7 a plural noun, eg *chairs, men*
8 a past tense form, eg *had, walked*
9 the name of a country, eg *Britain, Saudi Arabia*
10 an adjective for describing aspects of towns, eg *boring, safe*

2 (10 marks: 1 mark for each appropriate answer.)

1	do	6	smoke
2	play	7	wait
3	listen	8	stand
4	meet	9	have
5	come	10	wear

SECTION 2: GRAMMAR [30 marks]

3a (10 marks: 1 mark for each correct answer.)

1	c) went	6	c) isn't
2	a) cousin	7	c) two
3	b) an	8	c) to
4	c) so	9	a) because
5	b) got	10	c) didn't

b (10 marks: 1 mark for each appropriate answer.)
Possible answers

11	was	16	met
12	and/so	17	there
13	are	18	like/love
14	is	19	at
15	you	20	the

4 (10 marks: 1 mark for each correct question.)
Possible answers
1 How old are you?
2 What do you do?
3 How are you?
4 What time do you get up/ have breakfast/go to work?
5 Have you got any brothers and sisters?
6 Where do you go for your holidays?
7 What did you do yesterday evening/last night?
8 What are you doing/writing?
9 Is there a dining room in your house?
10 What would you like (to drink)?

SECTION 3: READING [20 marks]

5 (2 marks)
c

6 (10 marks: 2 marks for each correct answer.)
1 F
2 T
3 T
4 F
5 DS

7 (8 marks: 2 marks for each of the four things.)
1 Guests change their shoes in the *genkan*.
2 When guests leave, everyone comes to the *genkan* to say goodbye.
3 Guests usually sit in front of the *tokonoma*.
4 You never eat or entertain people in the kitchen.

SECTION 4: WRITING [20 marks]

8 (20 marks)
Tell students what you will take into consideration when marking their written work. Criteria should include:
- efficient communication of meaning (7 marks)
- grammatical accuracy (7 marks)
- coherence in the ordering or the information or ideas (3 marks)
- layout, capitalisation and punctuation (3 marks)

It is probably better not to use a rigid marking system with the written part of the test. If, for example, you always deduct a mark for a grammatical mistake, you may find that you are over-penalising students who write a lot or who take risks. Deduct marks if students haven't written the minimum number of sentences stated in the test.

swers Progress Test 2 Lessons 11–20

SECTION 1: VOCABULARY [30 marks]

1 (15 marks: 1 mark for each correct answer.)
- A jobs: actor, journalist, nurse
- B subjects: economics, history, science
- C clothes: suit, sweater, tights
- D town features: hospital, museum, stadium
- E things to eat: bread, onions, potatoes

b (5 marks: 1 mark for each appropriate answer.)

2 (10 marks: 1 mark for each appropriate answer.)

1	money	6	petrol/gas
2	packets	7	aspirin
3	row	8	penknife
4	beard	9	exhibition
5	trainers	10	passengers

SECTION 2: GRAMMAR [30 marks]

3a (10 marks: 1 mark for each correct answer.)

1	a) at	6	c) see
2	a) can't	7	a) best
3	b) in	8	a) does
4	b) next	9	b) much
5	c) on	10	a) I'll phone

b (10 marks: 1 mark for each appropriate answer.)

Possible answers

11	won't	16	far
12	if	17	takes
13	turn	18	like
14	on	19	at
15	to	20	your

4 (10 marks: 1 mark for each correct sentence.)
1 The library is opposite the post office.
2 I'm going to Switzerland so I bought a guidebook.
3 How do I get to the law courts?
4 Let's go to the opera next week.
5 He's more confident than she is.
6 The shirts are more expensive than the T-shirts. / The shirts aren't as cheap as the T-shirts.
7 The supermarket is in front of the chemist.
8 Chocolate isn't as good for you as fruit.
9 He hasn't got any hair.
10 She enjoys being at university.

SECTION 3: READING [20 marks]

5 (2 marks)
no

6 (8 marks: 1 mark for each correct answer.)

1	T	5	F
2	DS	6	T
3	T	7	F
4	F	8	DS

7 (10 marks: 2 marks for the notes under each heading.)
- a near Caracas: bananas, villages, coconut trees
- b near Valencia: orange trees, sugar cane
- c near Acarigna: farms, cowboys, roadside ovens
- d near Apartaderos: plants, flowers
- e near Mérida: villages, potato fields, rivers, hotels

SECTION 4: WRITING [20 marks]

8 (20 marks)
Tell students what you will take into consideration when marking their written work. Criteria should include:
- efficient communication of meaning (7 marks)
- grammatical accuracy (7 marks)
- coherence in the ordering or the information or ideas (3 marks)
- layout, capitalisation and punctuation (3 marks)

It is probably better not to use a rigid marking system with the written part of the test. If, for example, you always deduct a mark for a grammatical mistake, you may find that you are over-penalising students who write a lot or who take risks. Deduct marks if students haven't written the minimum number of sentences stated in the test.

Answers Progress Test 3 Lessons 21–30

SECTION 1: VOCABULARY [30 marks]

1a (10 marks: 1 mark for each correct answer.)
1	drink	6	poisoning
2	cold	7	book
3	pressure	8	attack
4	park	9	bite
5	player	10	parade

b (10 marks: 1 mark for each appropriate answer.)
Possible answers
11	crossing	16	school
12	minister	17	officer
13	line	18	inspector
14	shop	19	lights
15	bag	20	glass

2 (10 marks: 1 mark for each appropriate answer.)
1	speech	6	zoo
2	toes	7	hangover
3	tap	8	lift
4	soap	9	toys
5	pavement	10	machine

SECTION 2: GRAMMAR [30 marks]

3a (10 marks: 1 mark for each correct answer.)
1	c) get	6	a) found
2	a) for	7	a) where
3	c) has moved	8	a) can
4	c) lives	9	b) have
5	b) him	10	b) can't

b (10 marks: 1 mark for each appropriate answer.)
11	I	16	went
12	wasn't	17	ever
13	felt / was	18	like
14	didn't	19	his
15	lost	20	must

4 (10 marks: 1 mark for each correct sentence.)
1 It's for taking photos with.
2 We have to take some sun-tan lotion.
3 Are you able to run five kilometres?
4 You ought to go to the dentist.
5 I've got a sore throat.
6 Could you tell me the time?
7 You must never walk along a railway line.
8 Shall I carry your suitcase?
9 I wasn't able to read when I was five.
10 Can I borrow your car this evening?

SECTION 3: READING [20 marks]

5 (2 marks)
c

6 (10 marks: 1 mark for each correct answer.)
1	T	6	F
2	T	7	DS
3	F	8	T
4	T	9	F
5	DS	10	T

7 (8 marks: 1 mark for each of these eight things. Deduct marks if unimportant information is included.)
The most important points are:
(1) Adults and (2) children must wear seat belts in the (3) front of a car. (4) Adults and (5) children must wear seat belts in the (6) back of a car if they are (7) fitted and (8) available.

SECTION 4: WRITING [20 marks]

8 (20 marks)
Tell students what you will take into consideration when marking their written work. Criteria should include:
- efficient communication of meaning (7 marks)
- grammatical accuracy (7 marks)
- coherence in the ordering or the information or ideas (3 marks)
- layout, capitalisation and punctuation (3 marks)

It is probably better not to use a rigid marking system with the written part of the test. If, for example, you always deduct a mark for a grammatical mistake, you may find that you are over-penalising students who write a lot or who take risks. Deduct marks if students haven't written the minimum number of sentences stated in the test.

.TION 1: VOCABULARY [30 marks]

1 (5 marks: 1 mark for each correct answer.)
1 toast
2 spoon
3 storm
4 dry
5 did

b (5 marks: 1 mark for each appropriate answer.)
1 west
2 plate
3 winter
4 any noun associated with weather, eg rain, wind
5 any past participle, eg finished, said

2a (5 marks: 1 mark for each correct answer.)
business
harm
someone a favour
the cleaning
your best

b (5 marks: 1 mark for each appropriate answer.)
Possible answers

well	the shopping
the washing up	damage
the housework	the ironing
your homework	

3 (10 marks: 1 mark for each correct answer.)

1	on	6	for
2	into	7	away
3	down	8	up
4	in	9	to
5	after	10	off

SECTION 2: GRAMMAR [30 marks]

3a (10 marks: 1 mark for each correct answer.)

1	b) dangerous	6	c) were staying
2	b) had	7	c) understood
3	b) laughed	8	b) were
4	a) bad	9	b) might
5	a) never	10	b) I

b (10 marks: 1 mark for each appropriate answer.)

11	are	16	of
12	doesn't	17	didn't
13	have	18	had
14	to	19	went
15	when	20	good

4 (10 marks: 1 mark for each correct sentence.)
1 While I was drying some shirts, one of them caught fire.
2 There's been too much fog recently.
3 I'll take that pen from you if you don't give it to me.
4 Although we like pasta, we don't have it every day.
5 After they'd listened to the weather forecast, they set off.
6 Money is used for buying things.
7 It isn't windy enough to go hang-gliding.
8 There might not be much sun tomorrow.
9 The computer exploded when I plugged it in.
10 If a stranger asked me, I wouldn't tell him the time.

SECTION 3: READING [20 marks]

5 (6 marks)
a Paragraph 2
b Paragraph 1
c Paragraph 4

6 (9 marks: 1 mark for each correct answer.)

1	T	4	F	7	T
2	DS	5	F	8	T
3	F	6	DS	9	T

7 (5 marks: 1 mark for each of these five things.)
(1) They gathered fruit, nuts and insects to eat (2) and hunted kangaroos and emus (3). They moved from one area to another in their search for food. (4) Their ancestors' spirits lived on after death, sometimes in animals and plants. (5) Totems, which protected and gave groups their identity, were usually animals or plants.

SECTION 4: WRITING [20 marks]

8 (20 marks)
Tell students what you will take into consideration when marking their written work. Criteria should include:
– efficient communication of meaning (7 marks)
– grammatical accuracy (7 marks)
– coherence in the ordering or the information or ideas (3 marks)
– layout, capitalisation and punctuation (3 marks)

It is probably better not to use a rigid marking system with the written part of the test. If, for example, you always deduct a mark for a grammatical mistake, you may find that you are over-penalising students who write a lot or who take risks. Deduct marks if students haven't written the minimum number of sentences stated in the test.